LYNCHBURG COLLEGE
SYMPOSIUM READINGS

CLASSICAL SELECTIONS ON GREAT ISSUES

SERIES TWO
VOLUME II

SCIENCE, TECHNOLOGY AND SOCIETY

Descartes
Conant
Oppenheimer
Bush
Feibleman
Taylor
Muller
Mumford
Ellul
Mesthene
T. H. Huxley
Thoreau
Adams
A. Huxley
Wells

NIVERSITY
RESS OF
MERICA

LANHAM • NEW YORK • LONDON

SERIES TWO

SYMPOSIUM READINGS

Lynchburg College in Virginia

Compiled and Edited by the
following faculty members of Lynchburg College:

Kenneth E. Alrutz, A.M., University of Pennsylvania; Assistant Professor of English

Virginia B. Berger, M.A., Harvard University; Associate Professor of Music

Anne Marshall Huston, Ed.D., University of Virginia; Associate Professor of Education

James L. Campbell, Ph.D., University of Virginia; Associate Professor of English

Robert L. Frey, Ph.D., University of Minnesota; Professor of History

James A. Huston, Ph.D., New York University; Dean of the College, Professor of History and International Relations

Shannon McIntyre Jordan, Ph.D., University of Georgia; Instructor in Philosophy

Jan G. Linn, D.Min., Christian Theological Seminary; Associate Professor, College Chaplain

Peggy S. Pittas, M.A., Dalhousie University; Associate Professor of Psychology

Clifton W. Potter, Jr., Ph.D., University of Virginia; Professor of History

Julius A. Sigler, Ph.D., University of Virginia; Professor of Physics

Phillip H. Stump, Ph.D., University of California at Los Angeles; Assistant Professor of History

Thomas C. Tiller, Ph.D., Florida State University; Dean of Student Affairs, Professor of Education

University Press of America,™ Inc.

4720 Boston Way
Lanham, MD 20706

3 Henrietta Street
London WC2E 8LU England

Printed in the United States of America

Library of Congress Cataloging in Publication Data
Main entry under title:

Science, technology, and society.

(Classical selections on great issues; series 2, vol. 2)
1. Science–History–Sources. 2. Technology–History–Sources. 3. Science–Social aspects–History–Sources. 4. Technology–Social aspects–History–Sources. I. Descartes, René, 1596–1650. II. Lynchburg College. III. Series.

Q125.S43475 500 81–71947
ISBN 0–8191–2297–1 AACR2
ISBN 0–8191–2298–X (pbk.)

ACKNOWLEDGEMENTS

The following copyrighted materials have been used with the permission of the copyright holders:

From *Modern Science and Modern Man* by James Conant. c.1953, Columbia University Press.

From *Science and the Common Understanding* by J.Robert Oppenheimer. c.1953, Simon & Schuster.

From *The Children of Frankenstein* by Herbert J. Muller. c. 1970, Indiana University Press.

From *Philosophy and Technology,* "The Technological Order" by Jacques Ellul. c.1972, Wayne State University Press.

From *Brave New World* by Aldous Huxley. c.1932, Harper & Row.

The authors acknowledge with appreciation the permissions granted by these holders of the respective copyrights.

INTRODUCTION TO VOLUME II

Science and technology have altered society dramatically during the past century. Today we are accustomed to new ideas and a frequency of change which would have been devastating to society a century ago. One corollary to the phenomenal success of science has been man's increasing tendency to assume that all problems are correctable by the methods of science and technology. In recent years energy shortages, genetic engineering, and artificial intelligence have caused doubt about the direction in which science and technology are taking us. The readings in this unit present a variety of views toward science and technology. Perhaps no issue will be more significant in our lifetime than the relationship between science, technology, and society. Shall we be absorbed into a technological value system in a totally autocratic way as Ellul claims, or will science and technology allow man the freedom that Taylor forsees?

CONTENTS

1

Rene Descartes, DISCOURSE ON METHOD

1. What does Descartes mean by method?

2. What is the essence of Descartes' method?

3. What is Descartes' attitude toward the utility of mathematics in scientific reasoning?

4. What does Descartes' "Code of Morals" have to do with his method?

5. What are the implications of the famous Cartesian Cogito: "I think, hence I am"?

6. To what extent do you think the method of Descartes is useful in analyzing society?

7. How do you account for Descartes' ambivalence about publishing the results of his thinking?

Rene Descartes (1596-1650) has often been called the father of modern philosophy. A contemporary of Galileo, Descartes was preoccupied with the problem of intellectual creativity. He tried to develop a coherent system of thought based on principles which he determined to be necessarily true. He ground his work in these principles and made important contributions to both mathematics and philosophy. This selection from the *Discourse on Method* illustrates the optimism with which he views the analytical approach to life.

T. V. Smith, noted philosopher and teacher at the University of Chicago and Syracuse University, and one-time Congressman from Illinois, has related how the reading of Descartes' *Discourse on Method* changed his whole life. One long, winter night when, as a young lad, he was riding in a freight train caboose across Texas, he came across the book which the brakeman had left open on his table while he was out working the train. Here Smith found many of the questions, and many of the challenges to prevailing opinion which he often had felt, but never had been able to articulate.

DISCOURSE ON THE METHOD

OF RIGHTLY CONDUCTING THE REASON AND SEEKING THE TRUTH IN THE SCIENCES

By René Descartes

PART I

GOOD SENSE is, of all things among men, the most equally distributed; for every one thinks himself so abundantly provided with it, that those even who are the most difficult to satisfy in everything else, do not usually desire a larger measure of this quality than they already possess. And in this it is not likely that all are mistaken: the conviction is rather to be held as testifying that the power of judging aright and of distinguishing Truth from Error, which is properly what is called Good Sense or Reason, is by nature equal in all men; and that the diversity of our opinions, consequently, does not arise from some being endowed with a larger share of Reason than others, but solely from this, that we conduct our thoughts along different ways, and do not fix our attention on the same objects. For to be possessed of a vigorous mind is not enough; the prime requisite is rightly to apply it. The greatest minds, as they are capable of the highest excellencies, are open likewise to the greatest aberrations; and those who travel very slowly may yet make far greater progress, provided they keep always to the straight road, than those who, while they run, forsake it.

For myself, I have never fancied my mind to be in any respect more perfect than those of the generality; on the contrary, I have often wished that I were equal to some others in promptitude of thought, or in clearness and distinctness of imagination, or in fulness and readiness of memory. And besides these, I know of no other qualities that contribute to the perfection of the mind; for as to the Reason or

Sense, inasmuch as it is that alone which constitutes us men, and distinguishes us from the brutes, I am disposed to believe that it is to be found complete in each individual; and on this point to adopt the common opinion of philosophers, who say that the difference of greater and less holds only among the *accidents,* and not among the *forms* or *natures* of *individuals* of the same *species.*

I will not hesitate, however, to avow my belief that it has been my singular good fortune to have very early in life fallen in with certain tracks which have conducted me to considerations and maxims, of which I have formed a Method that gives me the means, as I think, of gradually augmenting my knowledge, and of raising it by little and little to the highest point which the mediocrity of my talents and the brief duration of my life will permit me to reach. For I have already reaped from it such fruits that, although I have been accustomed to think lowly enough of myself, and although when I look with the eye of a philosopher at the varied courses and pursuits of mankind at large, I find scarcely one which does not appear vain and useless, I nevertheless derive the highest satisfaction from the progress I conceive myself to have already made in the search after truth, and cannot help entertaining such expectations of the future as to believe that if, among the occupations of men as men, there is any one really excellent and important, it is that which I have chosen.

After all, it is possible I may be mistaken; and it is but a little copper and glass, perhaps, that I take for gold and diamonds. I know how very liable we are to delusion in what relates to ourselves, and also how much the judgments of our friends are to be suspected when given in our favour. But I shall endeavour in this Discourse to describe the paths I have followed, and to delineate my life as in a picture, in order that each one may be able to judge of them for himself, and that in the general opinion entertained of them, as gathered from current report, I myself may have a new help towards instruction to be added to those I have been in the habit of employing.

My present design, then, is not to teach the Method which each ought to follow for the right conduct of his Reason, but solely to describe the way in which I have endeavoured to conduct my own. They who set themselves to give precepts must of course regard themselves as possessed of greater skill than those to whom they prescribe; and if they err in the slightest particular, they subject themselves to censure. But as this Tract is put forth merely as a history, or, if you will, as a tale, in which, amid some examples worthy of imitation, there will be found, perhaps, as many more which it were advisable not to follow, I hope it will prove useful to some without being hurtful to any, and that my openness will find some favour with all.

From my childhood, I have been familiar with letters; and as I was given to believe that by their help a clear and certain knowledge of all that is useful in life might be acquired, I was ardently desirous of instruction. But as soon as I had finished the entire course of study, at the close of which it is customary to be admitted into the order of the learned, I completely changed my opinion. For I found myself involved in so many doubts and errors, that I was convinced I had advanced no farther in all my attempts at learning, than the discovery at every turn of my own ignorance. And yet I was studying in one of the most celebrated Schools in Europe, in which I thought there must be learned men, if such were anywhere to be found. I had been taught all that others learned there; and not contented with the sciences actually taught us, I had, in addition, read all the books that had fallen into my hands, treating of such branches as are esteemed the most curious and rare. I knew the judgment which others had formed of me; and I did not find that I was considered inferior to my fellows, although there were among them some who were already marked out to fill the places of our instructors. And, in fine, our age appears to me as flourishing, and as fertile in powerful minds as any preceding one. I was thus led to take the liberty of judging of all other men by myself, and of concluding that there was no science in existence that was of such a nature as I had previously been given to believe.

I still continued, however, to hold in esteem the studies of the Schools. I was aware that the Languages taught in them are necessary to the understanding of the writings of the ancients; that the grace of Fable stirs the mind; that the memorable deeds of History elevate it; and, if read with discretion, aid in forming the judgment; that the perusal of all excellent books is, as it were, to interview with the noblest men of past ages, who have written them, and even a studied interview, in which are discovered to us only their choicest thoughts; that Eloquence has incomparable force and beauty; that Poesy has its ravishing graces and delights; that in the Mathematics there are many refined discoveries eminently suited to gratify the inquisitive, as well as further all the arts and lessen the labour of man; that numerous highly useful precepts and exhortations to virtue are contained in treatises on Morals; that Theology points out the path to heaven; that Philosophy affords the means of discoursing with an appearance of truth on all matters, and commands the admiration of the more simple; that Jurisprudence, Medicine, and the other Sciences, secure for their cultivators honours and riches; and in fine, that it is useful to bestow some attention upon all, even upon those abounding the most in superstition and error, that we may be in a position to determine their real value, and guard against being deceived.

But I believed that I had already given sufficient time to Languages, and likewise to the reading of the writings of the ancients, to their Histories and Fables. For to hold converse with those of other ages and to travel, are almost the same thing. It is useful to know something of the manners of different nations, that we may be enabled to form a more correct judgment regarding our own, and be prevented from thinking that everything contrary to our customs is ridiculous and irrational,—a conclusion usually come to by those whose experience has been limited to their own country. On the other hand, when too much time is occupied in travelling, we become strangers to our native country; and the over curious in the customs of the past are generally ignorant of those of the present. Besides, fictitious narratives lead us to imagine the possibility of many events that are impossible; and even the most faithful histories, if they do not wholly misrepresent matters, or exaggerate their importance to render the account of them more worthy of perusal, omit, at least, almost always the meanest and least striking of the attendant circumstances; hence it happens that the remainder does not represent the truth, and that such as regulate their conduct by examples drawn from this source, are apt to fall into the extravagances of the knight-errants of Romance, and to entertain projects that exceed their powers.

I esteemed Eloquence highly, and was in raptures with Poesy; but I thought that both were gifts of nature rather than fruits of study. Those in whom the faculty of Reason is predominant, and who most skilfully dispose their thoughts with a view to render them clear and intelligible, are always the best able to persuade others of the truth of what they lay down, though they should speak only in the language of Lower Brittany, and be wholly ignorant of the rules of Rhetoric; and those whose minds are stored with the most agreeable fancies, and who can give expression to them with the greatest embellishment and harmony, are still the best poets, though unacquainted with the Art of Poetry.

I was especially delighted with the Mathematics, on account of the certitude and evidence of their reasonings: but I had not as yet a precise knowledge of their true use; and thinking that they but contributed to the advancement of the mechanical arts, I was astonished that foundations, so strong and solid, should have had no loftier superstructure reared on them. On the other hand, I compared the disquisitions of the ancient Moralists to very towering and magnificent palaces with no better foundation than sand and mud: they laud the virtues very highly, and exhibit them as estimable far above anything on earth; but they give us no adequate criterion of virtue, and frequently that which they designate with so fine a name is but apathy, or pride, or despair, or parricide.

I revered our Theology, and aspired as much as any one to reach heaven: but being given assuredly to understand that the way is not less open to the most ignorant than to the most learned, and that the revealed truths which lead to heaven are above our comprehension, I did not presume to subject them to the impotency of my Reason; and I thought that in order competently to undertake their examination, there was need of some special help from heaven, and of being more than man.

Of philosophy I will say nothing, except that when I saw that it had been cultivated for many ages by the most distinguished men, and that yet there is not a single matter within its sphere which is not still in dispute, and nothing, therefore, which is above doubt, I did not presume to anticipate that my success would be greater in it than that of others; and further, when I considered the number of conflicting opinions touching a single matter that may be upheld by learned men, while there can be but one true, I reckoned as well-nigh false all that was only probable.

As to the other Sciences, inasmuch as these borrow their principles from Philosophy, I judged that no solid superstructures could be reared on foundations so infirm; and neither the honour nor the gain held out by them was sufficient to determine me to their cultivation: for I was not, thank Heaven, in a condition which compelled me to make merchandise of Science for the bettering of my fortune; and though I might not profess to scorn glory as a Cynic, I yet made very slight account of that honour which I hoped to acquire only through fictitious titles. And, in fine, of false Sciences I thought I knew the worth sufficiently to escape being deceived by the professions of an alchemist, the predictions of an astrologer, the impostures of a magician, or by the artifices and boasting of any of those who profess to know things of which they are ignorant.

For these reasons, as soon as my age permitted me to pass from under the control of my instructors, I entirely abandoned the study of letters, and resolved no longer to seek any other science than the knowledge of myself, or of the great book of the world. I spent the remainder of my youth in travelling, in visiting courts and armies, in holding intercourse with men of different dispositions and ranks, in collecting varied experience, in proving myself in the different situations into which fortune threw me, and, above all, in making such reflection on the matter of my experience as to secure my improvement. For it occurred to me that I should find much more truth in the reasonings of each individual with reference to the affairs in which he is personally interested, and the issue of which must presently punish him if he has judged amiss, than in those conducted by a man of letters in his study, regarding speculative matters that are of no practical moment, and followed by no consequences to himself, farther, perhaps, than that they foster his vanity the better the more remote they are from common sense; requiring, as they must in this case, the exercise of greater ingenuity and art to render them probable. In addition, I had always a most earnest desire to know how to distinguish the true from the false, in order that I might be able clearly to discriminate the right path in life, and proceed in it with confidence.

It is true that, while busied only in considering the manners of other men, I found here, too, scarce any ground for settled conviction, and remarked hardly less contradiction among them than in the opinions of the philosophers. So that the greatest advantage I derived from the study consisted in this, that, observing many things which, however extravagant and ridiculous to our apprehension, are yet by common consent received and approved by other great nations, I learned to entertain too decided a belief in regard to nothing of the truth of which I had been persuaded merely by example and custom: and thus I gradually extricated myself from many errors powerful enough to darken our Natural Intelligence, and incapacitate us in great measure from listening to Reason. But after I had been occupied several years in thus studying the book of the world, and in essaying to gather some experience, I at length resolved to make myself an object of study, and to employ all the powers of my mind in choosing the paths I ought to follow; an undertaking which was accompanied with greater success than it would have been had I never quitted my country or my books.

PART II

I WAS then in Germany, attracted thither by the wars in that country, which have not yet been brought to a termination; and as I was returning to the army from the coronation of the Emperor, the setting in of winter arrested me in a locality where, as I found no society to interest me, and was besides fortunately undisturbed by any cares or passions, I remained the whole day in seclusion,[1] with full opportunity to occupy my attention with my own thoughts. Of these one of the very first that occurred to me was, that there is seldom so much perfection in works composed of many separate parts, upon which different hands have been employed, as in those completed by a single master. Thus it is observable that the buildings which a single architect has planned and executed, are generally more elegant and commodious than those which several have attempted to improve, by making old walls serve for purposes for which they were not originally built. Thus also, those ancient cities which, from being at first only villages, have become, in course of time, large towns, are usually but ill-laid out compared with the regularly constructed towns which a professional architect has freely planned on an open plain; so that although the several buildings of the former may often equal or surpass in beauty those of the latter, yet when one observes their indiscriminate juxtaposition, there a large one and here a small, and the consequent crookedness and irregularity of the streets, one is disposed to allege that chance rather than any human will guided by reason, must have led to such an arrangement. And if we consider that nevertheless there have been at all times certain officers whose duty it was to see that private buildings contributed to public ornament, the difficulty of reaching high perfection with but the materials of others to operate on, will be readily acknowledged. In the same way I fancied that those nations which, starting from a semi-barbarous state and advancing to civili-

[1] Literally, in a room heated by means of a stove.—*Tr.*

sation by slow degrees, have had their laws successively determined, and, as it were, forced upon them simply by experience of the hurtfulness of particular crimes and disputes, would by this process come to be possessed of less perfect institutions than those which, from the commencement of their association as communities, have followed the appointments of some wise legislator. It is thus quite certain that the constitution of the true religion, the ordinances of which are derived from God, must be incomparably superior to that of every other. And, to speak of human affairs, I believe that the past pre-eminence of Sparta was due not to the goodness of each of its laws in particular, for many of these were very strange, and even opposed to good morals, but to the circumstance that, originated by a single individual, they all tended to a single end. In the same way I thought that the sciences contained in books, (such of them at least as are made up of probable reasonings, without demonstrations,) composed as they are of the opinions of many different individuals massed together, are farther removed from truth than the simple inferences which a man of good sense using his natural and unprejudiced judgment draws respecting the matters of his experience. And because we have all to pass through a state of infancy to manhood, and have been of necessity, for a length of time, governed by our desires and preceptors, (whose dictates were frequently conflicting, while neither perhaps always counselled us for the best,) I farther concluded that it is almost impossible that our judgments can be so correct or solid as they would have been, had our Reason been mature from the moment of our birth, and had we always been guided by it alone.

It is true, however, that it is not customary to pull down all the houses of a town with the single design of rebuilding them differently, and thereby rendering the streets more handsome; but it often happens that a private individual takes down his own with the view of erecting it anew, and that people are even sometimes constrained to this when their houses are in danger of falling from age, or when the foundations are insecure. With this before me by way of example, I was persuaded that it would indeed be preposterous for a private individual to think of reforming a state by fundamentally changing it throughout, and overturning it in order to set it up amended; and the same I thought was true of any similar project for reforming the body of the Sciences, or the order of teaching them established in the Schools: but as for the opinions which up to that time I had embraced, I thought that I could not do better than resolve at once to sweep them wholly away, that I might afterwards be in a position to admit either others more correct, or even perhaps the same when they had undergone the scrutiny of Reason. I firmly believed that in this way I should much better succeed in the conduct of my life, than if I built only upon old foundations, and leant upon principles which, in my youth, I had taken upon trust. For although I recognised various difficulties in this undertaking, these were not, however, without remedy, nor once to be compared with such as attend the slightest reformation in public affairs. Large bodies, if once overthrown, are with great difficulty set up again, or even kept erect when once seriously shaken, and the fall of such is always disastrous. Then if there are any imperfections in the constitutions of states, (and that many such exist the diversity of constitutions is alone sufficient to assure us,) custom has without doubt materially smoothed their inconveniences, and has even managed to steer altogether clear of, or insensibly corrected a number which sagacity could not have provided against with equal effect; and, in fine, the defects are almost always more tolerable than the change necessary for their removal; in the same manner that highways which wind among mountains, by being much frequented, become gradually so smooth and commodious, that it is much better to follow them than to seek a straighter path by climbing over the tops of rocks and descending to the bottoms of precipices.

Hence it is that I cannot in any degree approve of those restless and busy meddlers who, called neither by birth nor fortune to take part in the management of public affairs, are yet always projecting reforms; and if I thought that this Tract contained aught which might justify the suspicion that I was a victim of such folly, I would by no means permit its publication. I have never contemplated anything higher than the reformation of my own opinions, and basing them on a foundation wholly my own. And although my own satisfaction with my work has led me to present here a draft of it, I do not by any means therefore recommend to every one else to make a similar attempt. Those whom God has endowed with a larger measure of genius will entertain, perhaps, designs still more exalted; but for the many I am much afraid lest even the present undertaking be more than they can safely venture to imitate. The single design to strip one's self of all past beliefs is one that ought not to be taken by every one. The majority of men is composed of two classes, for neither of which would this be at all a befitting resolution: in the *first* place, of those who with more than a due confidence in their own powers, are precipitate in their judgments and want the patience requisite for orderly and circumspect thinking; whence it happens, that if men of this class once take the liberty to doubt of their accustomed opinions, and quit the beaten highway, they will never be able to thread the byeway that would lead them by a shorter course, and will lose themselves and continue to wander for life; in the *second* place, of those who, possessed of sufficient sense or modesty to determine that there are others who excel them in the power of discriminating between truth and error, and by whom they may be instructed, ought rather to content themselves with the opinions of such than trust for more correct to their own Reason.

For my own part, I should doubtless have belonged to the latter class, had I received instruction from but one master, or had I never known the diversities of opinion that from time immemorial have prevailed among men of the greatest learning. But I had become aware, even so early as during my college life, that no opinion, however absurd and incredible, can be imagined, which has not been maintained by some one of the philosophers; and afterwards in the course of my travels I remarked that all those whose opinions are decidedly repugnant to ours are not on that account barbarians and savages, but on the contrary that many of these nations make an equally good, if not a better, use of their Reason than we do. I took into account also the very different character which a person brought up from infancy in France or Germany exhibits, from that which, with the same mind originally, this individual would have possessed had he lived always among the Chinese or with savages, and the circumstance that in dress itself the fashion which pleased us ten years ago, and which may again, perhaps, be received into favour before ten years have gone, appears to us at this moment extravagant and ridiculous. I was thus led to infer that the ground of our opinions is far more custom and example than any certain knowledge. And, finally, although such be the ground of our opinions, I remarked that a plurality of suffrages is no guarantee of truth where it is at all of difficult discovery, as in such cases it is much more likely that it will be found by one than by many. I could, however, select from the crowd no one whose opinions seemed worthy of preference, and thus I found myself constrained, as it were, to use my own Reason in the conduct of my life.

But like one walking alone and in the dark, I resolved to proceed so slowly and with such circumspection, that if I did not advance far, I would at least guard against falling. I did not even choose to dismiss summarily any of the opinions that had crept into my belief without having been introduced by Reason, but first of all took sufficient time carefully to satisfy myself of the general nature of the task I was setting myself, and ascertain the true Method by which to arrive at the knowledge of whatever lay within the compass of my powers.

Among the branches of Philosophy, I had, at an earlier period, given some attention to Logic, and among those of the Mathematics to Geometrical Analysis and Algebra,—three Arts or Sciences which ought, as I conceived, to contribute something to my design. But, on examination, I found that, as for Logic, its syllogisms and the majority of its other precepts are of avail rather in the communication of what we already know, or even as the Art of Lully, in speaking without judgment of things of which we are ignorant, than in the investigation of the unknown; and although this Science contains indeed a number of correct and very excellent precepts, there are, nevertheless, so many others, and these either injurious or superfluous, mingled with the former, that it is almost quite as difficult to effect a severance of the true from the false as it is to extract a Diana or a Minerva from a rough block of marble. Then as to the Analysis of the ancients and the Algebra of the moderns, besides that they embrace only matters highly abstract, and, to appearance, of no use, the former is so exclusively restricted to the consideration of figures, that it can exercise the Understanding only on condition of greatly fatiguing the Imagination;[2] and, in the latter, there is so complete a subjection to certain rules and formulas, that there results an art full of confusion and obscurity calculated to embarrass, instead of a science fitted to cultivate the mind. By these considerations I was induced to seek some other Method which would comprise the advantages of the three and be exempt from their defects. And as a multitude of laws often only hampers justice, so that a state is best governed when, with few laws, these are rigidly administered; in like manner, instead of the great number of precepts of which Logic is composed, I believed that the four following would prove perfectly sufficient for me, provided I took the firm and unwavering resolution never in a single instance to fail in observing them.

The *first* was never to accept anything for true which I did not clearly know to be such; that is to say, carefully to avoid precipitancy and prejudice, and to comprise nothing more in my judgment than what was presented to my mind so clearly and distinctly as to exclude all ground of doubt.

The *second,* to divide each of the difficulties under examination into as many parts as possible, and as might be necessary for its adequate solution.

The *third,* to conduct my thoughts in such order that, by commencing with objects the simplest and easiest to know, I might ascend by little and little, and, as it were, step by step, to the knowledge of the more complex; assigning in thought a certain order even to those objects which in their own nature do not stand in a relation of antecedence and sequence.

And the *last,* in every case to make enumerations so complete, and reviews so general, that I might be assured that nothing was omitted.

The long chains of simple and easy reasonings by means of which geometers are accustomed to reach the conclusions of their most difficult demonstrations, had led me to imagine that all things, to the knowledge of which man is competent, are mutually connected in the same way, and that there is nothing so far removed from us as to be beyond our reach, or so hidden that we cannot discover it, provided only we abstain from accepting the false for the true, and

[2] The Imagination must here be taken as equivalent simply to the Representative Faculty.—*Tr.*

always preserve in our thoughts the order necessary for the deduction of one truth from another. And I had little difficulty in determining the objects with which it was necessary to commence, for I was already persuaded that it must be with the simplest and easiest to know, and, considering that of all those who have hitherto sought truth in the Sciences, the mathematicians alone have been able to find any demonstrations, that is, any certain and evident reasons, I did not doubt but that such must have been the rule of their investigations. I resolved to commence, therefore, with the examination of the simplest objects, not anticipating, however, from this any other advantage than that to be found in accustoming my mind to the love and nourishment of truth, and to a distaste for all such reasonings as were unsound. But I had no intention on that account of attempting to master all the particular Sciences commonly denominated Mathematics: but observing that, however different their objects, they all agree in considering only the various relations or proportions subsisting among those objects, I thought it best for my purpose to consider these proportions in the most general form possible, without referring them to any objects in particular, except such as would most facilitate the knowledge of them, and without by any means restricting them to these, that afterwards I might thus be the better able to apply them to every other class of objects to which they are legitimately applicable. Perceiving further, that in order to understand these relations I should sometimes have to consider them one by one, and sometimes only to bear them in mind, or embrace them in the aggregate, I thought that, in order the better to consider them individually, I should view them as subsisting between straight lines, than which I could find no objects more simple, or capable of being more distinctly represented to my imagination and senses; and on the other hand, that in order to retain them in the memory, or embrace an aggregate of many, I should express them by certain characters the briefest possible. In this way I believed that I could borrow all that was best both in Geometrical Analysis and in Algebra, and correct all the defects of the one by help of the other.

And, in point of fact, the accurate observance of these few precepts gave me, I take the liberty of saying, such ease in unravelling all the questions embraced in these two sciences, that in the two or three months I devoted to their examination, not only did I reach solutions of questions I had formerly deemed exceedingly difficult, but even as regards questions of the solution of which I continued ignorant, I was enabled, as it appeared to me, to determine the means whereby, and the extent to which, a solution was possible; results attributable to the circumstance that I commenced with the simplest and most general truths, and that thus each truth discovered was a rule available in the discovery of subsequent ones. Nor in this perhaps shall I appear too vain if it be considered that, as the truth on any particular point is one, whoever apprehends the truth, knows all that on that point can be known. The child, for example, who has been instructed in the elements of Arithmetic, and has made a particular addition, according to rule, may be assured that he has found, with respect to the sum of the numbers before him, all that in this instance is within the reach of human genius. Now, in conclusion, the Method which teaches adherence to the true order, and an exact enumeration of all the conditions of the thing sought includes all that gives certitude to the rules of Arithmetic.

But the chief ground of my satisfaction with this Method, was the assurance I had of thereby exercising my reason in all matters, if not with absolute perfection, at least with the greatest attainable by me: besides, I was conscious that by its use my mind was becoming gradually habituated to clearer and more distinct conceptions of its objects; and I hoped also, from not having restricted this Method to any particular matter, to apply it to the difficulties of the other Sciences, with not less success than to those of Algebra. I should not, however, on this account have ventured at once on the examination of all the difficulties of the Sciences which presented themselves to me, for this would have been contrary to the order prescribed in the Method, but observing that the knowledge of such is dependent on principles borrowed from Philosophy, in which I found nothing certain, I thought it necessary first of all to endeavour to establish its principles. And because I observed, besides, that an inquiry of this kind was of all others of the greatest moment, and one in which precipitancy and anticipation in judgment were most to be dreaded, I thought that I ought not to approach it till I had reached a more mature age, (being at that time but twenty-three,) and had first of all employed much of my time in preparation for the work, as well by eradicating from my mind all the erroneous opinions I had up to that moment accepted, as by amassing variety of experience to afford materials for my reasonings, and by continually exercising myself in my chosen Method with a view to increased skill in its application.

PART III

AND, finally, as it is not enough, before commencing to rebuild the house in which we live, that it be pulled down, and materials and builders provided, or that we engage in the work ourselves, according to a plan which we have beforehand carefully drawn out, but as it is likewise necessary that we be furnished with some other house in which we may live commodiously during the operations, so that I might not remain irresolute in my actions, while my Reason compelled me to suspend my judgment, and that I might not be prevented from living thenceforward in the greatest possible felicity, I formed a provisory code of Morals, composed of three or four maxims, with which I am desirous to make you acquainted.

The *first* was to obey the laws and customs of my country, adhering firmly to the Faith in which, by the grace of God, I had been educated from my childhood, and regulating my conduct in every other matter according to the most moderate opinions, and the farthest removed from extremes, which should happen to be adopted in practice with general consent of the most judicious of those among whom I might be living. For, as I had from that time begun to hold my own opinions for nought because I wished to subject them all to examination, I was convinced that I could not do better than follow in the meantime the opinions of the most judicious; and although there are some perhaps among the Persians and Chinese as judicious as among ourselves, expediency seemed to dictate that I should regulate my practice conformably to the opinions of those with whom I should have to live; and it appeared to me that, in order to ascertain the real opinions of such, I ought rather to take cognizance of what they practised than of what they said, not only because, in the corruption of our manners, there are few disposed to speak exactly as they believe, but also because very many are not aware of what it is that they really believe; for, as the act of mind by which a thing is believed is different from that by which we know that we believe it, the one act is often found without the other. Also, amid many opinions held in equal repute, I chose always the most moderate, as much for the reason that these are always the most convenient for practice, and probably the best, (for all excess is generally vicious,) as that, in the event of my falling into error, I might be at less distance from the truth than if, having chosen one of the extremes, it should turn out to be the other which I ought to have adopted. And I placed in the class of extremes especially all promises by which somewhat of our freedom is abridged; not that I disapproved of the laws which, to provide against the instability of men of feeble resolution, when what is sought to be accomplished is some good, permit engagements by vows and contracts binding the parties to persevere in it, or even, for the security of commerce, sanction similar engagements where the purpose sought to be realized is indifferent: but because I did not find anything on earth which was wholly superior to change, and because, for myself in particular, I hoped gradually to perfect my judgments, and not to suffer them to deteriorate, I would have deemed it a grave sin against good sense, if, for the reason that I approved of something at a particular time, I therefore bound myself to hold it for good at a subsequent time, when perhaps it had ceased to be so, or I had ceased to esteem it such.

My *second* maxim was to be as firm and resolute in my actions as I was able, and not to adhere less steadfastly to the most doubtful opinions, when once adopted, than if they had been highly certain; imitating in this the example of travellers who, when they have lost their way in a forest, ought not to wander from side to side, far less remain in one place, but proceed constantly towards the same side in as straight a line as possible, without changing their direction for slight reasons, although perhaps it might be chance alone which at first determined the selection; for in this way, if they do not exactly reach the point they desire, they will come at least in the end to some place that will probably be preferable to the middle of a forest. In the same way, since in action it frequently happens that no delay is permissible, it is very certain that, when it is not in our power to determine what is true, we ought to act according to what is most probable; and even although we should not remark a greater probability in one opinion than in another, we ought notwithstanding to choose one or the other, and afterwards consider it, in so far as it relates to practice, as no longer dubious, but manifestly true and certain, since the reason by which our choice has been determined is itself possessed of these qualities. This principle was sufficient thenceforward to rid me of all those repentings and pangs of remorse that usually disturb the consciences of such feeble and uncertain minds as, destitute of any clear and determinate principle of choice, allow themselves one day to adopt a course of action as the best, which they abandon the next, as the opposite.

My *third* maxim was to endeavour always to conquer myself rather than fortune, and change my desires rather than the order of the world, and in general, accustom myself to the persuasion that, except our own thoughts, there is nothing absolutely in our power; so that when we have done our best in respect of things external to us, all wherein we fail of success is to be held, as regards us, absolutely impossible: and this single principle seemed to me sufficient to prevent me from desiring for the future anything which I could not obtain, and thus render me contented; for since our will naturally seeks those objects alone which the understanding represents as in some way possible of attainment, it is plain, that if we consider all external goods as equally beyond our power, we shall no more regret the absence of such goods as seem due to our birth, when deprived of them without any fault of ours, than our not possessing the kingdoms of China or Mexico; and thus making, so to speak, a virtue of necessity, we shall no more desire health in disease, or freedom in imprisonment, than we now do bodies incorruptible as diamonds, or the wings of birds to fly with. But I confess there is need of prolonged discipline and frequently repeated meditation to accustom the mind to view all objects in this light; and I believe that in this chiefly consisted the secret of the power of such philosophers as in former times were enabled to rise superior to the influence of fortune, and amid suffering and poverty, enjoy a happiness which their gods might have envied. For, occupied incessantly with the consideration of the limits prescribed to their power by nature, they became so entirely convinced that nothing was at their disposal except their own thoughts, that this conviction was of itself sufficient to prevent their entertaining any desire of other objects; and over their thoughts they acquired a sway so absolute, that they had some ground on this account for esteeming themselves more rich and more powerful, more free and more happy, than other men who, whatever be the favours heaped on them by nature and fortune, if destitute of this philosophy, can never command the realization of all their desires.

In fine, to conclude this code of Morals, I thought of reviewing the different occupations of men in this life, with the view of making choice of the best. And, without wishing to offer any remarks on the employments of others, I may state that it was my conviction that I could not do better than continue in that in which I was engaged, viz., in devoting my whole life to the culture of my Reason, and in making the greatest progress I was able in the knowledge of truth, on the principles of the Method which I had prescribed to myself. This Method, from the time I had begun to apply it, had been to me the source of satisfaction so intense as to lead me to believe that more perfect or more innocent could not be enjoyed in this life; and as by its means I daily discovered truths that appeared to me of some importance, and of which other men were generally ignorant, the gratification thence arising so occupied my mind that I was wholly indifferent to every other object. Besides, the three preceding maxims were founded singly on the design of continuing the work of self-instruction. For since God has endowed each of us with some Light of Reason by which to distinguish truth from error, I could not have believed that I ought for a single moment to rest satisfied with the opinions of another, unless I had resolved to exercise my own judgment in examining these whenever I should be duly qualified for the task. Nor could I have proceeded on such opinions without scruple, had I supposed that I should thereby forfeit any advantage for attaining still more accurate, should such exist. And, in fine, I could not have restrained my desires, nor remained satisfied, had I not followed a path in which I thought myself certain of attaining all the knowledge to the acquisition of which I was competent, as well as the largest amount of what is truly good which I could ever hope to secure. Inasmuch as we neither seek nor shun any object except in so far as our understanding represents it as good or bad, all that is necessary to right action is right judgment, and to the best action the most correct judgment,—that is, to the acquisition of all the virtues with all else that is truly valuable and within our reach; and the assurance of such an acquisition cannot fail to render us contented.

Having thus provided myself with these maxims, and having placed them in reserve along with the truths of Faith, which have ever occupied the first place in my belief, I came to the conclusion that I might with freedom set about ridding myself of what remained of my opinions. And, inasmuch as I hoped to be better able successfully to accomplish this work by holding intercourse with mankind, than by remaining longer shut up in the retirement where these thoughts had occurred to me, I betook me again to travelling before the winter was well ended. And, during the nine subsequent years, I did nothing but roam from one place to another, desirous of being a spectator rather than an actor in the plays exhibited on the theatre of the world; and, as I made it my business in each matter to reflect particularly upon what might fairly be doubted and prove a source of error, I gradually rooted out from my mind all the errors which had hitherto crept into it. Not that in this I imitated the Sceptics who doubt only that they may doubt, and seek nothing beyond uncertainty itself; for, on the contrary, my design was singly to find ground of assurance, and cast aside the loose earth and sand, that I might reach the rock or the clay. In this, as appears to me, I was successful enough; for, since I endeavoured to discover the falsehood or incertitude of the propositions I examined, not by feeble conjectures, but by clear and certain reasonings, I met with nothing so doubtful as not to yield some conclusion of adequate certainty, although this were merely the inference, that the matter in question contained nothing certain. And, just as in pulling down an old house, we usually reserve the ruins to contribute towards the erection, so, in destroying such of my opinions as I judged to be ill-founded, I made a variety of observations and acquired an amount of experience of which I availed myself in the establishment of more certain. And further, I continued to exercise myself in the Method I had prescribed; for, besides taking care in general to conduct all my thoughts according to its rules, I reserved some hours from time to time which I expressly devoted to the employment of the Method in the solution of Mathematical difficulties, or even in the solution likewise of some questions belonging to other Sciences, but which, by my having detached them from such principles of these Sciences as were of inadequate certainty, were rendered almost Mathematical: the truth of this will be manifest from the numerous examples contained in this volume.[1] And thus, without in appearance living otherwise than those who, with no other occupation than that of spending their lives agreeably and innocently, study to sever pleasure from vice, and who, that they may enjoy their leisure without ennui, have recourse to such pursuits as are honourable, I was nevertheless prosecuting my design, and making greater progress in the knowledge of truth, than I might, perhaps, have made had I been engaged in the perusal of books merely, or in holding converse with men of letters.

These nine years passed away, however, before I had come to any determinate judgment respecting the difficulties which form matter of dispute among the learned, or had commenced to seek the principles of any Philosophy more certain than the vulgar. And the examples of many men of the highest genius, who had, in former times, engaged in this inquiry, but, as appeared to me, without success, led me to imagine it to be a work of so much difficulty, that I would not perhaps have ventured on it so soon had I not heard it currently rumoured that I had already completed the inquiry. I know not what were the grounds of this opinion; and, if my conversation contributed in any measure to its rise, this must have happened rather from my having confessed my ignorance with greater freedom than those are accustomed to do who have studied a little, and expounded, perhaps, the reasons that led me to doubt of many of those things that by others are esteemed certain, than from my having boasted of any system of Philosophy. But, as I am of a disposition that makes me unwilling to be esteemed different from what I really am, I thought it necessary to endeavour by all means to render myself worthy of the reputation accorded to me; and it is now exactly eight years since this desire constrained me to remove from all those places where interruption from any of my acquaintances was possible, and

[1] The Discourse on Method was originally published along with the Dioptrics, the Meteorics, and the Geometry.—*Tr.*

betake myself to this country,[2] in which the long duration of the war has led to the establishment of such discipline, that the armies maintained seem to be of use only in enabling the inhabitants to enjoy more securely the blessings of peace; and where, in the midst of a great crowd actively engaged in business, and more careful of their own affairs than curious about those of others, I have been enabled to live without being deprived of any of the conveniences to be had in the most populous cities, and yet as solitary and as retired as in the midst of the most remote deserts.

[2] Holland; to which country he withdrew in 1629.—*Tr.*

PART IV

I AM in doubt as to the propriety of making my first meditations in the place above mentioned matter of discourse; for these are so metaphysical, and so uncommon, as not, perhaps, to be acceptable to every one. And yet, that it may be determined whether the foundations that I have laid are sufficiently secure, I find myself in a measure constrained to advert to them. I had long before remarked that, in (relation to) practice, it is sometimes necessary to adopt, as if above doubt, opinions which we discern to be highly uncertain, as has been already said; but as I then desired to give my attention solely to the search after truth, I thought that a procedure exactly the opposite was called for, and that I ought to reject as absolutely false all opinions in regard to which I could suppose the least ground for doubt, in order to ascertain whether after that there remained aught in my belief that was wholly indubitable. Accordingly, seeing that our senses sometimes deceive us, I was willing to suppose that there existed nothing really such as they presented to us; and because some men err in reasoning, and fall into paralogisms, even on the simplest matters of Geometry, I, convinced that I was as open to error as any other, rejected as false all the reasonings I had hitherto taken for demonstrations; and finally, when I considered that the very same thoughts (presentations) which we experience when awake may also be experienced when we are asleep, while there is at that time not one of them true, I supposed that all the objects (presentations) that had ever entered into my mind when awake, had in them no more truth than the illusions of my dreams. But immediately upon this I observed that, whilst I thus wished to think that all was false, it was absolutely necessary that I, who thus thought, should be somewhat; and as I observed that this truth, *I think, hence I am,* was so certain and of such evidence, that no ground of doubt, however extravagant, could be alleged by the Sceptics capable of shaking it, I concluded that I might, without scruple, accept it as the first principle of the Philosophy of which I was in search.

In the next place, I attentively examined what I was, and as I observed that I could suppose that I had no body, and that there was no world nor any place in which I might be; but that I could not therefore suppose that I was not; and that, on the contrary, from the very circumstance that I thought to doubt of the truth of other things, it most clearly and certainly followed that I was; while, on the other hand, if I had only ceased to think, although all the other objects which I had ever imagined had been in reality existent, I would have had no reason to believe that I existed; I thence concluded that I was a substance whose whole essence or nature consists only in thinking, and which, that it may exist, has need of no place, nor is dependent on any material thing; so that "I," that is to say, the mind by which I am what I am, is wholly distinct from the body, and is even more easily known than the latter, and is such, that although the latter were not, it would still continue to be all that it is.

After this I inquired in general into what is essential to the truth and certainty of a proposition; for since I had discovered one which I knew to be true, I thought that I must likewise be able to discover the ground of this certitude. And as I observed that in the words *I think, hence I am,* there is nothing at all which gives me assurance of their truth beyond this, that I see very clearly that in order to think it is necessary to exist, I concluded that I might take, as a general rule, the principle, that all the things which we very clearly and distinctly conceive are true, only observing, however, that there is some difficulty in rightly determining the objects which we distinctly conceive.

In the next place, from reflecting on the circumstance that I doubted, and that consequently my being was not wholly perfect, (for I clearly saw that it was a greater perfection to know than to doubt,) I was led to inquire whence I had learned to think of something more perfect than myself; and I clearly recognised that I must hold this notion from some Nature which in reality was more perfect. As for the thoughts of many other objects external to me, as of the sky, the earth, light, heat, and a thousand more, I was less at a loss to know whence these came; for since I remarked in them nothing which seemed to render them superior to myself, I could believe that, if these were true, they were dependencies on my own nature, in so far as it possessed a certain perfection, and, if they were false, that I held them from nothing, that is to say, that they were in me because of a certain imperfection of my nature. But this could not be the case with the idea of a Nature more perfect than myself; for to receive it from nothing was a thing manifestly impossible; and, because it is not less repugnant that the more perfect should be an effect of, and dependence on the less perfect, than that something should proceed from nothing, it was equally impossible that I could hold it from myself: accordingly, it but remained that it had been placed in me by a Nature which was in reality more perfect than mine, and which even possessed within itself all the perfections of which I could form any idea; that is to say, in a single word, which was God. And to this I added that, since I knew some perfections which I did not possess, I was not the only being in existence, (I will here, with your permission, freely use the terms of the schools); but, on the contrary, that there was of necessity some other more perfect Being upon whom I was dependent, and from whom I had received all that I possessed; for if I had existed alone, and independently of every other being, so as to have had from myself all the perfection, however little, which I actually possessed, I should have been able, for the same reason, to have had from myself the whole remainder of perfection, of the want of which I was conscious, and thus could of myself have become infinite, eternal, immutable, omniscient, all-powerful, and, in fine, have possessed all the perfection which I could recognise in God. For in order to know the nature of God, (whose existence has been established by the preceding reasonings,) as far as my own nature permitted, I had only to consider in reference to all the properties of which I found in my mind some idea, whether their possession was a mark of perfection; and I was assured that no one which indicated any imperfection was in him, and that none of the rest was awanting. Thus I perceived that doubt, inconstancy, sadness, and such like, could not be found in God, since I myself would have been happy to be free from them. Besides, I had ideas of many sensible and corporeal things; for although I might suppose that

I was dreaming, and that all which I saw or imagined was false, I could not, nevertheless, deny that the ideas were in reality in my thoughts. But, because I had already very clearly recognised in myself that the intelligent nature is distinct from the corporeal, and as I observed that all composition is an evidence of dependency, and that a state of dependency is manifestly a state of imperfection, I therefore determined that it could not be a perfection in God to be compounded of these two natures, and that consequently he was not so compounded; but that if there were any bodies in the world, or even any intelligences, or other natures that were not wholly perfect, their existence depended on his power in such a way that they could not subsist without him for a single moment.

I was disposed straightway to search for other truths; and when I had represented to myself the object of the geometers, which I conceived to be a continuous body, or a space indefinitely extended in length, breadth, and height or depth, divisible into divers parts which admit of different figures and sizes, and of being moved or transposed in all manner of ways, (for all this the geometers suppose to be in the object they contemplate,) I went over some of their simplest demonstrations. And, in the first place, I observed, that the great certitude which by common consent is accorded to these demonstrations, is founded solely upon this, that they are clearly conceived in accordance with the rules I have already laid down. In the next place, I perceived that there was nothing at all in these demonstrations which could assure me of the existence of their object: thus, for example, supposing a triangle to be given, I distinctly perceived that its three angles were necessarily equal to two right angles, but I did not on that account perceive anything which could assure me that any triangle existed: while, on the contrary, recurring to the examination of the idea of a Perfect Being, I found that the existence of the Being was comprised in the idea in the same way that the equality of its three angles to two right angles is comprised in the idea of a triangle, or as in the idea of a sphere, the equidistance of all points on its surface from the centre, or even still more clearly; and that consequently it is at least as certain that God, who is this Perfect Being, is, or exists, as any demonstration of Geometry can be.

But the reason which leads many to persuade themselves that there is a difficulty in knowing this truth, and even also in knowing what their mind really is, is that they never raise their thoughts above sensible objects, and are so accustomed to consider nothing except by way of imagination, which is a mode of thinking limited to material objects, that all that is not imaginable seems to them not intelligible. The truth of this is sufficiently manifest from the single circumstance, that the philosophers of the Schools accept as a maxim that there is nothing in the Understanding which was not previously in the Senses, in which however it is certain that the ideas of God and of the soul have never been; and it appears to me that they who make use of their imagination to comprehend these ideas do exactly the same thing as if, in order to hear sounds or smell odours, they strove to avail themselves of their eyes; unless indeed that there is this difference, that the sense of sight does not afford us an inferior assurance to those of smell or hearing; in place of which, neither our imagination nor our senses can give us assurance of anything unless our Understanding intervene.

Finally, if there be, still persons who are not sufficiently persuaded of the existence of God and of the soul, by the reasons I have adduced, I am desirous that they should know that all the other propositions, of the truth of which they deem themselves perhaps more assured, as that we have a body, and that there exist stars and an earth, and such like, are less certain; for, although we have a moral assurance of these things, which is so strong that there is an appearance of extravagance in doubting of their existence, yet at the same time no one, unless his intellect is impaired, can deny, when the question relates to a metaphysical certitude, that there is sufficient reason to exclude entire assurance, in the observation that when asleep we can in the same way imagine ourselves possessed of another body and that we see other stars and another earth, when there is nothing of the kind. For how do we know that the thoughts which occur in dreaming are false rather than those other which we experience when awake, since the former are often not less vivid and distinct than the latter? And though men of the highest genius study this question as long as they please, I do not believe that they will be able to give any reason which can be sufficient to remove this doubt, unless they presuppose the existence of God. For, in the first place, even the principle which I have already taken as a rule, viz., that all the things which we clearly and distinctly conceive are true, is certain only because God is or exists, and because he is a Perfect Being, and because all that we possess is derived from him: whence it follows that our ideas or notions, which to the extent of their clearness and distinctness are real, and proceed from God, must to that extent be true. Accordingly, whereas we not unfrequently have ideas or notions in which some falsity is contained, this can only be the case with such as are to some extent confused and obscure, and in this proceed from nothing, (participate of negation,) that is, exist in us thus confused because we are not wholly perfect. And it is evident that it is not less repugnant that falsity or imperfection, in so far as it is imperfection, should proceed from God, than that truth or perfection should proceed from nothing. But if we did not know that all which we possess of real and true proceeds from a Perfect and Infinite Being, however clear and distinct our ideas might be, we should have no ground on that account for the assurance that they possessed the perfection of being true.

But after the knowledge of God and of the soul has rendered us certain of this rule, we can easily understand that the truth of the thoughts we experience when awake, ought not in the slightest degree to be called in question on account of the illusions of our dreams. For if it happened that an individual, even when asleep, had some very distinct idea, as, for example, if a geometer should discover some new demonstration, the circumstance of his being asleep would not militate against its truth; and as for the most ordinary error of our dreams, which consists in their representing to us various objects in the same way as our external senses, this is not prejudicial, since it leads us very properly to suspect the truth of the ideas of sense; for we are not unfrequently deceived in the same manner when awake; as when persons in the jaundice see all objects yellow, or when the stars or bodies at a great distance appear to us much smaller than they are. For, in fine, whether awake or asleep, we ought never to allow ourselves to be persuaded of the truth of anything unless on the evidence of our Reason. And it must be noted that I say of our Reason, and not of our imagination or of our senses: thus, for example, although we very clearly see the sun, we ought not therefore to determine that it is only of the size which our sense of sight presents; and we may very distinctly imagine the head of a lion joined to the body of a goat, without being therefore shut up to the conclusion that a chimæra exists; for it is not a dictate of Reason that what we thus see or imagine is in reality existent; but it plainly tells us that all our ideas or notions contain in them some truth; for otherwise it could not be that God, who is wholly perfect and veracious, should have placed them in us. And because our reasonings are never so clear or so complete during sleep as when we are awake, although sometimes the acts of our imagination are then as lively and distinct, if not more so than in our waking moments, Reason further dictates that, since all our thoughts cannot be true because of our partial imperfection, those possessing truth must infallibly be found in the experience of our waking moments rather than in that of our dreams.

PART V

I WOULD here willingly have proceeded to exhibit the whole chain of truths which I deduced from these primary; but as with a view to this it would have been necessary now to treat of many questions in dispute among the learned, with whom I do not wish to be embroiled, I believe that it will be better for me to refrain from this exposition, and only mention in general what these truths are, that the more judicious may be able to determine whether a more special account of them would conduce to the public advantage. I have ever remained firm in my original resolution to suppose no other principle than that of which I have recently availed myself in demonstrating the existence of God and of the soul, and to accept as true nothing that did not appear to me more clear and certain than the demonstrations of the geometers had formerly appeared; and yet I venture to state that not only have I found means to satisfy myself in a short time on all the principal difficulties which are usually treated of in Philosophy, but I have also observed certain laws established in nature by God in such a manner, and of which he has impressed on our minds such notions, that after we have reflected sufficiently upon these, we cannot doubt that they are accurately observed in all that exists or takes place in the world: and farther, by considering the concatenation of these laws, it appears to me that I have discovered many truths more useful and more important than all I had before learned, or even had expected to learn.

But because I have essayed to expound the chief of these discoveries in a Treatise which certain considerations prevent me from publishing, I cannot make the results known more conveniently than by here giving a summary of the contents of this Treatise. It was my design to comprise in it all that, before I set myself to write it, I thought I knew of the nature of material objects. But like the painters who, finding themselves unable to represent equally well on a plain surface all the different faces of a solid body, select one of the chief, on which alone they make the light fall, and throwing the rest into the shade, allow them to appear only in so far as they can be seen while looking at the principal one; so, fearing lest I should not be able to comprise in my discourse all that was in my mind, I resolved to expound singly, though at considerable length, my opinions regarding light; then to take the opportunity of adding something on the sun and the fixed stars, since light almost wholly proceeds from them; on the heavens since they transmit it; on the planets, comets, and earth, since they reflect it; and particularly on all the bodies that are upon the earth, since they are either coloured, or transparent, or luminous; and finally on man, since he is the spectator of these objects. Further, to enable me to cast this variety of subjects somewhat into the shade, and to express my judgment regarding them with greater freedom, without being necessitated to adopt or refute the opinions of the learned, I resolved to leave all the people here to their disputes, and to speak only of what would happen in a new world, if God were now to create somewhere in the imaginary spaces matter sufficient to compose one, and were to agitate variously and confusedly the different parts of this matter, so that there resulted a chaos as disordered as the poets ever feigned, and after that did nothing more than lend his ordinary concurrence to nature, and allow her to act in accordance with the laws which he had established. On this supposition, I, in the first place, described this matter, and essayed to represent it in such a manner that to my mind there can be nothing clearer and more intelligible, except what has been recently said regarding God and the soul; for I even expressly supposed that it possessed none of those forms or qualities which are so debated in the Schools, nor in general anything the knowledge of which is not so natural to our minds that no one can so much as imagine himself ignorant of it. Besides, I have pointed out what are the laws of nature; and, with no other principle upon which to found my reasonings except the infinite perfection of God, I endeavoured to demonstrate all those about which there could be any room for doubt, and to prove that they are such, that even if God had created more worlds, there could have been none in which these laws were not observed.

Thereafter, I showed how the greatest part of the matter of this chaos must, in accordance with these laws, dispose and arrange itself in such a way as to present the appearance of heavens; how in the meantime some of its parts must compose an earth and some planets and comets, and others a sun and fixed stars. And, making a digression at this stage on the subject of light, I expounded at considerable length what the nature of that light must be which is found in the sun and the stars, and how thence in an instant of time it traverses the immense spaces of the heavens, and how from the planets and comets it is reflected towards the earth. To this I likewise added much respecting the substance, the situation, the motions, and all the different qualities of these heavens and stars; so that I thought I had said enough respecting them to show that there is nothing observable in the heavens or stars of our system that must not, or at least may not appear precisely alike in those of the system which I described. I came next to speak of the earth in particular, and to show how, even though I had expressly supposed that God had given no weight to the matter of which it is composed, this should not prevent all its parts from tending exactly to its centre; how with water and air on its surface, the disposition of the heavens and heavenly bodies, more especially of the moon, must cause a flow and ebb, like in all its circumstances to that observed in our seas, as also a certain current both of water and air from east to west, such as is likewise observed between the tropics; how the mountains, seas, fountains, and rivers might naturally be formed in it, and the metals produced in the mines, and the plants grow in the fields; and in general, how all the bodies which are commonly denominated mixed or composite might be generated: and, among other things in the discoveries alluded to, inasmuch as besides the stars, I knew nothing except fire which produces light, I spared no pains to set forth all that pertains to its nature,—the manner of its production and support, and to explain how heat is sometimes found without light, and light without heat; to show how it can induce various colours upon different bodies and other diverse qualities; how it reduces some to a liquid state and hardens others; how it can consume almost all bodies, or convert them into ashes and smoke; and finally, how from these ashes, by the mere intensity of its action, it forms glass: for as this transmutation of ashes into glass appeared to me as wonderful as any other in nature, I took a special pleasure in describing it.

I was not, however, disposed, from these circumstances, to conclude that this world had been created in the manner I described; for it is much more likely that God made it at the first such as it was to be. But this is certain, and an opinion commonly received among theologians, that the action by which he now sustains it is the same with that by which he originally created it; so that even although he had from the beginning given it no other form than that of chaos, provided only he had established certain laws of nature, and had lent it his concurrence to enable it to act as it is wont to do, it may be believed, without discredit to the miracle of creation, that, in this way alone, things purely material might, in course of time, have become such as we observe them at present; and their nature is much more easily conceived when they are beheld coming in this manner gradually into existence, than when they are only considered as produced at once in a finished and perfect state.

From the description of inanimate bodies and plants, I passed to animals, and particularly to man. But since I had not as yet sufficient knowledge to enable me to treat of these in the same manner as of the rest, that is to say, by deducing effects from their causes, and by showing from what elements and in what manner Nature must produce them, I remained satisfied with the supposition that God formed the body of man wholly like to one of ours, as well in the external shape of the members as in the internal conformation of the organs, of the same matter with that I had described, and at first placed in it no Rational Soul, nor any other principle, in room of the Vegetative or Sensitive Soul, beyond kindling in the heart one of those fires without light, such as I had already described, and which I thought was not different from the heat in hay that has been heaped together before it is dry, or that which causes fermentation in new wines before they are run clear of the fruit. For, when I examined the kind of functions which might, as consequences of this supposition, exist in this body, I found precisely all those which may exist in us independently of all power of thinking, and consequently without being in any measure owing to the soul; in other words, to that part of us which is distinct from the body, and of which it has been said above that the nature distinctively consists in thinking,—functions in which the animals void of Reason may be said wholly to resemble us; but among which I could not discover any of those that, as dependent on thought alone, belong to us as men, while, on the other hand, I did afterwards discover these as soon as I supposed God to have created a Rational Soul, and to have annexed it to this body in a particular manner which I described.

But, in order to show how I there handled this matter, I mean here to give the explication of the motion of the heart and arteries, which, as the first and most general motion observed in animals, will afford the means of readily determining what should be thought of all the rest. And that there may be less difficulty in understanding what I am about to say on this subject, I advise those who are not versed in Anatomy, before they commence the perusal of these observations, to take the trouble of getting dissected in their presence the heart of some large animal possessed of lungs, (for this is throughout sufficiently like the human,) and to have shewn to them its two ventricles or cavities: in the first place, that in the right side, with which correspond two very ample tubes, viz., the hollow vein, (*vena cava,*) which is the principal receptacle of the blood, and the trunk of the tree, as it were, of which all the other veins in the body are branches; and the arterial vein, (*vena arteriosa,*) inappropriately so denominated, since it is in truth only an artery, which, taking its rise in the heart, is divided, after passing out from it, into many branches which presently disperse themselves all over the lungs; in the second place, the cavity in the left side, with which correspond in the same manner two canals in size equal to or larger than the preceding, viz., the venous artery, (*arteria venosa,*) likewise inappropriately thus designated, because it is simply a vein which comes from the lungs, where it is divided into many branches, interlaced with those of the arterial vein, and those of the tube called the windpipe, through which the air we breathe enters; and the great artery which, issuing from the heart, sends it branches all over the body. I should wish also that such persons were carefully shewn the eleven pellicles which, like so many small valves, open and shut the four orifices that are in these two cavities, viz., three at the entrance of the hollow vein, where they are disposed in such a manner as by no means to prevent the blood which it contains from flowing into the right ventricle of the heart, and yet exactly to prevent its flowing out; three at the entrance to the arterial vein, which, arranged in a manner exactly the opposite of the former, readily permit the blood contained in this cavity to pass into the lungs, but hinder that contained in the lungs from returning to this cavity; and, in like manner, two others at the mouth of the venous artery, which allow the blood from the lungs to flow into the left cavity of the heart, but preclude its return; and three at the mouth of the great artery, which suffer the blood to flow from the heart, but prevent its reflux. Nor do we need to seek any other reason for the number of these pellicles beyond this that the orifice of the venous artery being of an oval shape from the nature of its situation, can be adequately closed with two, whereas the others being round are more conveniently closed with three. Besides, I wish such persons to observe that the grand artery and the arterial vein are of much harder and firmer texture than the venous artery and the hollow vein; and that the two last expand before entering the heart, and there form, as it were, two pouches denominated the auricles of the heart, which are composed of a substance similar to that of the heart itself; and that there is always more warmth in the heart than in any other part of the body; and, finally, that this heat is capable of causing any drop of blood that passes into the cavities rapidly to expand and dilate, just as all liquors do when allowed to fall drop by drop into a highly heated vessel.

For, after these things, it is not necessary for me to say anything more with a view to explain the motion of the heart, except that when its cavities are not full of blood, into these the blood of necessity flows,—from the hollow vein into the right, and from the venous artery into the left; because these two vessels are always full of blood, and their orifices, which are turned towards the heart, cannot then be closed. But as soon as two drops of blood have thus passed, one into each of the cavities, these drops which cannot but be very large, because the orifices through which they pass are wide, and the vessels from which they come full of blood, are immediately rarefied, and dilated by the heat they meet with. In this way they cause the whole heart to expand, and at the same time press home and shut the five small valves that are at the entrances of the two vessels from which they flow, and thus prevent any more blood from coming down into the heart, and becoming more and more rarefied, they push open the six small valves that are in the orifices of the other two vessels, through which they pass out, causing in this way all the branches of the arterial vein and of the grand artery to expand almost simultaneously with the heart—which immediately thereafter begins to contract, as do also the arteries, because the blood that has entered them has cooled, and the six small valves close, and the five of the hollow vein and of the venous artery open anew and allow a passage to other two drops of blood, which cause the heart and the arteries again to expand as before. And, because the blood which thus enters into the heart passes through these two pouches called auricles, it thence happens that their motion is the contrary of that of the heart, and that when it expands they contract. But lest those who are ignorant of the force of mathematical demonstrations, and who are not accustomed to distinguish true reasons from mere verisimilitudes, should venture, without examination, to deny what has been said, I wish it to be considered that the motion which I have now explained follows as necessarily from the very arrangement of the parts, which may be observed in the heart by the eye alone, and from the heat which may be felt with the fingers, and from the nature of the blood as learned from experience, as does the motion of a clock from the power, the situation, and shape of its counter-weights and wheels.

But if it be asked how it happens that the blood in the veins, flowing in this way continually into the heart, is not exhausted, and why the arteries do not become too full, since all the blood which passes through the heart flows into them, I need only mention in reply what has been written by a physician[1] of England, who has the honour of having broken the ice on this subject, and of having been the first to teach that there are many small passages at the extremities of the arteries, through which the blood received by them from the heart passes into the small branches of the veins, whence it again returns to the heart; so that its course amounts

[1] Harvey.

precisely to a perpetual circulation. Of this we have abundant proof in the ordinary experience of surgeons, who, by binding the arm with a tie of moderate straitness above the part where they open the vein, cause the blood to flow more copiously than it would have done without any ligature; whereas quite the contrary would happen were they to bind it below; that is, between the hand and the opening, or were to make the ligature above the opening very tight. For it is manifest that the tie, moderately straitened, while adequate to hinder the blood already in the arm from returning towards the heart by the veins, cannot on that account prevent new blood from coming forward through the arteries, because these are situated below the veins, and their coverings, from their greater consistency, are more difficult to compress; and also that the blood which comes from the heart tends to pass through them to the hand with greater force than it does to return from the hand to the heart through the veins. And since the latter current escapes from the arm by the opening made in one of the veins, there must of necessity be certain passages below the ligature, that is, towards the extremities of the arm through which it can come thither from the arteries. This physician likewise abundantly establishes what he has advanced respecting the motion of the blood, from the existence of certain pellicles, so disposed in various places along the course of the veins, in the manner of small valves, as not to permit the blood to pass from the middle of the body towards the extremities, but only to return from the extremities to the heart; and farther, from experience which shows that all the blood which is in the body may flow out of it in a very short time through a single artery that has been cut, even although this had been closely tied in the immediate neighbourhood of the heart, and cut between the heart and the ligature, so as to prevent the supposition that the blood flowing out of it could come from any other quarter than the heart.

But there are many other circumstances which evince that what I have alleged is the true cause of the motion of the blood: thus, in the first place, the difference that is observed between the blood which flows from the veins, and that from the arteries, can only arise from this, that being rarefied, and, as it were, distilled by passing through the heart, it is thinner, and more vivid, and warmer immediately after leaving the heart, in other words, when in the arteries, than it was a short time before passing into either, in other words, when it was in the veins; and if attention be given, it will be found that this difference is very marked only in the neighbourhood of the heart; and is not so evident in parts more remote from it. In the next place, the consistency of the coats of which the arterial vein and the great artery are composed, sufficiently shows that the blood is impelled against them with more force than against the veins. And why should the left cavity of the heart and the great artery be wider and larger than the right cavity and the arterial vein, were it not that the blood of the venous artery, having only been in the lungs after it has passed through the heart, is thinner, and rarefies more readily, and in a higher degree, than the blood which proceeds immediately from the hollow vein? And what can physicians conjecture from feeling the pulse unless they know that according as the blood changes its nature it can be rarefied by the warmth of the heart, in a higher or lower degree, and more or less quickly than before? And if it be inquired how this heat is communicated to the other members, must it not be admitted that this is effected by means of the blood, which, passing through the heart, is there heated anew, and thence diffused over all the body? Whence it happens, that if the blood be withdrawn from any part, the heat is likewise withdrawn by the same means; and although the heart were as hot as glowing iron, it would not be capable of warming the feet and hands as at present, unless it continually sent thither new blood. We likewise perceive from this, that the true use of respiration is to bring sufficient fresh air into the lungs, to cause the blood which flows into them from the right ventricle of the heart, where it has been rarefied and, as it were, changed into vapours, to become thick, and to convert it anew into blood, before it flows into the left cavity, without which process it would be unfit for the nourishment of the fire that is there. This receives confirmation from the circumstance, that it is observed of animals destitute of lungs that they have also but one cavity in the heart, and that in children who cannot use them while in the womb, there is a hole through which the blood flows from the hollow vein into the left cavity of the heart, and a tube through which it passes from the arterial vein into the grand artery without passing through the lung. In the next place, how could digestion be carried on in the stomach unless the heart communicated heat to it through the arteries, and along with this certain of the more fluid parts of the blood, which assists in the dissolution of the food that has been taken in? Is not also the operation which converts the juice of food into blood easily comprehended, when it is considered that it is distilled by passing and repassing through the heart perhaps more than one or two hundred times in a day? And what more need be adduced to explain nutrition, and the production of the different humours of the body, beyond saying, that the force with which the blood, in being rarefied, passes from the heart towards the extremities of the arteries, causes certain of its parts to remain in the members at which they arrive, and there occupy the place of some others expelled by them; and that according to the situation, shape, or smallness of the pores with which they meet, some rather than others flow into certain parts, in the same way that some sieves are observed to act, which, by being variously perforated, serve to separate different species of grain? And, in the last place, what above all is here worthy of observation, is the generation of the animal spirits, which are like a very subtle wind, or rather a very pure and vivid flame which, continually ascending in great abundance from the heart to the brain, thence penetrates through the nerves into the muscles, and gives motion to all the members; so that to account for other parts of the blood which, as most agitated and penetrating, are the fittest to compose these spirits, proceeding towards the brain, it is not necessary to suppose any other cause, than simply, that the arteries which carry them thither proceed from the heart in the most direct lines, and that, according to the rules of Mechanics, which are the same with those of Nature, when many objects tend at once to the same point where there is not sufficient room for all, (as is the case with the parts of the blood which flow forth from the left cavity of the heart and tend towards the brain,) the weaker and less agitated parts must necessarily be driven aside from that point by the stronger which alone in this way reach it.

I had expounded all these matters with sufficient minuteness in the Treatise which I formerly thought of publishing. And after these, I had shewn what must be the fabric of the nerves and muscles of the human body to give the animal spirits contained in it the power to move the members, as when we see heads shortly after they have been struck off still move and bite the earth, although no longer animated; what changes must take place in the brain to produce waking, sleep, and dreams; how light, sounds, odours, tastes, heat, and all the other qualities of external objects impress it with different ideas by means of the senses; how hunger, thirst, and the other internal affections can likewise impress upon it divers ideas; what must be understood by the common sense (*sensus communis*) in which these ideas are received, by the memory which retains them, by the fantasy which can change them in various ways, and out of them compose new ideas, and which, by the same means, distributing the animal spirits through the muscles, can cause the members of such a body to move in as many different ways, and in a manner as suited, whether to the objects that are presented to its senses or to its internal affections, as can take place in our own case apart from the guidance of the will. Nor will this appear at all strange to those who are acquainted with the variety of movements performed by the different automata, or moving machines fabricated by human industry, and that with help of but few pieces compared with the great multitude of bones, muscles, nerves, arteries, veins, and other parts that are found in the body of each animal. Such persons will look upon this body as a machine made by the hands of God, which is incomparably better arranged, and adequate to movements more admirable than is any machine of human invention. And here I specially stayed to show that, were there such machines exactly resembling in organs and outward form an ape or any other irrational animal, we could have no means of knowing that they were in any respect of a different nature from these animals; but if there were machines bearing the image of our bodies, and capable of imitating our actions as far as it is morally possible, there would still remain two most certain tests whereby to know that they were not therefore really men. Of these the first is that they could never use words or other signs arranged in such a manner as is competent to us in order to declare our thoughts to others: for we may easily conceive a machine to be so constructed that it emits vocables, and even that it emits some correspondent to the action upon it of external objects which cause a change in its organs; for example, if touched in a particular place it may demand what we wish to say to it; if in another it may cry out that it is hurt, and such like; but not that it should arrange them variously so as appositely to reply to what is said in its presence, as men of the lowest grade of intellect can do. The second test is, that although such machines might execute many things with equal or perhaps greater perfection than any of us, they would, without doubt, fail in certain others from which it could be discovered that they did not act from knowledge, but solely from the disposition of their organs: for while Reason is an universal instrument that is alike available on every occasion, these organs, on the contrary, need a particular arrangement for each particular action; whence it must be morally impossible that there should exist in any machine a diversity of organs sufficient to enable it to act in all the occurrences of life, in the way in which our reason enables us to act. Again, by means of these two tests we may likewise know the difference between men and brutes. For it is highly deserving of remark, that there are no men so dull and stupid, not even idiots, as to be incapable of joining together different words, and thereby constructing a declaration by which to make their thoughts understood; and that on the other hand, there is no other animal, however perfect or happily circumstanced which can do the like. Nor does this inability arise from want of organs: for we observe that magpies and parrots can utter words like ourselves, and are yet unable to speak as we do, that is, so as to show that they understand what they say; in place of which men born deaf and dumb, and thus not less, but rather more than the brutes, destitute of the organs which others use in speaking, are in the habit of spontaneously inventing certain signs by which they discover their thoughts to those who, being usually in their company, have leisure to learn their language. And this proves not only that the brutes have less Reason than man, but that they have none at all: for we see that very little is required to enable a person to speak; and since a certain inequality of capacity is observable among animals of the same species, as well as among men, and since some are more capable of being instructed than others, it is incredible that the most perfect ape or parrot of its species, should not in this be equal to the most stupid infant of its kind, or at least to one that was crack-brained, unless the soul of brutes were of a nature wholly different from ours. And we ought not to confound speech with the natural movements which indicate the passions, and can be imitated by machines as well as manifested by animals; nor must it be thought with certain of the ancients, that the brutes speak, although we do not understand their language. For if such were the case, since they are endowed with many organs analogous to ours, they could as easily communicate their thoughts to us as to their fellows. It is also very worthy of remark, that, though there are many animals which manifest more industry than we in certain of their actions, the same animals are yet observed to show none at all in many others: so that the circumstance that they do better than we does not prove that they are endowed with mind, for it would thence follow that they possessed greater Reason than any of us, and could surpass us in all things; on the contrary, it rather proves that they are destitute of Reason, and that it is Nature which acts in them according to the disposition of their organs: thus it is seen, that a clock composed only of wheels and weights can number the hours and measure time more exactly than we with all our skill.

I had after this described the Reasonable Soul, and shewn that it could by no means be educed from the power of matter, as the other things of which I had spoken but that it must be expressly created; and that it is not sufficient that it be lodged in the human body exactly like a pilot in a ship, unless perhaps to move its members, but that it is necessary for it to be joined and united more closely to the body, in order to have sensations and appetites similar to ours, and thus constitute a real man. I here entered, in conclusion, upon the subject of the soul at considerable length, because it is of the greatest moment: for after the error of those who deny the existence of God, an error which I think I have already sufficiently refuted, there is none that is more powerful in leading feeble minds astray from the straight path of virtue than the supposition that the soul of the brutes is of the same nature with our own; and consequently that after this life we have nothing to hope for or fear, more than flies and ants; in place of which, when we know how far they differ we much better comprehend the reasons which establish that the soul is of a nature wholly independent of the body, and that consequently it is not liable to die with the latter; and, finally, because no other causes are observed capable of destroying it, we are naturally led thence to judge that it is immortal.

PART VI

THREE years have now elapsed since I finished the Treatise containing all these matters; and I was beginning to revise it, with the view to put it into the hands of a printer, when I learned that persons to whom I greatly defer, and whose authority over my actions is hardly less influential than is my own Reason over my thoughts, had condemned a certain doctrine in Physics, published a short time previously by another individual,[1] to which I will not say that I adhered, but only that, previously to their censure, I had observed in it nothing which I could imagine to be prejudicial either to religion or to the state, and nothing therefore which would have prevented me from giving expression to it in writing, if Reason had persuaded me of its truth; and this led me to fear lest among my own doctrines likewise some one might be found in which I had departed from the truth, notwithstanding the great care I have always taken not to accord belief to new opinions of which I had not the most certain demonstrations, and not to give expression to aught that might tend to the hurt of any one. This has been sufficient to make me alter my purpose of publishing them; for although the reasons by which I had been induced to take this resolution were very strong, yet my inclination, which has always been hostile to writing books, enabled me immediately to discover other considerations sufficient to excuse me for not undertaking the task. And these reasons, on one side and the other, are such, that not only is it in some measure my interest here to state them, but that of the public, perhaps, to know them.

I have never made much account of what has proceeded from my own mind; and so long as I gathered no other advantage from the Method I employ beyond satisfying myself on some difficulties belonging to the speculative sciences, or endeavouring to regulate my actions according to the principles it taught me, I never thought myself bound to publish anything respecting it. For in what

[1] Galileo.—*Tr.*

regards manners, every one is so full of his own wisdom, that there might be found as many reformers as heads, if any were allowed to take upon themselves the task of mending them, except those whom God has constituted the supreme rulers of his people, or to whom he has given sufficient grace and zeal to be prophets; and although my speculations greatly pleased myself, I believed that others had theirs, which perhaps pleased them still more. But as soon as I had acquired some general notions respecting Physics, and beginning to make trial of them in various particular difficulties, had observed how far they can carry us, and how much they differ from the principles that have been employed up to the present time, I believed that I could not keep them concealed without sinning grievously against the law by which we are bound to promote, as far as in us lies, the general good of mankind. For by them I perceived it to be possible to arrive at knowledge highly useful in life; and in room of the Speculative Philosophy usually taught in the Schools, to discover a Practical, by means of which, knowing the force and action of fire, water, air, the stars, the heavens, and all the other bodies that surround us, as distinctly as we know the various crafts of our artizans, we might also apply them in the same way to all the uses to which they are adapted, and thus render ourselves the lords and possessors of nature. And this is a result to be desired, not only in order to the invention of an infinity of arts, by which we might be enabled to enjoy without any trouble the fruits of the earth, and all its comforts, but also and especially for the preservation of health, which is without doubt, of all the blessings of this life, the first and fundamental one; for the mind is so intimately dependent upon the condition and relation of the organs of the body, that if any means can ever be found to render men wiser and more ingenious than hitherto, I believe that it is in Medicine they must be sought for. It is true that the science of Medicine, as it now exists, contains few things whose utility is very remarkable: but without any wish to depreciate it, I am confident that there is no one, even among those whose profession it is, who does not admit that all at present known in it is almost nothing in comparison of what remains to be discovered; and that we could free ourselves from an infinity of maladies of body as well as of mind, and perhaps also even from the debility of age, if we had sufficiently ample knowledge of their causes, and of all the remedies provided for us by Nature. But since I designed to employ my whole life in the search after so necessary a Science, and since I had fallen in with a path which seems to me such, that if any one follow it he must inevitably reach the end desired, unless he be hindered either by the shortness of life or the want of experiments, I judged that there could be no more effectual provision against these two impediments than if I were faithfully to communicate to the public all the little I might myself have found, and incite men of superior genius to strive to proceed farther, by contributing, each according to his inclination and ability, to the experiments which it would be necessary to make, and also by informing the public of all they might discover, so that, by the last beginning where those before them had left off, and thus connecting the lives and labours of many, we might collectively proceed much farther than each by himself could do.

I remarked, moreover, with respect to experiments, that they become always more necessary the more one is advanced in knowledge; for, at the commencement, it is better to make use only of what is spontaneously presented to our senses, and of which we cannot remain ignorant, provided we bestow on it any reflection, however slight, than to concern ourselves about more uncommon and recondite phænomena: the reason of which is, that the more uncommon often only mislead us so long as the causes of the more ordinary are still unknown; and the circumstances upon which they depend are almost always so special and minute as to be highly difficult to detect. But in this I have adopted the following order: first, I have essayed to find in general the principles, or first causes of all that is or can be in the world, without taking into consideration for this end anything but God himself who has created it, and without educing them from any other source than from certain germs of truths naturally existing in our minds. In the second place, I examined what were the first and most ordinary effects that could be deduced from these causes; and it appears to me that, in this way I have found heavens, stars, an earth, and even on the earth, water, air, fire, minerals, and some other things of this kind, which of all others are the most common and simple, and hence the easiest to know. Afterwards, when I wished to descend to the more particular, so many diverse objects presented themselves to me, that I believed it to be impossible for the human mind to distinguish the forms or species of bodies that are upon the earth, from an infinity of others which might have been, if it had pleased God to place them there, or consequently to apply them to our use, unless we rise to causes through their effects, and avail ourselves of many particular experiments. Thereupon, turning over in my mind all the objects that had ever been presented to my senses, I freely venture to state that I have never observed any which I could not satisfactorily explain by the principles I had discovered. But it is necessary also to confess that the power of nature is so ample and vast, and these principles so simple and general, that I have hardly observed a single particular effect which I cannot at once recognize as capable of being deduced in many different modes from the principles, and that my greatest difficulty usually is to discover in which of these modes the effect is dependent upon them; for out of this difficulty I cannot otherwise extricate myself than by again seeking certain experiments, which may be such that their result is not the same, if it is in the one of these modes that we must explain it, as it would be if it were to be explained in the other. As to what remains, I am now in a position to discern, as I think, with sufficient clearness what course must be taken to make the majority of those experiments which may conduce to this end: but I perceive likewise that they are such and so numerous, that neither my hands nor my income, though it were a thousand times larger than it is, would be sufficient for them all; so that, according as henceforward I shall have the means of making more or fewer experiments, I shall in the same proportion make greater or less progress in the knowledge of nature. This was what I had hoped to make known by the Treatise I had written, and so clearly to exhibit the advantage that would thence accrue to the public, as to induce all who have the common good of man at heart, that is, all who are virtuous in truth, and not merely in appearance, or according to opinion, as well to communicate to me the experiments they had already made, as to assist me in those that remain to be made.

But since that time other reasons have occurred to me, by which I have been led to change my opinion, and to think that I ought indeed to go on committing to writing all the results which I deemed of any moment, as soon as I should have tested their truth, and to bestow the same care upon them as I would have done had it been my design to publish them. This course commended itself to me, as well because I thus afforded myself more ample inducement to examine them thoroughly, for doubtless that is always more narrowly scrutinized which we believe will be read by many, than that which is written merely for our private use, (and frequently what has seemed to me true when I first conceived it, has appeared false when I have set about committing it to writing;) as because I thus lost no opportunity of advancing the interests of the public, as far as in me lay, and since thus likewise, if my writings possess any value, those into whose hands they may fall after my death may be able to put them to what use they deem proper. But I resolved by no means to consent to their publication during my lifetime, lest either the oppositions or the controversies to which they might give rise, or even the reputation, such as it might be, which they would acquire for me, should be any occasion of my losing the time that I had set apart for my own improvement. For though it be true that every one is bound to promote to the extent of his ability the good of others, and that to be useful to no one is really to be worthless, yet it is likewise true that our cares ought to extend beyond the present; and it is good to omit doing what might perhaps bring some profit to the living, when we have in view the accomplishment of other ends that will be of much greater advantage to posterity. And in truth, I am quite willing it should be known that the little I have hitherto learned is almost nothing in comparison with that of which I am ignorant, and to the knowledge of which I do not despair of being able to attain; for it is much the same with those who gradually discover truth in the Sciences, as with those who when growing rich find less difficulty in making great acquisitions, than they formerly experienced when poor in making acquisitions of much smaller amount. Or they may be compared to the commanders of armies, whose forces usually increase in proportion to their victories, and who need greater prudence to keep together the residue of their troops after a defeat than after a victory to take towns and provinces. For he truly engages in battle who endeavors to surmount all the difficulties and errors which prevent him from reaching the knowledge of truth, and he is overcome in fight who admits a false opinion touching a matter of any generality and importance, and he requires thereafter much more skill to recover his former position than to make great advances when once in possession of thoroughly ascertained principles. As for myself, if I have succeeded in discovering any truths in the Sciences, (and I trust that what is contained in this volume will show that I have found some,) I can declare that they are but the consequences and results of five or six principal difficulties which I have surmounted, and my encounters with which I reckoned as battles in which victory declared for me. I will not hesitate even to avow my belief that nothing further is wanting to enable me fully to realize my designs than to gain two or three similar victories; and that I am not so far advanced in years but that, according to the ordinary course of nature, I may still have sufficient leisure for this end. But I conceive myself the more bound to husband the time that remains, the greater my expectation of being able to employ it aright, and I should doubtless have much to rob me of it, were I to publish the principles of my Physics: for although they are almost all so evident that to assent to them no more is needed than simply to understand them, and although there is not one of them of which I do not expect to be able to give demonstration, yet, as it is impossible that they can be in accordance with all the diverse opinions of others, I foresee that I should frequently be turned aside from my grand design, on occasion of the opposition which they would be sure to awaken.

It may be said, that these oppositions would be useful both in making me aware of my errors, and, if my speculations contain anything of value, in bringing others to a fuller understanding of it; and still farther, as many can see better than one, in leading others who are now beginning to avail themselves of my principles, to assist me in turn with their discoveries. But though I recognise my extreme liability to error, and scarce ever trust to the first thoughts which occur to me, yet the experience I have had of possible objections to my views prevents me from anticipating any profit from them. For I have already had frequent proof of the judgments, as well of those I esteemed friends, as of some others to whom I thought I was an object of indifference, and even of some whose malignity and envy would, I knew, determine them to endeavour to discover what partiality concealed from the eyes of my friends. But it has rarely happened that anything has been objected to me which I had myself altogether overlooked, unless it were something far removed from the subject: so that I have never met with a single critic of my opinions who did not appear to me either less rigorous or less equitable than myself. And further, I have never observed that any truth before unknown has been brought to light by the disputations that are practised in the Schools; for while each strives for the victory, each is much more occupied in making the best of mere verisimilitude, than in weighing the reasons on both sides of the question; and those who have been long good advocates are not afterwards on that account the better judges.

As for the advantage that others would derive from the communication of my thoughts, it could not be very great; because I have not yet so far prosecuted them as that much does not remain to be added before they can be applied to practice. And I think I may say without vanity, that if there is any one who can carry them out that length, it must be myself rather than another: not that there may not be in the world many minds incomparably superior to mine, but because one cannot so well seize a thing and make it one's own, when it has been learned from another, as when one has himself discovered it. And so true is this of the present subject that, though I have often explained some of my opinions to persons of much acuteness, who, whilst I was speaking, appeared to understand them very distinctly, yet, when they repeated them, I have observed that they almost always changed them to such an extent that I could no longer acknowledge them as mine. I am glad, by the way, to take this opportunity of requesting posterity never to believe on hearsay that anything has proceeded from me which has not been published by myself; and I am not at all astonished at the extravagances attributed to those ancient philosophers whose own writings we do not possess; whose thoughts, however, I do not on that account suppose to have been really absurd, seeing they were among the ablest men of their times, but only that these have been falsely represented to us. It is observable, accordingly, that scarcely in a single instance has any one of their disciples surpassed them; and I am quite sure that the most devoted of the present followers of Aristotle would think themselves happy if they had as much knowledge of nature as he possessed, were it even under the condition that they should never afterwards attain to higher. In this respect they are like the ivy which never strives to rise above the tree that sustains it, and which frequently even returns downwards when it has reached the top; for it seems to me that they also sink, in other words, render themselves less wise than they would be if they gave up study, who, not contented with knowing all that is intelligibly explained in their author, desire in addition to find in him the solution of many difficulties of which he says not a word, and never perhaps so much as thought. Their fashion of philosophizing, however, is well suited to persons whose abilities fall below mediocrity; for the obscurity of the distinctions and principles of which they make use enables them to speak of all things with as much confidence as if they really knew them, and to defend all that they say on any subject against the most subtle and skilful, without its being possible for any one to convict them of error. In this they seem to me to be like a blind man, who, in order to fight on equal terms with a person that sees, should have made him descend to the bottom of an intensely dark cave: and I may say that such persons have an interest in my refraining from publishing the principles of the Philosophy of which I make use; for, since these are of a kind the simplest and most evident, I should, by publishing them, do much the same as if I were to throw open the windows, and allow the light of day to enter the cave into which the combatants had descended. But even superior men have no reason for any great anxiety to know these principles, for if what they desire is to be able to speak of all things, and to acquire a reputation for learning, they will gain their end more easily by remaining satisfied with the appearance of truth, which can be found without much difficulty in all sorts of matters, than by seeking the truth itself which unfolds itself but slowly and that only in some departments, while it obliges us, when we have to speak of others, freely to confess our ignorance. If, however, they prefer the knowledge of some few truths to the vanity of appearing ignorant of none, as such knowledge is undoubtedly much to be preferred, and, if they choose to follow a course similar to mine, they do not require for this that I should say anything more than I have already said in this Discourse. For if they are capable of making greater advancement than I have made, they will much more be able of themselves to discover all that I believe myself to have found; since as I have never examined aught except in order, it is certain that what yet remains to be discovered is in itself more difficult and recondite, than that which I have already been enabled to find, and the gratification would be much less in learning it from me than in discovering it for themselves. Besides this, the habit which they will acquire, by seeking first what is easy, and then passing onward slowly and step by step to the more difficult, will benefit them more than all my instructions. Thus, in my own case, I am persuaded that if I had been taught from my youth all the truths of which I have since sought out demonstrations, and had thus learned them without labour, I should never, perhaps, have known any beyond these; at least, I should never have acquired the habit and the facility which I think I possess in always discovering new truths in proportion as I give myself to the search. And, in a single word, if there is any work in the world which cannot be so well finished by another as by him who has commenced it, it is that at which I labour.

It is true, indeed, as regards the experiments which may conduce to this end, that one man is not equal to the task of making them all; but yet he can advantageously avail himself, in this work, of no hands besides his own, unless those of artisans, or parties of the same kind, whom he could pay, and whom the hope of gain (a means of great efficacy) might stimulate to accuracy in the performance of what was prescribed to them. For as to those who, through curiosity or a desire of learning, of their own accord, perhaps, offer him their services, besides that in general their promises exceed their performance, and that they sketch out fine designs of which not one is ever realized, they will, without doubt, expect to be compensated for their trouble by the explication of some difficulties, or, at least, by compliments and useless speeches, in which he cannot spend any portion of his time without loss to himself. And as for the experiments that others have already made, even although these parties should be willing of themselves to communicate them to him, (which is what those who esteem them secrets will never do,) the experiments are, for the most part, accompanied with so many circumstances and superfluous elements, as to make it exceedingly difficult to disentangle the truth from its adjuncts; besides, he will find almost all of them so ill described, or even so false, (because those who made them have wished to see in them only such facts as they deemed conformable to their principles,) that, if in the entire number there should be some of a nature suited to his purpose, still their value could not compensate for the time that would be necessary to make the selection. So that if there existed any one whom we assuredly knew to be capable of making discoveries of the highest kind, and of the greatest possible utility to the public; and if all other men were therefore eager by all means to assist him in successfully prosecuting his designs, I do not see that they could do aught else for him beyond contributing to defray the expenses of the experiments that might be necessary; and for the rest, prevent his being deprived of his leisure by the unseasonable interruptions of any one. But besides that I neither have so high an opinion of myself as to be willing to make promise of anything extraordinary, nor feed on imaginations so vain as to fancy that the public must be much interested in my designs; I do not, on the other hand, own a soul so mean as to be capable of accepting from any one a favour of which it could be supposed that I was unworthy.

These considerations taken together were the reason why, for the last three years, I have been unwilling to publish the Treatise I had on hand, and why I even resolved to give publicity during my life to no other that was so general, or by which the principles of my Physics might be understood. But since then, two other reasons have come into operation that have determined me here to subjoin some particular specimens, and give the public some account of my doings and designs. Of these considerations, the first is, that if I failed to do so, many who were cognizant of my previous intention to publish some writings, might have imagined that the reasons which induced me to refrain from so doing, were less to my credit than they really are; for although I am not immoderately desirous of glory, or even, if I may venture so to say, although I am averse from it in so far as I deem it hostile to repose which I hold in greater account than aught else, yet, at the same time, I have never sought to conceal my actions as if they were crimes, nor made use of many precautions that I might remain unknown; and this partly because I should have thought such a course of conduct a wrong against myself, and partly because it would have occasioned me some sort of uneasiness which would again have been contrary to the perfect mental tranquillity which I court. And forasmuch as, while thus indifferent to the thought alike of fame or of forgetfulness, I have yet been unable to prevent myself from acquiring some sort of reputation, I have thought it incumbent on me to do my best to save myself at least from being ill-spoken of. The other reason that has determined me to commit to writing these specimens of philosophy is, that I am becoming daily more and more alive to the delay which my design of self-instruction suffers, for want of the infinity of experiments I require, and which it is impossible for me to make without the assistance of others; and, without flattering myself so much as to expect the public to take a large share in my interests, I am yet unwilling to be found so far wanting in the duty I owe to myself as to give occasion to those who shall survive me to make it matter of reproach against me some day, that I might have left them many things in a much more perfect state than I have done, had I not too much neglected to make them aware of the ways in which they could have promoted the accomplishment of my designs.

And I thought that it was easy for me to select some matters which should neither be obnoxious to much controversy, nor should compel me to expound more of my principles than I desired, and which should yet be sufficient clearly to exhibit what I can or cannot accomplish in the Sciences. Whether or not I have succeeded in this it is not for me to say; and I do not wish to forestall the judgments of others by speaking myself of my writings; but it will gratify me if they be examined, and, to afford the greater inducement to this, I request all who may have any objections to make to them, to take the trouble of forwarding these to my publisher, who will give me notice of them, that I may endeavour to subjoin at the same time my reply; and in this way readers seeing both at once will more easily determine where the truth lies; for I do not engage in any case to make prolix replies, but only with perfect frankness to avow my errors if I am convinced of them, or if I cannot perceive them, simply to state what I think is required for defence of the matters I have written, adding thereto no explication of any new matter that it may not be necessary to pass without end from one thing to another.

If some of the matters of which I have spoken in the beginning of the Dioptrics and Meteorics should offend at first sight, because I call them hypotheses and seem indifferent about giving proof of them, I request a patient and attentive reading of the whole, from which I hope those hesitating will derive satisfaction; for it appears to me that the reasonings are so mutually connected in these Treatises, that, as the last are demonstrated by the first which are their causes, the first are in their turn demonstrated by the last which are their effects. Nor must it be imagined that I here commit the fallacy which the logicians call a circle; for since experience renders the majority of these effects most certain, the causes from which I deduce them do not serve so much to establish their reality as to explain their existence; but on the contrary, the reality of the causes is established by the reality of the effects. Nor have I called them hypotheses with any other end in view except that it may be known that I think I am able to deduce them from those first truths which I have already expounded; and yet that I have expressly determined not to do so, to prevent a certain class of minds from thence taking occasion to build some extravagant Philosophy upon what they may take to be my principles, and my being blamed for it. I refer to those who imagine that they can master in a day all that another has taken twenty years to think out, as soon as he has spoken two or three words to them on the subject; or who are the more liable to error and the less capable of perceiving truth in very proportion as they are more subtle and lively. As to the opinions which are truly and wholly mine, I offer no apology for them as new,—persuaded as I am that if their reasons be well considered they will be found to be so simple and so conformed to common sense as to appear less extraordinary and less paradoxical than any others which can be held on the same subjects; nor do I even boast of being the earliest discoverer of any of them, but only of having adopted them, neither because they had nor because they had not been held by others, but solely because Reason has convinced me of their truth.

Though artisans may not be able at once to execute the invention which is explained in the Dioptrics, I do not think that any one on that account is entitled to condemn it; for since address and practice are required in order so to make and adjust the machines described by me as not to overlook the smallest particular, I should not be less astonished if they succeeded on the first attempt than if a person were in one day to become an accomplished performer on the guitar, by merely having excellent sheets of music set up before him. And if I write in French, which is the language of my country, in preference to Latin, which is that of my preceptors, it is because I expect that those who make use of their unprejudiced natural Reason will be better judges of my opinions than those who give heed to the writings of the ancients only; and as for those who unite good sense with habits of study, whom alone I desire for judges, they will not, I feel assured, be so partial to Latin as to refuse to listen to my reasonings merely because I expound them in the vulgar Tongue.

In conclusion, I am unwilling here to say anything very specific of the progress which I expect to make for the future in the Sciences, or to bind myself to the public by any promise which I am not certain of being able to fulfil; but this only will I say, that I have resolved to devote what time I may still have to live to no other occupation than that of endeavouring to acquire some knowledge of Nature, which shall be of such a kind as to enable us therefrom to deduce rules in Medicine of greater certainty than those at present in use; and that my inclination is so much opposed to all other pursuits, especially to such as cannot be useful to some without being hurtful to others, that if, by any circumstances, I had been constrained to engage in such, I do not believe that I should have been able to succeed. Of this I here make a public declaration, though well aware that it cannot serve to procure for me any consideration in the world, which, however, I do not in the least affect; and I shall always hold myself more obliged to those through whose favour I am permitted to enjoy my retirement without interruption than to any who might offer me the highest earthly preferments.

2

James Conant, MODERN SCIENCE AND MODERN MAN

J. Robert Oppenheimer, SCIENCE AND COMMON UNDERSTANDING

1. Explain what Conant sees as the greatest threat to the advance of science.

2. Describe how the public concept of scientists changed during the first half of the twentieth-century?

3. What sort of relation does Conant see between advances in the pure sciences and what may be done in the practical arts?

4. Compare Conant's attitude toward science with that of Descartes.

5. Compare Oppenheimer's view of science with that of Conant. Would Oppenheimer's view of the nature of man be more like that of Hobbs or Locke? Why?

6. What does "The greatest of the changes that science has brought is the acuity of change...," imply to one? Have the years since this was written altered the validity of this statement?

7. One of the most common criticisms of science is that it lacks humanity. Do you find any support for this charge in Oppenheimer's writings?

8. After World War II, Oppenheimer was a powerful opponent of secrecy in science. His security clearance was revoked partly because of this attitude and partly because of his opposition to building the hydrogen bomb. Do scientists have a moral and professional responsibility to oppose what they consider inappropriate uses of research?

James B. Conant (1893-) was an organic chemist who served as president of Harvard University for two decades. As a member of the highest circle of scientific advisers during World War II, he was uniquely situated to see the impact of an ever-growing scientific establishment on a nation recovering from war. *Modern Science and Modern Man,* published in 1953, is a provocative discussion of the role of science in society.

MODERN SCIENCE AND MODERN MAN

Science and Technology in the Last Decade

A FEW WORDS may be in order at the outset to explain why I have chosen the title "Modern Science and Modern Man" for this series of four lectures. When I was honored by the invitation to be the Bampton lecturer for 1952, President Eisenhower expressed on behalf of the committee the hope that I would undertake to provide "some understanding of the significance of recent developments in the physical sciences." On my inquiring as to the nature of the audience, I was assured that professional philosophers and scientists would be conspicuous by their absence. My exposition, if not aimed at the proverbial man-on-the-street, was to be directed to the equally proverbial college graduate—the hypothetical individual whom college presidents welcome each commencement to the fellowship of educated men. Being thus assured that I was not expected either to give an appraisal of the impact of physics on metaphysics or a technical account of the inner workings of the atom, I gratefully accepted the privilege of being a guest lecturer at Columbia University.

Under the terms of reference thus graciously arranged, I was still left great latitude as to my choice of topics. In my endeavors to narrow down a vast array of subjects to a few that could be handled in a general fashion in four lectures, I asked myself, what are the recent developments of physical science that are significant to those young men and women now graduating from our colleges and universities? To ask this question is to come up with the trite answer in three words: the atomic bomb. Or if you are more optimistic, with a variant, atomic energy. Now admittedly it is extremely difficult today to make an address without referring in some way to atomic bombs; particularly so if the speaker, as in my case, has been to some degree concerned with the development of these weapons. Yet an exposition of the elementary facts about nuclear physics or an appraisal of the military possibilities of atomic bombs, or an estimate of the peaceful uses of atomic energy—surely this is not what is expected from a Bampton lecturer in 1952. The technological implications of recent developments in physics and chemistry are certainly of the utmost importance, but a popular exposition of them would not go to the heart of the question that I raised. For advances in technology are not a new phenomenon, and the addition of one more source of energy, when viewed in the cool light of engineering analysis, appears unlikely to be revolutionary as long as coal and oil last. As for warfare, that is a special subject; it would certainly be out of place for me to give here my standard address on "The Necessity for the Defense of Europe in the Atomic Age."

I propose to examine the cultural significance of what has been going on in science since, say, 1935. And by "cultural" I have in mind the whole social pattern of Western civilization. I have in mind far more than the modes of production, distribution, and communication; I wish to include in my survey the impact of modern science on the philosophic presuppositions of the average enlightened citizen of a modern democracy—on his ambitions, his hopes, his fears, his outlook on the world. So in a sense I shall relate physics to philosophy, but only by handling both subjects in a general fashion and viewing each from the point of view of a deeply troubled modern man.

During the last decade and a little more, the attitude of the American citizen towards science and scientists has undergone a profound change. At least that is the thesis of this first lecture. I have just spoken of modern man as being "deeply troubled." I judge few would challenge this statement nor the validity of the overtones of an alternative title I thought of suggesting for these lectures: science and the predicament of the intelligent citizen. Now the awesome nature of modern weapons, the deep division of the world, and the continued high tension in international situations may be the only causes of the trouble that disturbs the dream of the intelligent citizen. Certainly they account for a large share of his apprehension. The changed attitude of society to science may be basically caused by fear of the atomic bomb, but its manifestations ramify in many directions and appear in surprising ways. So much so that if, as I sincerely hope, ten or fifteen years from now this fear can be enormously diminished by international action, the old relations of society and science will *not* automatically reappear. Let me make it plain—I wish they would: I do not like the atomic age or any of its consequences. To learn to adjust to these consequences with charity and sanity is the chief spiritual problem of our time. Therefore, in considering what I have to say, please erase from your minds, if you can, the knowledge that we are assembled on a prime target and that a potential enemy is every year increasing his stockpile of atomic bombs and may well be improving his methods of delivery.

When you have obliterated all thoughts of war and destruction, let me ask you to contrast public reaction to the development of the atomic bomb with public reaction to the steam engine, the electric light and motor, the telephone, the automobile, and the airplane or, in our time, to radio and television. According to my reading of history, modern science first reached the walking and talking stage about 1700, and only about 1780 did it start growing into manhood. Therefore, an appreciation of the changes wrought by science and invention could hardly be older than the industrial revolution of the eighteenth century. But that it is certainly so old is indicated by a statement of Thomas Jefferson in 1818 which he wrote as chairman of the Commissioners for the University of Virginia. "It cannot be but that each generation succeeding to the knowledge acquired by all those who preceded it, adding to it their own acquisitions and discoveries, and handing the mass down for successive and constant accumulation, must advance the knowledge and well-being of mankind, not *infinitely*, as some have said, but *indefinitely*, and to a term which no one can fix and foresee." (Now comes the significant statement.) "Indeed, we need look back half a century, to times which many now living remember well, and see the wonderful advances in the sciences and arts which have been made within that period. Some of these have rendered the elements themselves subservient to the purposes of man, have harnessed them to the yoke of his labours, and effected the great blessings of moderating his own, of accomplishing what was beyond his feeble force, and extending the comforts of life to a much enlarged circle, to those who had before known its necessaries only. That these are not the vain dreams of sanguine hope, we have before our eyes real and living examples." [1]

Such praises and such hopes have been repeated every generation in the Western world since Jefferson's time. But popular credit throughout the nineteenth century went largely to the inventor, not to the scientist. This was particularly true in the United States but not much less so in Great Britain. Imperial Germany was the only nation where the physical scientist as scientist received something like his due. The fact that from 1800 to the present, various writers were continually attempting to right the balance by pointing out the significance of science as contrasted to invention is evidence for the correctness of my generalization. By and large, the scientist in

[1] *Report of the Rockfish Gap Commissioners.*

the nineteenth century was supposed to be concerned only with discovering nature's laws; the inventor was taking advantage of these discoveries for practical ends. The attitude of James Clerk-Maxwell, the founder of the electromagnetic theory of light, towards the inventor Alexander Graham Bell was one of patronizing condescension. (Clerk-Maxwell referred to Bell as "a speaker, who to gain his private ends, has become an electrician.") Professor Rowland of the Johns Hopkins University, addressing his fellow physicists in 1879, said, "He who makes two blades of grass grow where one grew before is the benefactor of mankind; but he who obscurely worked to find the laws of such growth is the intellectual superior as well as the greater benefactor of the two." The scientist looked down upon the inventor and the inventor in turn was a bit contemptuous of the scientist; so too were some of the men of business who backed the inventors in successful enterprises.

In World War I, President Wilson appointed a consulting board to assist the Navy. Thomas Edison was the chairman; his appointment was widely acclaimed by the press—the best brains would now be available for the application of science to naval problems. The solitary physicist on the board owed his appointment to the fact that Edison in choosing his fellow board members had said to the President, "We might have one mathematical fellow in case we have to calculate something out."

Another story illustrating the popular attitude towards science and invention in 1916 concerns chemists, not mathematicians or physicists. At the time of our entry into World War I, a representative of the American Chemical Society called on the Secretary of War, Newton Baker, and offered the service of the chemists in the conflict. He was thanked and asked to come back the next day. On so doing, he was told by the Secretary of War that while he appreciated the offer of the chemists, he found that it was unnecessary as he had looked into the matter and found the War Department already had *a* chemist.

By 1940 the scene had completely altered: science had moved into industry and, belatedly, even in the United States, industry had moved into science. The electrical, the chemical, and the pharmaceutical industries may be said to have led the way. The great research laboratories of the General Electric Company, the Bell Telephone Laboratory, and the DuPont Company may be mentioned as examples. Therefore, when in World War II it came to mobilizing technologists to assist the military forces, there could be no question but that the scientists as scientists would be called on by the government.

I have oversimplified the history of the last forty years; the National Research Council was established during World War I, and by this action the foundation was laid for a closer relation between science and the national government. Indeed, the National Academy of Science had been chartered by Congress during the War between the States. Science was not an outcast in this country, but even as late as 1917 it was primarily the inventor, not the scientist, who was looked to by the general public as being the prime mover of technology; he was the man who had changed our habits and made possible our new comforts; in other words, the inventor had conquered nature and put it to man's use.

From the end of World War I to the beginning of World War II, the relation of science to industry in the United States was changing at great speed. The development of the atomic bomb only demonstrated to the public what was already known to many industrialists, namely that the scientists had now become inventors. The fact that so many *scientists*, and I emphasize the word, had been concerned in this development that had both brought about an extraordinary new weapon and kindled hopes for future industrial revolutions impressed the people of the United

States; not only the people of the United States but those of the entire world. The transition was complete; the scientist was no longer thought of as a man in an ivory tower, gradually unravelling the secrets of nature for his own spiritual satisfaction, but as a miracle-worker who like Watt or Edison before him could bring about tremendous transformations of man's relation to his material surroundings.

The changed status of the scientist, to my mind, is one of the major significant developments of the last decade. Another is the altered relation of many scientists to organized society, that is, the government. The development of the first atomic bomb introduced a novel and highly significant element into the complex relations between science and society. This was in part due to the revolutionary military potential of the weapon, to be sure; but the essence of the novelty, to my mind, is only secondarily a result of the physical consequences of Einstein's famous equation $E=mc^2$

To me, the fact of prime importance is this: in the period 1940 to 1945 a whole army of specialists was engaged in advancing science in a spectacular fashion and at the same time was developing a new weapon of great military power. This combined activity was a new social phenomenon; we are still struggling with its consequences.

To illustrate what I have in mind, let me contrast two enterprises about which I can speak from personal experience, the manufacture of a new poison "gas" in World War I and the production of the first atomic bomb. In the first case, even if the "gas" had been as effective a weapon as we who were concerned with its development then believed, the impact of this war work in 1917–1918 on chemistry would have been of no significance. Why? Because as in all the other cases of the application of science to war until 1945, the technical area of the new knowledge was small and easily circumscribed. Actually, the "gas" was shown by later animal experimentation to be of doubtful value, but even if it had had all the terrible characteristics we then dreamed of—high killing power, ability to penetrate all masks, no odor, complete stability—even then this discovery would have been easily accommodated within the nineteenth-century framework of peaceful international science. Only if in the process of discovering the new "gas" the chemists involved had unlocked a safe full of exciting new chemical facts would this military application of science have caused an appreciable reorientation of scientific attitudes or the attitude of society towards science.

Chemical warfare, radar, proximity fuses, underwater warfare, jet planes, new mines and missiles—all these developments depended on the special application of the publicly known facts and principles of physics and chemistry; what new scientific knowledge was gained in these military developments had no revolutionary consequences in relation to the forward march of science. How completely otherwise it was in the manufacture of the atomic bomb! No one could be sure that the phenomenon known as the "critical mass" was an experimental reality until a large-scale manufacturing operation was put into motion. In 1940 the physicists were in possession of the results of certain experiments with microscopic amounts of materials; they likewise held in their hands powerful theoretical concepts in the new field of nuclear physics and chemistry. By an enormous extrapolation they predicted the operation of atomic piles and the explosion of an atomic bomb. Most of these predictions, however, could not be tested by any large-scale laboratory experiments or even by the erection of small-scale pilot plants. The flowering of this whole new field of science was dependent on the expenditure of a large sum of taxpayers' money; this expenditure could be justified in the 1940's only in terms of the destructive power of a weapon required in a desperate global struggle. Thus a vast scientific activity was born as a war necessity; it was marked at its birth with a word hitherto the anathema of scientists—secrecy.

After the war was over, as much as was deemed impossible to keep secret was published in the Smyth report; over all the rest a ban of complete censorship was placed. The difficulties of such a situation need not be recounted here; nor the struggles to decide what was vital and what was not; nor the inevitable conflict between the presuppositions of the scientist and the government official. Who can judge the wisdom or lack of wisdom of those who made decisions about security regulations in the days immediately following the surrender of Japan? Few could then foresee the tensions of the divided world in which we now must live as best we can. In retrospect it all seems clearer now; one whole area of science was inherently different from the rest; applied nuclear physics and chemistry were wards of the national governments of every industrialized nation; here, at least, science, technology, and politics merged, though often the confluence was a turbulent and, at times, a muddy stream.

If some nineteenth-century scientist could come to earth, he would certainly be amazed at what has now occurred. Michael Faraday was presumably in a cynical mood, rare for him, when he replied to Gladstone's skeptical comment about his rudimentary electric motor —the little wire whirling around the magnet— "Someday you can tax it." But he could never have imagined a time when the governments of free nations would exercise a complete monopoly over similar embryonic bits of science or would prohibit publication of new advances in these sciences—in short, a time when the government would regard the whole matter as a national secret. If he had envisaged any such social phenomenon, he would have realized the consequences that were bound to follow, namely, complete failure of communication between the scientists involved in the secret undertaking and the general public, complete failure of scientists and engineers to discuss new proposals, and failure to have adequate appraisals of the significance of each year's advances.

The future of atomic energy for military and industrial purposes has now become a matter not only of national pride but also a matter of pride for politicians. Moscow announces that the Soviet Union is using atomic bombs to move mountains; few if any informed persons believe these stories; within Russia, of course, no one can even discuss such matters. From Argentina come strange stories of new sources of atomic energy; no one credits them, few discuss them; within that country, no skeptics appear to have raised their voices. All that you may attribute to the authoritarian rule of dictators, to the consequences of a totalitarian state, but the situation is not far different in North America or in Great Britain. We read in the press about the alleged first demonstration of atomic energy for industrial purposes, about future applications of atomic energy for military aircraft or naval vessels; we are told that this or that development is right around the corner. The officials of one nation make claims about the future prospects of industrial atomic power; those of another nation, about hydrogen bombs. The significant, new, and alarming fact is this: in the face of all such statements those who are competent to discuss such matters cannot do so because of security regulations. Those who have scientific ability and access to the technical facts must remain silent. Consequently the public is largely informed through statements by politicians only partially aware of their own distortion of the facts, certainly unconscious of the degree of uncertainty in their predictions. At times half-truths and necessarily ambiguous reports "leak" into newspaper columns: these are some of the methods by which the public is informed of the progress of applied nuclear physics. I can underline what I have been saying by making one bold statement based on twelve years of experience behind the veil of secrecy: It is impossible today or in the foreseeable future to have a frank, rational, searching discussion of the industrial uses of atomic energy. The general public might just as well stop reading anything in the papers about atomic energy or atomic bombs. By the nature of the case it is almost certain to be misleading.[2]

[2] To illustrate what I have in mind, suppose that in the next few years some one of the four national governments now admittedly developing atomic energy should announce the completion of a plant for the production of power for civilian use. Such an announcement would be front-page news with some such headlines as "First Atomic Energy Power Plant —Government of —— Wins the Race." This headline would be misleading. Any time in the last seven years the United States Atomic Energy Commission could have generated electric power by coupling an atomic pile to a steam plant operating with low temperature steam and coupling this in turn to a turbine and generator. Such an outfit would have produced electric power from atomic energy but at a fantastic cost. Whether or not an atomic power plant is producing power at a cost competitive with a coal- or oil-burning plant can only be determined by an engineering and cost-accounting analysis; these analyses are always subject to argument and, to be trustworthy, must be public. This would require the pub-

There is no use in wringing one's hands about the situation. There is no use in trying to find a scapegoat for what has happened or in denouncing all secrecy and security regulations. The world being what it is today and is likely to be for a long time to come, secrecy and applied nuclear physics are words that must be joined together. But it is of the utmost importance that the general public understand the consequences of this union. Furthermore, it is important that the necessary curtain of secrecy not be extended any further than is required; above all, that it not be extended into branches of science far distant from nu-

lication of such secret information as the cost of the nuclear fuel, the cost of the plant, and the details of the operation, including the handling of the radioactive waste products. That any government can make such information public in the foreseeable future is extremely doubtful. Without such information, any news story on atomic power would be at best meaningless and in all probability would be entirely misleading. (For a discussion of the factors affecting the development of atomic power for industrial purposes, see United States Atomic Energy Commission, *Report*, July, 1948, pp. 43–46.)

clear physics. Advances in science are difficult within a secret national monopoly because necessarily its research is guarded from all but a few branches of the government. Secrecy and science are fundamentally antithetic propositions.

Let me contrast what I have just been saying with two other major developments in the application of science within the last ten years. Consider the revolution in fabrics that is taking place, and the use of the so-called "miracle drugs" in surgery and medicine. The facts in both cases are so well known that I need only remind you of them here. Starting with a more or less accidental discovery made by a remarkable man, Wallace Carruthers, in the course of his investigation of the structure of substances of high molecular weight, the DuPont Company produced nylon. During the development the matter was a secret, but as soon as the fabric was on the market, essentially all the basic scientific facts were made available and protected by patents; I do not doubt that to a large degree the so-called "know how" was made known as well.

A new lead had been opened up by Carruthers' work, and it was not long before other fabrics were on the market and competing companies were in the business. Indeed, the whole development of synthetic plastics and synthetic fabrics in the last twenty or thirty years makes a fascinating story in itself. But there is nothing revolutionary in this chapter in applied science. If you will, it is a further application of the procedures used by the German chemists nearly a century ago when they manufactured the first synthetic dyes and, later, synthetic drugs.

How is all this different from what has gone on with atomic energy, you may ask. First, all the developments I have just described have been in private hands; second, the work was not kept secret (by its nature it could not be); and third, various competing companies have taken up the ideas. Indeed, I venture the opinion that if there had been an attempt at a government monopoly and secrecy, some of the new fabrics which are now coming on the market would have been produced very much later, if at all.

Let me, if I may, just refer to the story of the production of the first so-called miracle drug, penicillin. Here again there were no secrets except for a short time. During the war, the work on penicillin was as highly classified as that on the atomic bomb. But notice that the traditions of the medical sciences and the whole feeling of the world essentially made it impossible for any one nation to keep this secret. Therefore, almost as soon as the war was over, people all over the world began to work on this problem in many different ways. As a consequence, a series of antibiotics has been developed and enormous progress has been made in their application. This is an example of free international science. It impresses the public, but notice how rarely it is referred to by people speaking about science in general terms. Why? Because there is essentially no novelty in this procedure. For several generations now, people have come to expect medical scientists to produce better drugs and better surgical procedures. Public health and the advance of medicine are almost taken as a matter of course. Along with the public's feeling that the scientist has now turned inventor —an inventor of strange weapons—there goes the reassuring thought that the scientist has continued to help the physician to improve the art of healing. Of the importance of underlining this aspect of scientific work, I shall have more to say in my concluding lecture.

Now I should like to examine a little more closely the procedures employed by the scientists today in physics, in chemistry, in biology, and in the application of science for practical ends in the whole area of the natural sciences. If in so doing, I appear to be underlining the obvious, I hope I shall be forgiven; for although a discussion of the so-called scientific method is almost a standard topic in a university, what I have to say is so antithetical to much of the current doctrine that I venture to present it in some detail here.

It would be my thesis that those historians of science, and I might add philosophers as well, who emphasize that there is no such thing as *the* scientific method are doing a public service. To my mind, some of the oversimplified accounts of science and its workings to be found in the elementary texts in high schools, for example, are based on a fallacious reading of the history of physics, chemistry, and biology. I will not attempt to trace back the sources of what seems to some of us a set of erroneous conclusions. Let me rather present my own analysis of the nature of science, if I may.

In order to produce a straw man that I may knock down, let me quote a definition of the scientific method written some years ago in an elementary text of chemistry of which I am a joint author: "The scientific way of thinking requires the habit of facing reality quite unprejudiced by any earlier conceptions. Accurate observation and dependence upon experiments are the guiding principles. The watchword is not 'what does the book say about this or that, but let's try to find out for ourselves.'" Then follows an account of the steps in the process of finding out. Such accounts in many textbooks, including the one I helped to write, usually run as follows: "Scientists collect their facts by carefully observing what is happening. They group them and try to interpret them in the light of other facts that are already known. Then a scientist sets up a theory or picture that will explain the newly discovered facts, and finally he tests out his theory by getting more data of a similar kind and comparing them with the facts he got through the earlier experiments. When his theory does not quite fit the facts, he must modify it and at the same time verify the facts by getting more data."

Another and more sophisticated account by a biologist is entitled "Steps in the Scientific Method." Note the use of the word "the" rather than the plural "scientific methods." Here the steps are listed: "Recognize that an indeterminate situation exists. This is a conflicting or obscure situation demanding inquiry. Two, state the problem in specific terms. Three, formulate a working hypothesis. Four, devise a controlled method of investigation by observation . . . or by experimentation or both. Five, gather and record the testimony or 'raw data.' Six, transform these raw data into a statement having meaning and significance. Seven, arrive at an assertion which appears to be warranted. If the assertion is correct, predictions may be made from it. Eight, unify the warranted assertion, if it proves to be new knowledge in science, with the body of knowledge already established."

The simplified account from my own elementary textbook and the more sophisticated version of the biologist, I have become convinced, serve only to confuse a layman. The basic fault is a failure to distinguish between two closely related activities which together have made the history of science possible. As I now read the story of the advances that have been made in the natural sciences since the time of Galileo, what has happened is essentially this. An age-old process of inquiry by which artisans and skillful workers improved the methods of handling inanimate nature became gradually associated with the type of thinking up to then characteristic of mathematics. Another way of putting it is to say that two streams of human activity, separated until the sixteenth century, gradually came together. These were abstract reasoning, as represented by Euclidian geometry, and experimentation, as represented by the work of the metallurgists who over the generations had improved the methods of winning metals from the ores.

Closely related to abstract reasoning were the broad speculative ideas of philosophers about the nature of the universe: of these the Aristotelian picture was predominant in the Middle Ages. A rival conceptual scheme associated with the name of Democritus was never wholly lost to view. Those who followed this line of thought imagined that the world was composed of corpuscles or atoms.

The usual descriptions of "the scientific method" are descriptions actually of the very limited procedure by which a person can improve a particular practical art. Indeed, if we take any one of these descriptions, we can apply it to almost any practical problems that arise in this mechanized age at home. Take the eight steps so carefully set forth by the biologist and apply them to the failure of an electric light to go on in one's country house. Certainly there is an indeterminate situation —something is wrong. The problem can be stated in specific terms, namely, when I turn the switch, the light does not work. What can be found that can be changed in order to make the light work when I turn it on? As to the working hypotheses, the first one will perhaps be that the bulb is burned out. The method of investigation is obvious: you try replacing the bulb with another—and the controlled experiment here is to see that the other bulb does light in a proper socket—you arrive at an assertion that is warranted, and the new knowledge easily fits into what you had before. Or you know by experience that it is more likely that the fuse is blown out, in which case you formulate another working hypothesis and make the test by turning on another light on the same circuit. In short, much of the so-called scientific method consists of rational experimentation or well-ordered inquiry directed towards solving a specific practical problem.

Now more than one teacher has pointed out to his or her class that, after all, the scientific method can be applied to everyday life and has used examples such as I have given. But I think this form of exposition is really putting the cart before the horse. The point is that what the scientist does in his laboratory is simply to carry over, into another frame of reference, habits that go back to the caveman.

We can imagine, for example, that one of our early ancestors tried the impact of fire on various kinds of minerals, probably starting from a chance observation, and so gradually evolved a process of making a metal from an ore.

All this information is conveniently called "empirical," which means essentially cookbook information. John Tyndall, in a famous address [3] of a popular nature on fermentation, wrote, "Hitherto the art and practice of the brewer have resembled those of the physician, both being founded on empirical observation." (And I am giving this quotation to define my use of the word "empirical.") "By this is meant the observation of facts apart from the principles which explain them, and which give the mind an intelligent mastery over them. The brewer learned from long experience the conditions, not the reasons, of suc-

[3] Published in *Essays on the Floating-Matter of the Air in Relation to Putrefaction and Infection* (London: Longmans and Co., 1881), p. 238.

cess. But he had to contend, and has still to contend, against unexplained perplexities."

At the time Tyndall was writing one could say that the art of wine making and beer manufacture was empirical. The work of Pasteur and subsequent microbiologists and chemists has greatly lowered the degree of empiricism. Still, even today, there are many procedures in these industries which are simply based on experience and cannot be related to the concepts or theories of chemistry or biology. It is convenient to characterize a given practical art or a branch of science by assigning to it a degree of empiricism. If one wants to find an activity where the degree of empiricism is very low, I suggest turning to the work of the surveyor. Long ago the science of optics was developed so that it is possible to calculate by mathematical formulas the shapes of mirrors and lenses and to construct the optical part of a surveyor's instruments. Furthermore, Euclidian geometry provides a mathematical framework for the observations of the surveyor. Therefore, one can say that the surveyor's work represents an applied science in which the degree of empiricism is essentially zero. At the other end of the scale I would put the labors of any excellent cook, for in spite of all our knowledge of the chemistry of proteins, fats, and carbohydrates, the recipe for a good sauce or a good dessert is still entirely empirical.

I shall use this idea of the degree of empiricism repeatedly throughout these lectures. I introduce it at this time in order to tie together the work of the scientist and the inventor. In the heyday of the inventor, which I regard as being the nineteenth century, the elements of science at his disposal were relatively rudimentary and he was able, operating with those and with the aid of very little mathematics or elaborate theory, to apply the new knowledge by a series of largely empirical procedures.

Earlier I was dwelling on the new and revolutionary aspects of the relation between society and science. I pointed out that the man in the street—the public at large—had come to see that the scientist is today taking the place of the inventor; that the "long-haired" professors who were elaborating highly abstruse mathematical theories had been able to play an important part in the extraordinary development of the atomic bomb. Also, people are beginning to realize that in other fields as well, the lowering of the degree of empiricism of which I have just spoken, that is, the application of theory to practice, has paid enormous dividends in terms of actual control of inanimate nature. Many hours could be spent relating the story of modern chemistry or our triumphs in electricity which follow this same pattern. The discovery of Wallace Carruthers which led to nylon fabric, the story of the development of various electronic devices, including the new solid material as a substitute for a vacuum tube—all these would illustrate the same point.

In emphasizing the distinction between

developments in which new theories have played an important part and those based on essentially empirical procedures, I have perhaps done less than justice to one change which has been going on during the whole period of the growth of modern science. We have learned that there are ways in which empirical procedures can be improved without the introduction of theory. We have, as it were, empirically improved empiricism. The inventor in the last century working in his attic might more or less erratically try first this procedure and then that in order to get a better way of carrying out a step in his process, for there was no one to criticize him; if his methods were far from being effective, that was no one's business but his own. But when invention was replaced by the work of scientists in laboratories, then self-criticism was replaced by group criticism. People learned that if a specific problem is to be solved as rapidly as possible, accurate observations of each experiment should be kept; furthermore, even in trial-and-error procedures there is a difference between well-ordered and disordered empiricism. Take the simple homely example of finding out what is wrong with the electric light; we can all imagine procedures which systematically test one limited working hypothesis and then another or helter-skelter methods of operating where the same trial is repeated several times to no purpose.

What is often defined as the scientific method is nothing more or less than an approximate description of a well-ordered, systematized empirical[4] inquiry. Now, systematized or well-ordered empirical inquiries are one element in the advance of science; the other element is the use of new concepts, new conceptual schemes that serve as working hypotheses on a grand scale. Only by the introduction of a theoretical element

[4] Any philosopher who happens to read these lectures will note that throughout I am using "empirical" and "empiricism" in a sense other than that to which he is accustomed; I am using these words as they are commonly employed by scientists

can the degree of empiricism be reduced. Only by the use of new ideas of broad significance has science advanced—such ideas as those embodied in Newton's laws, as the notion that the earth is surrounded by a sea of air which exerts pressure, that light is a vibratory motion in an all-pervading ether, that matter is composed of atoms that unite in definite proportions to form compounds. The essential element in the advance of modern science has been the curious interplay between such theoretical notions and the experimentation of the artisan; through such an interplay scientists have built up a fabric of interconnected concepts and conceptual schemes.

Without unduly laboring the obvious, let me repeat: by well-ordered empirical procedures it has been possible to make great progress in the practical arts. Such procedures are still being used in almost all branches of applied science. In metallurgy and organic chemistry, for example, the degree of empiricism is still high, but the development of new concepts and wide conceptual schemes was essential to the progress of physics, chemistry, and biology in the last three hundred and fifty years. As these sciences became equipped with more and more satisfactory theories, the degree of empiricism in the arts related to these sciences diminished. As a consequence, in these practical endeavors it became more and more possible to attain the accuracy of prediction of such undertakings as those of the surveyor. The practical significance of the advance in theoretical science lies right here. The history of the last hundred and fifty years in particular shows what occurs when advances are made in pure science. New principles evolve which can be related to empirical observations; at that point it becomes possible to control with far greater accuracy than before what one is doing in the practical arts and to predict the outcome of a large-scale operation directed towards making a commercial product.

No better illustration of this can be found than in applied biology, the very field that John Tyndall was discussing. I referred to Tyndall's words which were concerned with the practical art of brewing. Explaining why the brewer could not proceed as long as his art was founded entirely upon empirical observations, he said, "But he had to contend and still has to contend against unexplained perplexities. Over and over again his care has been rendered nugatory; his beer has fallen into acidity or rottenness and disastrous losses have been sustained of which he has been unable to assign the cause. It is the hidden enemies against which the physician and the brewer have hitherto contended that recent researches are dragging into the light of day, thus preparing the way for their final extermination." These researches had been instituted by Louis Pasteur in his study of fermentation. He had shown that living microorganisms were the "hidden enemies" with which the brewer had contended without knowing it. This knowledge was the consequence of a bold working hypothesis on a grand scale by Pasteur, namely, that fermentation and putrefaction were the consequences of the growth of microorganisms. Armed with his theory, Pasteur reduced the degree of empiricism in the whole range of fermentation industries. We accept the working hypothesis of Pasteur today as an established principle; we base all our procedures for handling food on the assumption of its correctness. Furthermore, this theory or principle is basic to medicine and public health. One need elaborate no further on this example of what theory can do to practice.

Let me give a reverse case where the failure to have any satisfactory theory has led to surprising consequences. I refer to the question of the action of chemicals on living organisms. In spite of an enormous amount of experimentation by chemists in making new substances and pharmacologists in testing them on animals and on men, one can say that it is almost impossible to predict the action of a chemical substance of a given structure on a human being or on a microorganism. An adequate theory would enable one to write down the molecular architecture of a substance and from this structure predict the effect of the compound on a living organism. Such predictions are possible today only within very narrow limits and with very special classes of substances. And even here the correlation is essentially empirical—we have no broad, overall theory of drug action. Only within the last half a dozen years has some glimmering of hope appeared that chemotherapy can become a science.

To recapitulate, the inventor was largely an empiricist; he continued the tradition of the ingenious artisan. His place today has largely been taken by teams of scientists and engineers. To a considerable degree they likewise operate empirically; but their procedures are well ordered and systematized; they are disciplined by experience. In almost every field into which theory has been introduced, the degree of empiricism has been lessened. The teams of scientists in industry and in medicine are engaged both in reducing further the degree of empiricism and in applying the theories of science as they now stand. All this has become an organized social undertaking. As a sociological phenomenon modern science deserves careful study. Agriculture, medicine, public health, and the production of raw materials and their processing in industry are all interpenetrated heavily by well-trained empiricists using modern instruments to assist their trial-and-error experimentations. For the most part their success in advancing science, that is to say, developing new theories and testing these theories by experiment, are publicly recorded. Scientific societies and scientific journals make this intercommunication possible. Destroy the social nature of scientific research in the sense of destroying the intercommunication of scientists, and the advance of science would almost cease. Recognizing this fact, one must ponder on the consequences of the vast sums of money now being spent on secret military research and development undertakings. One cannot help wondering how long a large fraction of our scientific manpower can be employed in this atypical scientific work without threatening the traditions that have made science possible.

The history of science demonstrates beyond a doubt that the really revolutionary and significant advances come not from empiricism but from new theories. The development of these theories, in turn, has in the past depended on free discussion of their consequences. How much can be accomplished behind a wall of secrecy remains to be determined. This wall represents public policy. I do not question its necessity in these grim years of rearmament. But it is essential that the public in free countries be aware constantly of its existence and understand the special conditions now imposed on some phases of an activity that industrialized societies can ill afford to damage: the advance of science. If this be fully realized, then the significance of one phase of the "recent developments in the physical sciences" can be better understood.

J. Robert Oppenheimer (1904-1967) was a theoretical nuclear physicist who during World War II was the director of the Manhattan Project's Los Alamos National Laboratory. The force of his scientific insight and personality was crucial in the drive to develop the atomic bomb. After the war, he was named director of the Institute for Advanced Studies at Princeton.

SCIENCE AND THE COMMON UNDERSTANDING

THE SCIENCES AND MAN'S COMMUNITY

FOR some moments during these lectures we have looked together into one of the rooms of the house called "science." This is a relatively quiet room that we know as quantum theory or atomic theory. The great girders which frame it, the lights and shadows and vast windows—these were the work of a generation our predecessor more than two decades ago. It is not wholly quiet. Young people visit it and study in it and pass on to other chambers; and from time to time someone rearranges a piece of the furniture to make the whole more harmonious; and many, as we have done, peer through its windows or walk through it as sight-seers. It is not so old but that one can hear the sound of the new wings being built nearby, where men walk high in the air to erect new scaffoldings, not unconscious of how far they may fall. All about there are busy workshops where the builders are active, and very near indeed are those of us who, learning more of the primordial structure of matter, hope some day for chambers as fair and lovely as that in which we have spent the years of our youth and our prime.

It is a vast house indeed. It does not appear to have been built upon any plan but to have grown as a great city grows. There is no central chamber, no one corridor from which all others debouch. All about the periphery men are at work studying the vast reaches of space and the state of affairs billions of years ago; studying the intricate and subtle but wonderfully meet mechanisms by which life proliferates, alters, and endures; studying the reach of the mind and its ways of learning; digging deep into the atoms and the atoms within atoms and their unfathomed order. It is a house so vast that none of us know it, and even the most fortunate have seen most rooms only from the outside or by a fleeting passage, as in a king's palace open to visitors. It is a house so vast that there is not and need not be complete concurrence on where its chambers stop and those of the neighboring mansions begin.

It is not arranged in a line nor a square nor a circle nor a pyramid, but with a wonderful randomness suggestive of unending growth and improvisation. Not many people live in the house, relatively speaking—perhaps if we count all its chambers and take residence requirements quite lightly, one tenth of one per cent, of all the people in this world—probably, by any reasonable definition, far fewer. And even those who live here live elsewhere also, live in houses where the rooms are not labelled atomic theory or genetics or the internal constitution of the stars, but quite different names like power and production and evil and beauty and history and children and the word of God.

We go in and out; even the most assiduous of us is not bound to this vast structure. One thing we find throughout the house: there are no locks; there are no shut doors; wherever we go there are the signs and usually the words of welcome. It is an open house, open to all comers.

The discoveries of science, the new rooms in this great house, have changed the way men think of things outside its walls. We have some glimmering now of the depth in time and the vastness in space of the physical world we live in. An awareness of how long our history and how immense our cosmos touches us even in simple earthly deliberations. We have learned from the natural history of the earth and from the story of evolution to have a sense of history, of time and change. We learn to talk of ourselves, and of the nature of the world and its reality as not wholly fixed in a silent quiet moment, but as unfolding with novelty and alteration, decay and new growth. We have understood something of the inner harmony and beauty of strange primitive cultures, and through this see the qualities of our own life in an altered perspective, and recognize its accidents as well as its inherent necessities. We are, I should think, not patriots less but patriots very differently for loving what is ours and understanding a little of the love of others for their lands and ways. We have begun to understand that it is not only in his rational life that man's psyche is intelligible, that even in what may appear to be his least rational actions and sentiments we may discover a new order. We have the beginnings of an understanding of what it is in man, and more in simple organisms, that is truly heritable, and rudimentary clues as to how the inheritance occurs. We know, in surprising detail, what is the physical counterpart of the act of vision and of other modes of perception. Not one of these new ideas and new insights is so little, or has so short a reach in its bearing on the common understanding but that it alone could make a proper theme for "Science and the Common Understanding." Yet we have been, bearing in mind my limited area of experience, in that one room of the part of the house where physics is, in which I have for some years worked and taught.

In that one room—in that relatively quiet room where we have been together—we have found things quite strange for those who have not been there before, yet reminiscent of what we have seen in other houses and known in other days. We have seen that in the atomic world we have been led by experience to use descriptions and ideas that apply to the large-scale world of matter, to the familiar world of our schoolday physics; ideas like the position of a body and its acceleration and its impulse and the forces acting on it; ideas like wave and interference; ideas like cause and probability. But what is new, what was not anticipated a half-century ago, is that, though to an atomic system there is a potential applicability of one or another of these ideas, in any real situation only some of these ways of description can be actual. This is because we need to take into account not merely the atomic system we are studying, but the means we use in observing it, and the fitness of these experimental means for defining and measuring selected properties of the system. All such ways of observing are needed for the whole experience of the atomic world; all but one are excluded in any actual experience. In the specific instance, there is a proper and consistent way to describe what the experience is; what it implies; what it predicts and thus how to deal with its consequences. But any such specific instance excludes by its existence the application of other ideas, other modes of prediction, other consequences. They are, we say, complementary to one another; atomic theory is in part an account of these descriptions and in part an understanding of the circumstances to which one applies, or another or another.

And so it is with man's life. He may be any of a number of things; he will not be all of them. He may be well versed, he may be a poet, he may be a creator in one or more than one science; he will not be all kinds of man or all kinds of scientist; and he will be lucky if he has a bit of familiarity outside the room in which he works.

So it is with the great antinomies that through the ages have organized and yet disunited man's experience: the antinomy between the ceaseless change and wonderful novelty and the perishing of all earthly things, and the eternity which inheres in every happening; in the antinomy between growth and order, between the spontaneous and changing and irregular and the symmetrical and balanced; in the related antinomy between freedom and necessity; between action, the life of the will, and observation and analysis and the life of reason; between the question "how?" and the questions "why?" and "to what end?"; between the causes that derive from natural law, from unvarying regularities in the natural world, and those other causes that express purposes and define goals and ends.

So it is in the antinomy between the individual and the community; man who is an end in himself and man whose tradition, whose culture, whose works, whose words have meaning in terms of other men and his relations to them. All our experience has shown that we can neither think, nor in any true sense live, without reference to these antinomic modes. We cannot in any sense be both the observers and the actors in any specific instance, or we shall fail properly to be either one or the other; yet we know that our life is built of these two modes, is part free and part inevitable, is part creation and part discipline, is part acceptance and part effort. We have no written rules that assign us to these ways; but we know that only folly and death of the spirit results when we deny one or the other, when we erect one as total and absolute and make the others derivative and secondary. We recognize this when we live as men. We talk to one another; we philosophize; we admire great men and their moments of greatness; we read; we study; we recognize and love in a particular act that happy union of the generally incompatible. With all of this we learn to use some reasonable part of the full register of man's resources.

We are, of course, an ignorant lot; even the best of us knows how to do only a very few things well; and of what is available in knowledge of fact, whether of science or of history, only the smallest part is in any one man's knowing.

The greatest of the changes that science has brought is the acuity of change; the greatest novelty the extent of novelty. Short of rare times of great disaster, civilizations have not known such rapid alteration in the conditions of their life, such rapid flowering of many varied sciences, such rapid changes in the ideas we have about the world and one another. What has been true in the days of a great disaster or great military defeat for one people at one time is true for all of us now, in the sense that our ends have little in common with our beginnings. Within a lifetime what we learned at school has been rendered inadequate by new discoveries and new inventions; the ways that we learn in childhood are only very meagerly adequate to the issues that we must meet in maturity.

In fact, of course, the notion of universal knowledge has always been an illusion; but it is an illusion fostered by the monistic view of the world in which a few great central truths determine in all its wonderful and amazing proliferation everything else that is true. We are not today tempted to search for these keys that unlock the whole of human knowledge and of man's experience. We know that we are ignorant; we are well taught it, and the more surely and deeply we know our own job the better able we are to appreciate the full measure of our pervasive ignorance. We know that these are inherent limits, compounded, no doubt, and exaggerated by that sloth and that complacency without which we would not be men at all.

But knowledge rests on knowledge; what is new is meaningful because it departs slightly from what was known before: this is a world of frontiers, where even the liveliest of actors or observers will be absent most of the time from most of them. Perhaps this sense was not so sharp in the village—that village which we have learned a little about but probably do not understand too well—the village of slow change and isolation and fixed culture which evokes our nostalgia even if not our full comprehension. Perhaps in the villages men were not so lonely; perhaps they found in each other a fixed community, a fixed and only slowly growing store of knowledge—a single world. Even that we may doubt, for there seem to be always in the culture of such times and places vast domains of mystery, if not unknowable, then imperfectly known, endless and open.

As for ourselves in these times of change, of ever-increasing knowledge, of collective power and individual impotence, of heroism and of drudgery, of progress and of tragedy, we too are brothers. And if we, who are the inheritors of two millennia of Christian tradition, understand that for us we have come to be brothers second by being children first, we know that in vast parts of the world where there has been no Christian tradition, and with men who never have been and never may be Christian in faith there is nevertheless a bond of brotherhood. We know this not only because of the almost universal ideal of human brotherhood and human community; we know it at first hand from the more modest, more diverse, more fleeting associations which are the substance of our life. The ideal of brotherhood, the ideal of fraternity in which all men, wicked and virtuous, wretched and fortunate, are banded together has its counterpart in the experience of communities, not ideal, not universal, imperfect, impermanent, as different from the ideal and as reminiscent of it as are the ramified branches of science from the ideal of a unitary, all-encompassing science of the eighteenth century.

Each of us knows from his own life how much even a casual and limited association of men goes beyond him in knowledge, in understanding, in humanity, and in power. Each of us, from a friend or a book or by concerting of the little we know with what others know, has broken the iron circle of his frustration. Each of us has asked help and been given it, and within our measure each of us has offered it. Each of us knows the great new freedom sensed almost as a miracle, that men banded together for some finite purpose experience from the power of their common effort. We are likely to remember the times of the last war, where the common danger brought forth in soldier, in worker, in scientist, and engineer a host of new experiences of the power and the comfort in even bleak undertakings, of common, concerted, co-operative life. Each of us knows how much he has been transcended by the group of which he has been or is a part; each of us has felt the solace of other men's knowledge to stay his own ignorance, of other men's wisdom to stay his folly, of other men's courage to answer his doubts or his weakness.

These are the fluid communities, some of long duration when circumstances favored—like the political party or many a trade union—some fleeting and vivid, encompassing in the time of their duration a moment only of the member's life; and in our world at least they are ramified and improvised, living and dying, growing and falling off almost as a form of life itself. This may be more true of the United States than of any other country. Certainly the bizarre and comical aspects impressed de Tocqueville more than a century ago when he visited our land and commented on the readiness with which men would band together: to improve the planting of a town, or for political reform, or for the pursuit or inter-exchange of knowledge, or just for the sake of banding together, because they liked one another or disliked someone else. Circumstances may have exaggerated the role of the societies, of the fluid and yet intense communities in the United States; yet these form a common pattern for our civilization. It brought men together in the Royal Society and in the French Academy and in the Philosophical Society that Franklin founded, in family, in platoon, on a ship, in the laboratory, in almost everything but a really proper club.

If we err today—and I think we do—it is in expecting too much of knowledge from the individual and too much of synthesis from the community. We tend to think of these communities, no less than of the larger brotherhood of man, as made up of individuals, as composed of them as an atom is of its ingredients. We think similarly of general laws and broad ideas as made up of the instances which illustrate them, and from an observation of which we may have learned them.

Yet this is not the whole. The individual event, the act, goes far beyond the general law. It is a sort of intersection of many generalities, harmonizing them in one instance as they cannot be harmonized in general. And we as men are not only the ingredients of our communities; we are their intersection, making a harmony which does not exist between the communities except as we, the individual men, may create it and reveal it. So much of what we think, our acts, our judgments of beauty and of right and wrong, come to us from our fellow men that what would be left were we to take all this away would be neither recognizable nor human. We are men because we are part of, but not because only part of, communities; and the attempt to understand man's brotherhood in terms only of the individual man is as little likely to describe our world as is the attempt to describe general laws as the summary of their instances. These are indeed two complementary views, neither reducible to the other, no more reducible than is the electron as wave to the electron as particle.

And this is the mitigant of our ignorance. It is true that

none of us will know very much; and most of us will see the end of our days without understanding in all its detail and beauty the wonders uncovered even in a single branch of a single science. Most of us will not even know, as a member of any intimate circle, anyone who has such knowledge; but it is also true that, although we are sure not to know everything and rather likely not to know very much, we can know anything that is known to man, and may, with luck and sweat, even find out some things that have not before been known to him. This possibility, which, as a universal condition of man's life is new, represents today a high and determined hope, not yet a reality; it is for us in England and in the United States not wholly remote or unfamiliar. It is one of the manifestations of our belief in equality, that belief which could perhaps better be described as a commitment to unparalleled diversity and unevenness in the distribution of attainments, knowledge, talent, and power.

This open access to knowledge, these unlocked doors and signs of welcome, are a mark of a freedom as fundamental as any. They give a freedom to resolve difference by converse, and, where converse does not unite, to let tolerance compose diversity. This would appear to be a freedom barely compatible with modern political tyranny. The multitude of communities, the free association for converse or for common purpose, are acts of creation. It is not merely that without them the individual is the poorer; without them a part of human life, not more nor less fundamental than the individual, is foreclosed. It is a cruel and humorless sort of pun that so powerful a present form of modern tyranny should call itself by the very name of a belief in community, by a word "communism" which in other times evoked memories of villages and village inns and of artisans concerting their skills, and of men of learning content with anonymity. But perhaps only a malignant end can follow the systematic belief that all communities are one community: that all truth is one truth; that all experience is compatible with all other; that total knowledge is possible; that all that is potential can exist as actual. This is not man's fate; this is not his path; to force him on it makes him resemble not that divine image of the all-knowing and all-powerful but the helpless, iron-bound prisoner of a dying world. The open society, the unrestricted access to knowledge, the unplanned and uninhibited association of men for its furtherance—these are what may make a vast, complex, ever-growing, ever-changing, ever more specialized and expert technological world nevertheless a world of human community.

So it is with the unity of science—that unity that is far more a unity of comparable dedication than a unity of common total understanding. This heartening phrase, "the unity of science," often tends to evoke a wholly false picture, a picture of a few basic truths, a few critical techniques, methods, and ideas, from which all discoveries and understanding of science derive; a sort of central exchange, access to which will illuminate the atoms and the galaxies, the genes and the sense organs. The unity of science is based rather on just such a community as I have described. All parts of it are open to all of us, and this is no merely formal invitation. The history of science is rich in example of the fruitfulness of bringing two sets of techniques, two sets of ideas, developed in separate contexts for the pursuit of new truth, into touch with one another. The sciences fertilize each other; they grow by contact and by common enterprise. Once again, this means that the scientist may profit from learning about any other science; it does not mean that he must learn about them all. It means that the unity is a potential unity, the unity of the things that might be brought together and might throw light one on the other. It is not global or total or hierarchical.

Even in science, and even without visiting the room in its house called atomic theory, we are again and again reminded of the complementary traits in our own life, even in our own professional life. We are nothing without the work of others our predecessors, others our teachers, others our contemporaries. Even when, in the measure of our adequacy and our fullness, new insight and new order are created, we are still nothing without others. Yet we are more.

There is a similar duality in our relations to wider society. For society our work means many things: pleasure, we hope, for those who follow it; instruction for those who perhaps need it; but also and far more widely, it means a common power, a power to achieve that which could not be achieved without knowledge. It means the cure of illness and the alleviation of suffering; it means the easing of labor and the widening of the readily accessible frontiers of experience, of communication, and of instruction. It means, in an earthy way, the power of betterment—that riddled word. We are today anxiously aware that the power to change is not always necessarily good.

As new instruments of war, of newly massive terror, add to the ferocity and totality of warfare, we understand that it is a special mark and problem of our age that man's ever-present preoccupation with improving his lot, with alleviating hunger and poverty and exploitation, must be brought into harmony with the over-riding need to limit and largely to eliminate resort to organized violence between nation and nation. The increasingly expert destruction of man's spirit by the power of police, more wicked if not more awful than the ravages of nature's own hand, is another such power, good only if never to be used.

We regard it as proper and just that the patronage of science by society is in large measure based on the increased power which knowledge gives. If we are anxious that the power so given and so obtained be used with wisdom and with love of humanity, that is an anxiety we share with almost everyone. But we also know how little of the deep new knowledge which has altered the face of the world, which has changed—and increasingly and ever more profoundly must change—man's views of the world, resulted from a quest for practical ends or an interest in exercising the power that knowledge gives. For most of us, in most of those moments when we were most free of corruption, it has been the beauty of the world of nature and the strange and compelling harmony of its order, that has sustained, inspirited, and led us. That also is as it should be. And if the forms in which society provides and exercises its patronage leave these incentives strong and secure, new knowledge will never stop as long as there are men.

We know that our work is rightly both an instrument and an end. A great discovery is a thing of beauty; and our faith—our binding, quiet faith—is that knowledge is good and good in itself. It is also an instrument; it is an instrument for our successors, who will use it to probe elsewhere and more deeply; it is an instrument for technology, for the practical arts, and for man's affairs. So it is with us as scientists; so it is with us as men. We are at once instrument and end, discoverers and teachers, actors and observers. We understand, as we hope others understand, that in this there is a harmony between knowledge in the sense of science, that specialized and general knowledge which it is our purpose to uncover, and the community of man. We, like all men, are among those who bring a little light to the vast unending darkness of man's life and world. For us as for all men, change and eternity, specialization and unity, instrument and final purpose, community and individual man alone, complementary each to the other, both require and define our bonds and our freedom.

3

Vannevar Bush, SCIENCE - THE ENDLESS FRONTIER

James K. Feibleman, PURE SCIENCE, APPLIED SCIENCE, AND TECHNOLOGY: AN ATTEMPT AT DEFINITIONS

Frederick W. Taylor, THE PRINCIPLES OF SCIENTIFIC MANAGEMENT

1. Why did Bush feel that it is necessary for government to support major projects of scientific research?

2. What advantages accrue to society generally from scientific research?

3. How does scientific research help or hurt the basic economy according to Bush?

4. Define pure science.

5. Define applied science. How does applied science differ from pure science?

6. Define technology. How does technology differ from applied science?

7. Why does Taylor think efficiency is so important?

8. What will result from improving efficiency from the point of view of the worker?

9. What are the basic steps or principles of Taylor's scientific management?

10. What are the basic causes of inefficiency according to Taylor?

11. Do you object to thinking of the management of people as a science? Why or why not?

Vannevar Bush (1890-1974), an electrical engineer, became the first director of the Office of Scientific Research and Development during World War II. In this position, he was responsible for mobilizing and coordinating the scientific aspects of the United States' war effort, a task that included the development of radar and the atomic bomb. After the war, as the excerpt from *Science - The Endless Frontier: A Report to the President* demonstrates, he helped provide the impetus for the federal government's encouragement of science and basic research.

SCIENCE: THE ENDLESS FRONTIER

Science can be effective in the national welfare only as a member of a team, whether the conditions be peace or war. But without scientific progress no amount of achievement in other directions can insure our health, prosperity, and security as a nation in the modern world.

For the War Against Disease

We have taken great strides in the war against disease. The death rate for all diseases in the Army, including overseas forces, has been reduced from 14.1 per thousand in the last war to 0.6 per thousand in this war. In the last 40 years life expectancy has increased from 49 to 65 years, largely as a consequence of the reduction in the death rates of infants and children. But we are far from the goal. The annual deaths from one or two diseases far exceed the total number of American lives lost in battle during this war. A large fraction of these deaths in our civilian population cut short the useful lives of our citizens. Approximately 7,000,000 persons in the United States are mentally ill and their care costs the public over $175,000,000 a year. Clearly much illness remains for which adequate means of prevention and cure are not yet known.

The responsibility for basic research in medicine and the underlying sciences, so essential to progress in the war against disease, falls primarily upon the medical schools and universities. Yet we find that the traditional sources of support for medical research in the medical schools and universities, largely endowment income, foundation grants, and private donations, are diminishing and there is no immediate prospect of a change in this trend. Meanwhile, the cost of medical research has been rising. If we are to maintain the progress in medicine which has marked the last 25 years, the Government should extend financial support to basic medical research in the medical schools and in universities.

For Our National Security

The bitter and dangerous battle against the U-boat was a battle of scientific techniques—and our margin of success was dangerously small. The new eyes which radar has supplied can sometimes be blinded by new scientific developments. V–2 was countered only by capture of the launching sites.

We cannot again rely on our allies to hold off the enemy while we struggle to catch up. There must be more—and more adequate—military research in peacetime. It is essential that the civilian scientists continue in peacetime some portion of those contributions to national security which they have made so effectively during the war. This can best be done through a civilian-controlled organization with close liaison with the Army and Navy, but with funds direct from Congress, and the clear power to initiate military research which will supplement and strengthen that carried on directly under the control of the Army and Navy.

And for the Public Welfare

One of our hopes is that after the war there will be full employment. To reach that goal the full creative and productive energies of the American people must be released. To create more jobs we must make new and better and cheaper products. We want plenty of new, vigorous enterprises. But new products and processes are not born full-grown. They are founded on new principles and new conceptions which in turn result from basic scientific research. Basic scientific research is scientific capital. Moreover, we cannot any longer depend upon Europe as a major source of this scientific capital. Clearly, more and better scientific research is one essential to the achievement of our goal of full employment.

How do we increase this scientific capital? First, we must have plenty of men and women trained in science, for upon them depends both the creation of new knowledge and its application to practical purposes. Second, we must strengthen the centers of basic research which are principally the colleges, universities, and research institutes. These institutions provide the environment which is most conducive to the creation of new scientific knowledge and least under pressure for immediate, tangible results. With some notable exceptions, most research in industry and in Government involves application of existing scientific knowledge to practical problems. It is only the colleges, universities, and a few research institutes that devote most of their research efforts to expanding the frontiers of knowledge.

Expenditures for scientific research by industry and Government increased from $140,000,000 in 1930 to $309,000,000 in 1940. Those for the colleges and universities increased from $20,000,000 to $31,000,000, while those for research institutes declined from $5,200,000 to $4,500,000 during the same period. If the colleges, universities, and research institutes are to meet the rapidly increasing demands of industry and Government for new scientific knowledge, their basic research should be strengthened by use of public funds.

For science to serve as a powerful factor in our national welfare, applied research both in Government and in industry must be vigorous. To improve the quality of scientific research within the Government, steps should be taken to modify the procedures for recruiting, classifying, and compensating scientific personnel in order to reduce the present handicap of governmental scientific bureaus in competing with industry and the universities for top-grade scientific talent. To provide coordination of the common scientific activities of these governmental agencies as to policies and budgets, a permanent Science Advisory Board should be created to advise the executive and legislative branches of Government on these matters.

The most important ways in which the Government can promote industrial research are to increase the flow of new scientific knowledge through support of basic research, and to aid in the development of scientific talent. In addition, the Government should provide suitable incentives to industry to conduct research (*a*) by clarification of present uncertainties in the Internal Revenue Code in regard to the deductibility of research and development expenditures as current charges against net income, and (*b*) by strengthening the patent system so as to eliminate uncertainties which now bear heavily on small industries and so as to prevent abuses which reflect discredit upon a basically sound system. In addition, ways should be found to cause the benefits of basic research to reach industries which do not now utilize new scientific knowledge.

We Must Renew Our Scientific Talent

The responsibility for the creation of new scientific knowledge—and for most of its application—rests on that small body of men and women who understand the fundamental laws of nature and are skilled in the techniques of scientific research. We shall have rapid or slow advance on any scientific frontier depending on the number of highly qualified and trained scientists exploring it.

The deficit of science and technology students who, but for the war, would have received bachelor's degrees is about 150,000. It is estimated that the deficit of those obtaining advanced degrees in these fields will amount in 1955 to about 17,000—for it takes at least 6 years from college entry to achieve a doctor's degree or its equivalent in science or engineering. The real ceiling on our productivity of new scientific knowledge and its application in the war against disease, and the development of new products and new industries, is the number of trained scientists available.

The training of a scientist is a long and expensive process. Studies clearly show that there are talented individuals in every part of the population, but with few exceptions, those without the means of buying higher education go without it. If ability, and not the circumstance of family fortune, determines who shall receive higher education in science, then we shall be assured of constantly improving quality at every level of scientific activity. The Government should provide a reasonable number of undergraduate scholarships and graduate fellowships in order to develop scientific talent in American youth. The plans should be designed to attract into science only that proportion of youthful talent appropriate to the needs of science in relation to the other needs of the Nation for high abilities.

Including Those in Uniform

The most immediate prospect of making up the deficit in scientific personnel is to develop the scientific talent in the generation now in uniform. Even if we should start now to train the current crop of high-school graduates none would complete graduate studies before 1951. The Armed Services should comb their records for men who, prior to or during the war, have given evidence of talent for science, and make prompt arrangements, consistent with current discharge plans, for ordering those who remain in uniform, as soon as militarily possible, to duty at institutions here and overseas where they can continue their scientific education. Moreover, the Services should see that those who study overseas have the benefit of the latest scientific information resulting from research during the war.

The Lid Must Be Lifted

While most of the war research has involved the application of existing scientific knowledge to the problems of war, rather than basic research, there has been accumulated a vast amount of information relating to the application of science to particular problems. Much of this can be used by industry. It is also needed for teaching in the colleges and universities here and in the Armed Forces Institutes overseas. Some of this information must remain secret, but most of it should be made public as soon as there is ground for belief that the enemy will not be able to turn it against us in this war. To select that portion which should be made public, to coordinate its release, and definitely to encourage its publication, a Board composed of Army, Navy, and civilian scientific members should be promptly established.

A Program for Action

The Government should accept new responsibilities for promoting the flow of new scientific knowledge and the development of scientific talent in our youth. These responsibilities are the proper concern of the Government, for they vitally affect our health, our jobs, and our national security. It is in keeping also with basic United States policy that the Government should foster the opening of new frontiers and this is the modern way to do it. For many years the Government has wisely supported research in the agricultural colleges and the benefits have been great. The time has come when such support should be extended to other fields.

The effective discharge of these new responsibilities will require the full attention of some over-all agency devoted to that purpose. There is not now in the permanent governmental structure receiving its funds from Congress an agency adapted to supplementing the support of basic research in the colleges, universities, and research institutes, both in medicine and the natural sciences, adapted to supporting research on new weapons for both Services, or adapted to administering a program of science scholarships and fellowships.

Therefore I recommend that a new agency for these purposes be established. Such an agency should be composed of persons of broad interest and experience, having an understanding of the peculiarities of scientific research and scientific education. It should have stability of funds so that long-range programs may be undertaken. It should recognize that freedom of inquiry must be preserved and should leave internal control of policy, personnel, and the method and scope of research to the institutions in which it is carried on. It should be fully responsible to the President and through him to the Congress for its program.

Early action on these recommendations is imperative if this Nation is to meet the challenge of science in the crucial years ahead. On the wisdom with which we bring science to bear in the war against disease, in the creation of new industries, and in the strengthening of our Armed Forces depends in large measure our future as a nation.

Scientific Progress Is Essential

We all know how much the new drug, penicillin, has meant to our grievously wounded men on the grim battlefronts of this war—the countless lives it has saved—the incalculable suffering which its use has prevented. Science and the great practical genius of this Nation made this achievement possible.

Some of us know the vital role which radar has played in bringing the Allied Nations to victory over Nazi Germany and in driving the Japanese steadily back from their island bastions. Again it was painstaking scientific research over many years that made radar possible.

What we often forget are the millions of pay envelopes on a peacetime Saturday night which are filled because new products and new industries have provided jobs for countless Americans. Science made that possible, too.

In 1939 millions of people were employed in industries which did not even exist at the close of the last war—radio, air conditioning, rayon and other synthetic fibers, and plastics are examples of the products of these industries. But these things do not mark the end of progress—they are but the beginning if we make full use of our scientific resources. New manufacturing industries can be started and many older industries greatly strengthened and expanded if we continue to study nature's laws and apply new knowledge to practical purposes.

Great advances in agriculture are also based upon scientific research. Plants which are more resistant to disease and are adapted to short growing seasons, the prevention and cure of livestock diseases, the control of our insect enemies, better fertilizers, and improved agricultural practices, all stem from painstaking scientific research.

Advances in science when put to practical use mean more jobs, higher wages, shorter hours, more abundant crops, more leisure for recreation, for study, for learning how to live without the deadening drudgery which has been the burden of the common man for ages past. Advances in science will also bring higher standards of living, will lead to the prevention or cure of diseases, will promote conservation of our limited national resources, and will assure means of defense against aggression. But to achieve these objectives—to secure a high level of employment, to maintain a position of world leadership—the flow of new scientific knowledge must be both continuous and substantial.

Our population increased from 75 million to 130 million between 1900 and 1940. In some countries comparable increases have been accompanied by famine. In this country the increase has been accompanied by more abundant food supply, better living, more leisure, longer life, and better health. This is, largely, the product of three factors—the free play of initiative of a vigorous people under democracy, the heritage of great natural wealth, and the advance of science and its application.

Science, by itself, provides no panacea for individual, social, and economic ills. It can be effective in the national welfare only as a member of a team, whether the conditions be peace or war. But without scientific progress no amount of achievement in other directions can insure our health, prosperity, and security as a nation in the modern world.

Science Is a Proper Concern of Government

It has been basic United States policy that Government should foster the opening of new frontiers. It opened the seas to clipper ships and furnished land for pioneers. Although these frontiers have more or less disappeared, the frontier of science remains. It is in keeping with the American tradition —one which has made the United States great—that new frontiers shall be made accessible for development by all American citizens.

Moreover, since health, well-being, and security are proper concerns of Government, scientific progress is, and must be, of vital interest to Government. Without scientific progress the national health would deteriorate; without scientific progress we could not hope for improvement in our standard of living or for an increased number of jobs for our citizens; and without scientific progress we could not have maintained our liberties against tyranny.

Government Relations to Science—Past and Future

From early days the Government has taken an active interest in scientific matters. During the nineteenth century the Coast and Geodetic Survey, the Naval Observatory, the Department of Agriculture, and the Geological Survey were established. Through the Land Grant College Acts the Government has supported research in state institutions for more than 80 years on a gradually increasing scale. Since 1900 a large number of scientific agencies have been established within the Federal Government, until in 1939 they numbered more than 40.

Much of the scientific research done by Government agencies is intermediate in character between the two types of work commonly referred to as basic and applied research. Almost all Government scientific work has ultimate practical objectives but, in many fields of broad national concern, it commonly involves long-term investigation of a fundamental nature. Generally speaking, the scientific agencies of Government are not so concerned with immediate practical objectives as are the laboratories of industry nor, on the other hand, are they as free to explore any natural phenomena without regard to possible economic applications as are the educational and private research institutions. Government scientific agencies have splendid records of achievement, but they are limited in function.

We have no national policy for science. The Government has only begun to utilize science in the Nation's welfare. There is no body within the Government charged with formulating or executing a national science policy. There are no standing committees of the Congress devoted to this important subject. Science has been in the wings. It should be brought to the center of the stage—for in it lies much of our hope for the future.

There are areas of science in which the public interest is acute but which are likely to be cultivated inadequately if left without more support than will come from private sources. These areas—such as research on military problems, agriculture, housing, public health, certain medical research, and research involving expensive capital facilities beyond the capacity of private institutions—should be advanced by active Government support. To date, with the exception of the intensive war research conducted by the Office of Scientific Research and Development, such support has been meager and intermittent.

For reasons presented in this report we are entering a period when science needs and deserves increased support from public funds.

Freedom of Inquiry Must Be Preserved

The publicly and privately supported colleges, universities, and research institutes are the centers of basic research. They are the wellsprings of knowledge and understanding. As long as they are vigorous and healthy and their scientists are free to pursue the truth wherever it may lead, there will be a flow of new scientific knowledge to those who can apply it to practical problems in Government, in industry, or elsewhere.

Many of the lessons learned in the war-time application of science under Government can be profitably applied in peace. The Government is peculiarly fitted to perform certain functions, such as the coordination and support of broad programs on problems of great national importance. But we must proceed with caution in carrying over the methods which work in wartime to the very different conditions of peace. We must remove the rigid controls which we have had to impose, and recover freedom of inquiry and that healthy competitive scientific spirit so necessary for expansion of the frontiers of scientific knowledge.

Scientific progress on a broad front results from the free play of free intellects, working on subjects of their own choice, in the manner dictated by their curiosity for exploration of the unknown. Freedom of inquiry must be preserved under any plan for Government support of science in accordance with the Five Fundamentals listed on page 32.

The study of the momentous questions presented in President Roosevelt's letter has been made by able committees working diligently. This report presents conclusions and recommendations based upon the studies of these committees which appear in full as the appendices. Only in the creation of one over-all mechanism rather than several does this report depart from the specific recommendations of the committees. The members of the committees have reviewed the recommendations in regard to the single mechanism and have found this plan thoroughly acceptable.

SCIENCE AND THE PUBLIC WELFARE

Relation to National Security

In this war it has become clear beyond all doubt that scientific research is absolutely essential to national security. The bitter and dangerous battle against the U-boat was a battle of scientific techniques—and our margin of success was dangerously small. The new eyes which radar supplied to our fighting forces quickly evoked the development of scientific countermeasures which could often blind them. This again represents the ever continuing battle of techniques. The V–1 attack on London was finally defeated by three devices developed during this war and used superbly in the field. V–2 was countered only by capture of the launching sites.

The Secretaries of War and Navy recently stated in a joint letter to the National Academy of Sciences:

> This war emphasizes three facts of supreme importance to national security: (1) Powerful new tactics of defense and offense are developed around new weapons created by scientific and engineering research; (2) the competitive time element in developing those weapons and tactics may be decisive; (3) war is increasingly total war, in which the armed services must be supplemented by active participation of every element of civilian population.
>
> To insure continued preparedness along farsighted technical lines, the research scientists of the country must be called upon to continue in peacetime some substantial portion of those types of contribution to national security which they have made so effectively during the stress of the present war * * *.

There must be more—and more adequate—military research during peacetime. We cannot again rely on our allies to hold off the enemy while we struggle to catch up. Further, it is clear that only the Government can undertake military research; for it must be carried on in secret, much of it has no commercial value, and it is expensive. The obligation of Government to support research on military problems is inescapable.

Modern war requires the use of the most advanced scientific techniques. Many of the leaders in the development of radar are scientists who before the war had been exploring the nucleus of the atom. While there must be increased emphasis on science in the future training of officers for both the Army and Navy, such men cannot be expected to be specialists in scientific research. Therefore, a professional partnership between the officers in the Services and civilian scientists is needed.

The Army and Navy should continue to carry on research and development on the improvement of current weapons. For many years the National Advisory Committee for Aeronautics has supplemented the work of the Army and Navy by conducting basic research on the problems of flight. There should now be permanent civilian activity to supplement the research work of the Services in other scientific fields so as to carry on in time of peace some part of the activities of the emergency wartime Office of Scientific Research and Development.

Military preparedness requires a permanent independent, civilian-controlled organization, having close liaison with the Army and Navy, but with funds directly from Congress and with the clear power to initiate military research which will supplement and strengthen that carried on directly under the control of the Army and Navy.

Science and Jobs

One of our hopes is that after the war there will be full employment, and that the production of goods and services will serve to raise our standard of living. We do not know yet how we shall reach that goal, but it is certain that it can be achieved only by releasing the full creative and productive energies of the American people.

Surely we will not get there by standing still, merely by making the same things we made before and selling them at the same or higher prices. We will not get ahead in international trade unless we offer new and more attractive and cheaper products.

Where will these new products come from? How will we find ways to make better products at lower cost? The answer is clear. There must be a stream of new scientific knowledge to turn the wheels of private and public enterprise. There must be plenty of men and women trained in science and technology for upon them depend both the creation of new knowledge and its application to practical purposes.

More and better scientific research is essential to the achievement of our goal of full employment.

The Importance of Basic Research

Basic research is performed without thought of practical ends. It results in general knowledge and an understanding of nature and its laws. This general knowledge provides the means of answering a large number of important practical problems, though it may not give a complete specific answer to any one of them. The function of applied research is to provide such complete answers. The scientist doing basic research may not be at all interested in the practical applications of his work, yet the further progress of industrial development would eventually stagnate if basic scientific research were long neglected.

One of the peculiarities of basic science is the variety of paths which lead to productive advance. Many of the most important discoveries have come as a result of experiments undertaken with very different purposes in mind. Statistically it is certain that important and highly useful discoveries will result from some fraction of the undertakings in basic science; but the results of any one particular investigation cannot be predicted with accuracy.

Basic research leads to new knowledge. It provides scientific capital. It creates the fund from which the practical applications of knowledge must be drawn. New products and new processes do not appear full-grown. They are founded on new principles and new conceptions, which in turn are painstakingly developed by research in the purest realms of science.

Today, it is truer than ever that basic research is the pacemaker of technological progress. In the nineteenth century, Yankee mechanical ingenuity, building largely upon the basic discoveries of European scientists, could greatly advance the technical arts. Now the situation is different.

A nation which depends upon others for its new basic scientific knowledge will be slow in its industrial progress and weak in its competitive position in world trade, regardless of its mechanical skill.

Centers of Basic Research

Publicly and privately supported colleges and universities and the endowed research institutes must furnish both the new scientific knowledge and the trained research workers. These institutions are uniquely qualified by tradition and by their special characteristics to carry on basic research. They are charged with the responsibility of conserving the knowledge accumulated by the past, imparting that knowledge to students, and contributing new knowledge of all kinds. It is chiefly in these institutions that scientists may work in an atmosphere which is relatively free from the adverse pressure of convention, prejudice, or commercial necessity. At their best they provide the scientific worker with a strong sense of solidarity and security, as well as a substantial degree of personal intellectual freedom. All of these factors are of great importance in the development of new knowledge, since much of new knowledge is certain to arouse opposition because of its tendency to challenge current beliefs or practice.

Industry is generally inhibited by preconceived goals, by its own clearly defined standards, and by the constant pressure of commercial necessity. Satisfactory progress in basic science seldom occurs under conditions prevailing in the normal industrial laboratory. There are some notable exceptions, it is true, but even in such cases it is rarely possible to match the universities in respect to the freedom which is so important to scientific discovery.

To serve effectively as the centers of basic research these institutions must be strong and healthy. They must attract our best scientists as teachers and investigators. They must offer research opportunities and sufficient compensation to enable them to compete with industry and government for the cream of scientific talent.

During the past 25 years there has been a great increase in industrial research involving the application of scientific knowledge to a multitude of practical purposes—thus providing new products, new industries, new investment opportunities, and millions of jobs. During the same period research within Government—again largely applied research—has also been greatly expanded. In the decade from 1930 to 1940 expenditures for industrial research increased from $116,000,000 to $240,000,000 and those for scientific research in Government rose from $24,000,000 to $69,000,000. During the same period expenditures for scientific research in the colleges and universities increased from $20,000,000 to $31,000,000, while those in the endowed research institutes declined from $5,200,000 to $4,500,000. These are the best estimates available. The figures have been taken from a variety of sources and arbitrary definitions have necessarily been applied, but it is believed that they may be accepted as indicating the following trends:

(a) Expenditures for scientific research by industry and Government—almost entirely applied research—have more than doubled between 1930 and 1940. Whereas in 1930 they were six times as large as the research expenditures of the colleges, universities, and research institutes, by 1940 they were nearly ten times as large.

(b) While expenditures for scientific research in the colleges and universities increased by one-half during this period, those for the endowed research institutes have slowly declined.

If the colleges, universities, and research institutes are to meet the rapidly increasing demands of industry and Government for new scientific knowledge, their basic research should be strengthened by use of public funds.

Research Within the Government

Although there are some notable exceptions, most research conducted within governmental laboratories is of an applied nature. This has always been true and is likely to remain so. Hence Government, like industry, is dependent upon the colleges, universities, and research institutes to expand the basic scientific frontiers and to furnish trained scientific investigators.

Research within the Government represents an important part of our total research activity and needs to be strengthened and expanded after the war. Such expansion should be directed to fields of inquiry and service which are of public importance and are not adequately carried on by private organizations.

The most important single factor in scientific and technical work is the quality of personnel employed. The procedures currently followed within the Government for recruiting, classifying and compensating such personnel place the Government under a severe handicap in competing with industry and the universities for first-class scientific talent. Steps should be taken to reduce that handicap.

In the Government the arrangement whereby the numerous scientific agencies form parts of large departments has both advantages and disadvantages. But the present pattern is firmly established and there is much to be said for it. There is, however, a very real need for some measure of coordination of the common scientific activities of these agencies, both as to policies and budgets, and at present no such means exist.

A permanent Science Advisory Board should be created to consult with these scientific bureaus and to advise the executive and legislative branches of Government as to the policies and budgets of Government agencies engaged in scientific research.

This board should be composed of disinterested scientists who have no connection with the affairs of any Government agency.

Industrial Research

The simplest and most effective way in which the Government can strengthen industrial research is to support basic research and to develop scientific talent.

The benefits of basic research do not reach all industries equally or at the same speed. Some small enterprises never receive any of the benefits. It has been suggested that the benefits might be better utilized if "research clinics" for such enterprises were to be established. Businessmen would thus be able to make more use of research than they now do. This proposal is certainly worthy of further study.

One of the most important factors affecting the amount of industrial research is the income-tax law. Government action in respect to this subject will affect the rate of technical progress in industry. Uncertainties as to the attitude of the Bureau of Internal Revenue regarding the deduction of research and development expenses are a deterrent to research expenditure. These uncertainties arise from lack of clarity of the tax law as to the proper treatment of such costs.

The Internal Revenue Code should be amended to remove present uncertainties in regard to the deductibility of research and development expenditures as current charges against net income.

Research is also affected by the patent laws. They stimulate new invention and they make it possible for new industries to be built around new devices or new processes. These industries generate new jobs and new products, all of which contribute to the welfare and the strength of the country.

Yet, uncertainties in the operation of the patent laws have impaired the ability of small industries to translate new ideas into processes and products of value to the Nation. These uncertainties are, in part, attributable to the difficulties and expense incident to the operation of the patent system as it presently exists. These uncertainties are also attributable to the existence of certain abuses which have appeared in the use of patents. The abuses should be corrected. They have led to extravagantly critical attacks which tend to discredit a basically sound system.

It is important that the patent system continue to serve the country in the manner intended by the Constitution, for it has been a vital element in the industrial vigor which has distinguished this Nation.

The National Patent Planning Commission has reported on this subject. In addition, a detailed study, with recommendations concerning the extent to which modifications should be made in our patent laws is currently being made under the leadership of the Secretary of Commerce. It is recommended, therefore, that specific action with regard to the patent laws be withheld pending the submission of the report devoted exclusively to that subject.

International Exchange of Scientific Information

International exchange of scientific information is of growing importance. Increasing specialization of science will make it more important than ever that scientists in this country keep continually abreast of developments abroad. In addition, a flow of scientific information constitutes one facet of general international accord which should be cultivated.

The Government can accomplish significant results in several ways: by aiding in the arrangement of international science congresses, in the official accrediting of American scientists to such gatherings, in the official reception of foreign scientists of standing in this country, in making possible a rapid flow of technical information, including translation service, and possibly in the provision of international fellowships. Private foundations and other groups partially fulfill some of these functions at present, but their scope is incomplete and inadequate.

The Government should take an active role in promoting the international flow of scientific information.

The Special Need for Federal Support

We can no longer count on ravaged Europe as a source of fundamental knowledge. In the past we have devoted much of our best efforts to the application of such knowledge which has been discovered abroad. In the future we must pay increased attention to discovering this knowledge for ourselves particularly since the scientific applications of the future will be more than ever dependent upon such basic knowledge.

New impetus must be given to research in our country. Such new impetus can come promptly only from the Government. Expenditures for research in the colleges, universities, and research institutes will otherwise not be able to meet the additional demands of increased public need for research.

Further, we cannot expect industry adequately to fill the gap. Industry will fully rise to the challenge of applying new knowledge to new products. The commercial incentive can be relied upon for that. But basic research is essentially noncommercial in nature. It will not receive the attention it requires if left to industry.

For many years the Government has wisely supported research in the agricultural colleges and the benefits have been great. The time has come when such support should be extended to other fields.

In providing Government support, however, we must endeavor to preserve as far as possible the private support of research both in industry and in the colleges, universities, and research institutes. These private sources should continue to carry their share of the financial burden.

The Cost of a Program

It is estimated that an adequate program for Federal support of basic research in the colleges, universities, and research institutes and for financing important applied research in the public interest, will cost about 10 million dollars at the outset and may rise to about 50 million dollars annually when fully underway at the end of perhaps 5 years.

THE MEANS TO THE END

New Responsibilities for Government

One lesson is clear from the reports of the several committees attached as appendices. The Federal Government should accept new responsibilities for promoting the creation of new scientific knowledge and the development of scientific talent in our youth.

The extent and nature of these new responsibilities are set forth in detail in the reports of the committees whose recommendations in this regard are fully endorsed.

In discharging these responsibilities Federal funds should be made available. We have given much thought to the question of how plans for the use of Federal funds may be arranged so that such funds will not drive out of the picture funds from local governments, foundations, and private donors. We believe that our proposals will minimize that effect, but we do not think that it can be completely avoided. We submit, however, that the Nation's need for more and better scientific research is such that the risk must be accepted.

It is also clear that the effective discharge of these responsibilities will require the full attention of some over-all agency devoted to that purpose. There should be a focal point within the Government for a concerted program of assisting scientific research conducted outside of Government. Such an agency should furnish the funds needed to support basic research in the colleges and universities, should coordinate where possible research programs on matters of utmost importance to the national welfare, should formulate a national policy for the Government toward science, should sponsor the interchange of scientific information among scientists and laboratories both in this country and abroad, and should ensure that the incentives to research in industry and the universities are maintained. All of the committees advising on these matters agree on the necessity for such an agency.

The Mechanism

There are within Government departments many groups whose interests are primarily those of scientific research. Notable examples are found within the Departments of Agriculture, Commerce, Interior, and the Federal Security Agency. These groups are concerned with science as collateral and peripheral to the major problems of those Departments. These groups should remain where they are, and continue to perform their present functions, including the support of agricultural research by grants to the land grant colleges and experimental stations, since their largest contribution lies in applying fundamental knowledge to the special problems of the Departments within which they are established.

By the same token these groups cannot be made the repository of the new and large responsibilities in science which belong to the Government and which the Government should accept. The recommendations in this report which relate to research within the Government, to the release of scientific information, to clarification of the tax laws, and to the recovery and development of our scientific talent now in uniform can be implemented by action within the existing structure of the Government. But nowhere in the governmental structure receiving its funds from Congress is there an agency adapted to supplementing the support of basic research in the universities, both in medicine and the natural sciences; adapted to supporting research on new weapons for both Services; or adapted to administering a program of science scholarships and fellowships.

A new agency should be established, therefore, by the Congress for the purpose. Such an agency, moreover, should be an independent agency devoted to the support of scientific research and advanced scientific education alone. Industry learned many years ago that basic research cannot often be fruitfully conducted as an adjunct to or a subdivision of an operating agency or department. Operating agencies have immediate operating goals and are under constant pressure to produce in a tangible way, for that is the test of their value. None of these conditions is favorable to basic research. Research is the exploration of the unknown and is necessarily speculative. It is inhibited by conventional approaches, traditions, and standards. It cannot be satisfactorily conducted in an atmosphere where it is gauged and tested by operating or production standards. Basic scientific research should not, therefore, be placed under an operating agency whose paramount concern is anything other than research. Research will always suffer when put in competition with operations. The decision that there should be a new and independent agency was reached by each of the committees advising in these matters.

I am convinced that these new functions should be centered in one agency. Science is fundamentally a unitary thing. The number of independent agencies should be kept to a minimum. Much medical progress, for example, will come from fundamental advances in chemistry. Separation of the sciences in tight compartments, as would occur if more than one agency were involved, would retard and not advance scientific knowledge as a whole.

Five Fundamentals

There are certain basic principles which must underlie the program of Government support for scientific research and education if such support is to be effective and if it is to avoid impairing the very things we seek to foster. These principles are as follows:

(1) Whatever the extent of support may be, there must be stability of funds over a period of years so that long-range programs may be undertaken.

(2) The agency to administer such funds should be composed of citizens selected only on the basis of their interest in and capacity to promote the work of the agency. They should be persons of broad interest in and understanding of the peculiarities of scientific research and education.

(3) The agency should promote research through contracts or grants to organizations outside the Federal Government. It should not operate any laboratories of its own.

(4) Support of basic research in the public and private colleges, universities, and research institutes must leave the internal control of policy, personnel, and the method and scope of the research to the institutions themselves. This is of the utmost importance.

(5) While assuring complete independence and freedom for the nature, scope, and methodology of research carried on in the institutions receiving public funds, and while retaining discretion in the allocation of funds among such institutions, the Foundation proposed herein must be responsible to the President and the Congress. Only through such responsibility can we maintain the proper relationship between science and other aspects of a democratic system. The usual controls of audits, reports, budgeting, and the like, should, of course, apply to the administrative and fiscal operations of the Foundation, subject, however, to such adjustments in procedure as are necessary to meet the special requirements of research.

Basic research is a long-term process—it ceases to be basic if immediate results are expected on short-term support. Methods should therefore be found which will permit the agency to make commitments of funds from current appropriations for programs of five years duration or longer. Continuity and stability of the program and its support may be expected (*a*) from the growing realization by the Congress of the benefits to the public from scientific research, and (*b*) from the conviction which will grow among those who conduct research under the auspices of the agency that good quality work will be followed by continuing support.

Military Research

As stated earlier in this report, military preparedness requires a permanent, independent, civilian-controlled organization, having close liaison with the Army and Navy, but with funds direct from Congress and the clear power to initiate military research which will supplement and strengthen that carried on directly under the control of the Army and Navy. As a temporary measure the National Academy of Sciences has established the Research Board for National Security at the request of the Secretary of War and the Secretary of the Navy. This is highly desirable in order that there may be no interruption in the relations between scientists and military men after the emergency wartime Office of Scientific Research and Development goes out of existence. The Congress is now considering legislation to provide funds for this Board by direct appropriation.

I believe that, as a permanent measure, it would be appropriate to add to the agency needed to perform the other functions recommended in this report the responsibilities for civilian-initiated and civilian-controlled military research. The function of such a civilian group would be primarily to conduct long-range scientific research on military problems—leaving to the Services research on the improvement of existing weapons.

Some research on military problems should be conducted, in time of peace as well as in war, by civilians independently of the military establishment. It is the primary responsibility of the Army and Navy to train the men, make available the weapons, and employ the strategy that will bring victory in combat. The Armed Services cannot be expected to be experts in all of the complicated fields which make it possible for a great nation to fight successfully in total war. There are certain kinds of research—such as research on the improvement of existing weapons—which can best be done within the military establishment. However, the job of long-range research involving application of the newest scientific discoveries to military needs should be the responsibility of those civilian scientists in the universities and in industry who are best trained to discharge it thoroughly and successfully. It is essential that both kinds of research go forward and that there be the closest liaison between the two groups.

Placing the civilian military research function in the proposed agency would bring it into close relationship with a broad program of basic research in both the natural sciences and medicine. A balance between military and other research could thus readily be maintained.

The establishment of the new agency, including a civilian military research group, should not be delayed by the existence of the Research Board for National Security, which is a temporary measure. Nor should the creation of the new agency be delayed by uncertainties in regard to the postwar organization of our military departments themselves. Clearly, the new agency, including a civilian military research group within it, can remain sufficiently flexible to adapt its operations to whatever may be the final organization of the military departments.

National Research Foundation

It is my judgment that the national interest in scientific research and scientific education can best be promoted by the creation of a National Research Foundation.

I. *Purposes*

The National Research Foundation should develop and promote a national policy for scientific research and scientific education, should support basic research in nonprofit organizations, should develop scientific talent in American youth by means of scholarships and fellowships, and should by contract and otherwise support long-range research on military matters.

James K. Feibleman (1904-), Professor Emeritus of Philosophy at Tulane University, is a versatile thinker who has written in many areas of history and philosophy.

Pure Science, Applied Science, and Technology: An Attempt at Definitions

James K. Feibleman

It is not the business of scientists to investigate just what the business of science is. Yet the business of science is in need of investigation. If we are to consider the relations between science and engineering, the relation between pure and applied science will have first to be made very clear; and for this purpose we shall need working definitions. Once stated, these definitions may seem an elaboration of the obvious and an over-simplification. But the elaboration often seems obvious only *after* it has been stated, and the definitions may have to be simple in order to bring out the necessary distinctions.

By *pure science* or *basic research* is meant a method of investigating nature by the experimental method in an attempt to satisfy the need to know. Many activities in pure science are not experimental, as, for instance, biological taxonomy; but it can always be shown that in such cases the activities are ancillary to experiment. In the case of biological taxonomy the classifications are of experimental material. Taxonomy is practiced in other areas where it is not scientific, such as in the operation of libraries.

By *applied science* is meant the use of pure science for some practical human purpose.

Thus science serves two human purposes: to know and to do. The former is a matter of understanding, the latter a matter of action. Technology, which began as the attempt to satisfy a practical need without the use of science, will receive a fuller treatment in a later section.

Applied science, then, is simply pure science applied. But scientific method has more than one end; it leads to explanation and application. It achieves explanation in the discovery of laws, and the laws can be applied. Thus both pure science and applied science have aims and results. Pure science has as its aim the understanding of nature; it seeks explanation. Applied science has as its aim the control of nature; it has the task of employing the findings of pure science to get practical tasks done. Pure science has as a result the furnishing of laws for application in applied science. And, as we shall learn later in this essay, applied science has as a result the stimulation of discovery in pure science.

Applied science puts to practical human uses the discoveries made in pure science. Whether there would be such a thing as pure science alone is hard to say; there are reasons for thinking that there would be, for pure science has a long history and, as we have noted, another justification. There could be technology without science; for millennia, in fact, there was. But surely there could be no applied science without pure science: applied science means just what it says, namely, the application of science, and so without pure science there would be nothing to apply.

Logically, pure science pursued in disregard of applied science seems to be the *sine qua non* of applied science, while historically the problems toward which applied science is directed came before pure science.

It has been asserted, for instance, that Greek geometry, which is certainly pure, arose out of the interest in land-surveying problems in Egypt, where the annual overflow of the Nile obliterated all conventional boundaries. Certainly it is true that the same concept of infinity is necessary for the understanding of Euclidean geometry and for the division of farms. Be that as it may, it yet remains true that the relations between pure and applied science are often varied and subtle, and will require exploration.

Let us propose the hypothesis that all pure science is applicable.

No proof exists for such an hypothesis; all that can be offered is evidence in favor of it. This evidence consists of two parts, the first logical and the second historical.

The logical evidence in favor of the hypothesis is contained in the very nature of pure science itself. Any discovery in pure science that gets itself established will have gained the support of experimental data. Thus there must be a connection between the world of fact—the actual world, in other words, which corresponds to sense experience—and the laws of pure science. It is not too difficult to take the next step, and so to suppose that the laws, which were suggested by facts in the world corresponding to sense experience, could be applied back to that world.

The second part of the evidence for the applicability of all of the laws of pure science is contained in the record of those laws which have been applied. The modern Western cultures have been altered by applied science, and now Soviet Russia and China are following their lead. Indeed, so prevalent are the effects of applied science, and so concealed the leadership of pure science, that those whose understanding of science is limited are apt to identify all of science with applied science and even to assume that science itself means technology or heavy industry.

One argument against the position advocated here would be based upon the number of pure theories in science for which no application has yet been found. But this is no argument at all; for to have any weight, it would have to show not only that there had been no application but also why there could be none. Yet a time lag between the discovery of a theory and its application to practice is not uncommon. How many years elapsed between Faraday's discovery of the dynamo and its general manufacture and use in industry? Conic sections were discovered by Apollonius of Perga in the third century B.C., when they were of intellectual interest only, and they were applied to the problems of engineering only in the seventeenth century. Non-Euclidean geometry, worked out by Riemann as an essay in pure mathematics early in the nineteenth century, was used by Einstein in his theory of relativity in the twentieth century. The coordinate geometry of Descartes made possible the study of curves by means of quadratic equations. It was in Descartes' time that the application of conic sections to the orbit of planets was first noticed; later, the same curves were used in the analysis of the paths of projectiles, in searchlight reflectors, and in the cables of suspension bridges. Chlorinated diphenylethane was synthesized in 1874. Its value as insecticide (DDT) was not recognized until 1939 when a systematic search for moth repellants was undertaken for the military. The photoelectric cell was used in pure science, notably by George E. Hale on observations of the sun's corona in 1894. Twenty-five years later it was found employed in making motion pictures. It often happens that the discovery of a useful material is not sufficient; it is necessary also to discover a use, to connect the material to some function in which it could prove advantageous. Paracelsus discovered ether and even observed its anesthetic properties, and Valerius Cordus gave the formula for its preparation as early as the sixteenth century. But it was many centuries before ether was used as an anesthetic.

Presumably, then, pure scientific formulations which have not been applied are merely those for which as yet no applications have been found. In the effort to extend knowledge it is not strategically wise to hamper investigation with antecedent assurances of utility. Many of the scientific discoveries which later proved most advantageous in industry had not been self-evidently applicable. This is certainly true of Gibbs's phase rule in chemistry, for instance. It often happens that for the most abstract theories new acts of discovery are necessary in order to put them to practical use.

It should be observed at the outset that applications are matters of relevance. The line between pure and applied science is a thin one; they are distinct in their differences, but one fades into the other. For instance, the use of crystallography in the packing industry is an application, but so is the use of the mathematical theory of groups in pure crystallography and in quantum mechanics. The employment of mathematics in pure science means application from the point of view of mathematics but remoteness from application from the point of view of experiment. Some branches of mathematics have been so widely employed that we have come to think of them as practical affairs. This is the case with probability or differential equations. Both branches, however, considered in their mathematical aspects and not at all in relation to the various experimental sciences in which they have proved so useful, are theoretical disciplines with a status of their own which in no wise depends upon the uses to which they may be put.

Procedurally, the practitioner introduces into his problem the facilitation afforded by some abstract but relevant theory from either mathematics or pure science. The statistical theory of extreme values is a branch of mathematics, yet it has application to studies of metal fatigue and to such meteorological phenomena as annual droughts, atmospheric pressure, and temperature, snowfalls, and rainfalls. The discovery of the Salk vaccine against polio virus was an achievement of applied science. Yet Pasteur's principle of pure science, that dead or attenuated organisms could induce the production of antibodies within blood serum, was assumed by Salk; so the immensely important practical applications would not have been possible without the previous theoretical work.

It is clear, then, that we need three separate and distinct kinds of pursuits, and, perforce, three types of interest to accompany them. The first is pure science. Pure theoretical sciences are concerned with the discovery of natural law and the description of nature, and with nothing else. These sciences are conducted by men whose chief desire is to know, and this requires a detached inquiry—which Einstein has somewhere called "the holy curiosity of inquiry" and which Emerson declared to be perpetual. Such a detachment and such a pursuit are comparable in their high seriousness of purpose only to religion and art.

The second type is applied science, in which are included all applications of the experimental pure sciences. These are concerned with the improvement of human means and ends and with nothing else. They are conducted by men whose chief desires are practical: either the improvement of human conditions or profit, or both. Temperamentally, the applied scientists are not the same as the pure scientists: their sights while valid are lower; they are apt to be men of greater skill but of lesser imagination; what they lack in loftiness they gain in humanity. It would be a poor view which in all respects held either variety secondary. Yet there is a scale of order to human enterprises, even when we are sure we could dispense with none of them; and so we turn with some measure of dependence to the type of leadership which a preoccupation with detached inquiry is able to provide.

The third type is the *modus operandi* level, which is represented by the scientist with an interest in the solution of the problems presented by the task of getting from theory to practice. As Whitehead said, a short but concentrated interval for the development of imaginative design lies between them. Consider for example the role of the discovery of Hertzian waves, which not only led to the development of radio but also brilliantly confirmed Maxwell's model of an electromagnetic field, specifically the existence of electromagnetic waves, with a constant representing the velocity of light in two of the four equations.

The conception of science as exclusively pure or utterly applied is erroneous; the situation is no longer so absolute. When scientific theories were not too abstract, it was possible for practical-minded men to address themselves both to a knowledge of the theory and to the business of applying it in practice. The nineteenth century saw the rise of the "inventor", the technologist who employed the results of the theoretical scientist in the discovery of devices, or instruments, of new techniques in electromagnetics, in chemistry, and in many other fields. Earlier scientists like Maxwell had prepared the way for inventors like Edison. In some sciences, notably physics, however, this simple situation no longer prevails. The theories discovered there are of such a degree of mathematical abstraction that an intermediate type of interest and activity is now required. The theories which are discovered in the physicists' laboratories and published as journal articles take some time to make their way into engineering handbooks and contract practices. Some intermediate theory is necessary for getting from theory to practice.

A good example of the *modus operandi* level is furnished by the activities and the scientists concerned with making the first atomic bombs. Hahn and Strasmann discovered in 1938 that neutrons could split the nuclei of uranium. Einstein and Planck had earlier produced the requisite theories, but it was Enrico Fermi, Lise Meitner, and others who worked out the method of getting from relativity and quantum mechanics to bombs which could be made to explode by atomic fission.

Technology

There has been some misunderstanding of the distinction between applied science and technology; and understandably so, for the terms have not been clearly distinguished. Primarily the difference is one of type of approach. The applied scientist as such is concerned with the task of discovering applications for pure theory. The technologist has a problem which lies a little nearer to practice. Both applied scientist and technologist employ experiment; but in the former case guided by hypotheses deduced from theory, while in the latter case employing trial and error or skilled approaches derived from concrete experience. The theoretical biochemist is a pure scientist, working for the most part with the carbon compounds. The biochemist is an applied scientist when he explores the physiological effects of some new drug, perhaps trying it out to begin with on laboratory animals, then perhaps on himself or on volunteers from his laboratory or from the charity ward of some hospital. The doctor or practicing physician is a technologist when he prescribes it for some of his patients.

Speaking historically, the achievements of technology are those which developed without benefit of science; they arose empirically either by accident or as a matter of common experience. The use of certain biochemicals in the practice of medicine antedates the development of science: notably, ephedrine, cocaine, curare, and quinine. This is true also of the prescientific forms of certain industrial processes, such as cheese-making, fermentation, and tanning.

The applied scientist fits a case under a class; the technologist takes it from there and works it out, so to speak, *in situ*. Applied science consists in a system of concrete interpretations of scientific propositions directed to some end useful for human life. Technology might now be described as a further step in applied science by means of the improvement of instruments. In this last sense, technology has always been with us; it was vastly accelerated in efficiency by having been brought under applied science as a branch.

Technology is more apt to develop empirical laws than theoretical laws, laws which are generalizations from practice rather than laws which are intuited and then applied to practice. Empirical procedures like empirical laws are often the product of technological practice without benefit of theory. Since 1938, when Cerletti and Bini began to use electrically induced convulsions in the treatment of schizophrenia, the technique of electroshock therapy has been widespread in psychiatric practice. Yet there is no agreement as to what precisely occurs or how the improvement is produced; a theory to explain the practice is entirely wanting.

Like applied science, technology has its ideals. Let us consider the technological problem of improving the airplane, for instance. For a number of decades now, the problem to which airplane designers have addressed themselves is how to increase the speed and the pay load of airplanes. This means cutting down on the weight of the empty airplane in proportion to its carrying capacity while increasing its effective speed. If we look back at what has been accomplished in this direction, then extrapolate our findings into the future in order to discern the outlines of the ideal, we shall be surprised to discover that what the designers have been working toward is an airplane that will carry an infinite amount of pay load at an infinite speed while itself weighing nothing at all! This of course is a limit and, like all such limits, is an ideal intended to be increasingly approached without ever being absolutely reached.

Conception of the ideal is evidently of the utmost practicality and cannot be escaped in applied formulations. Yet the existence of such a thing as a technical ideal is fairly recent and is peculiar to Western culture. The ideal of a general character envisaged in this connection is that of fitness of purpose and of economy; no material or energy is to be wasted. Roman engineers built bridges designed to support loads far in excess of anything that might be carried over them; their procedures would be regarded as bad engineering today. The modern engineer builds his bridge to carry exactly the load that will be put upon it plus a small margin for safety, but no more; he must not waste structural steel nor use more rivets than necessary, and labor must be held to minimum. The ideal of technology is efficiency.

Although technologists work in terms of ideals, they are nevertheless more bound down to materials than is the applied scientist, just as the applied scientist in his turn is more bound down to materials than is the pure scientist. Since the technologist is limited by what is available, when he increases the going availability it is usually at the material level. The environment with which a society reacts is the available environment, not the entire or total environment; and the available environment is that part of the environment which is placed within reach of the society by its knowledge and techniques. These are laid out for it and increased by the pure and the applied sciences. Only when these limits are set can the technologist go to work. For example, discovery of the internal combustion engine which required gasoline as fuel turned men's attention to possible sources of oil. The applied scientist found ways of locating oil in the ground, while the petroleum technologist made the actual discoveries. In the hundred years since Edwin Drake drilled the world's first oil well near Titusville, Pennsylvania, in 1859, the technologists have taken this discovery very far. Oil is now a natural resource, a part of the available environment; but it can hardly be said to have been so a hundred years ago, although it was just as much in the ground then as now.

Another kind of technologist is the engineer; engineering is the most down-to-earth of all scientific work that can justify the name of science at all. In engineering the solutions of the technologists are applied to particular cases. The building of bridges, the medical treatment of patients, the designing of instruments, all improvements in model constructions of already existing tools—these are the work of the engineer. But the theories upon which such work rests, such as studies in the flow or "creep" of metals, the physics of lubrication, the characteristics of surface tensions of liquids—these belong to the applied scientist.

The industrial scientific laboratory is devoted to the range of applied science, from "fundamental research," by which is meant long-range work designed to produce or improve practical technology, to immediate technological gains from which manufacturing returns are expected, for instance, the testing of materials and of manufactured products. Technological laboratories have been established in the most important of the giant industrial companies, such as DuPont, General Electric, Eastman Kodak, and Bell Telephone. The work of such laboratories is cumulative and convergent; applied science in such an institution is directed toward eventual technological improvement, the range of applied science and technology being employed as a series of connecting links to tie up pure science with manufacturing. University and foundation laboratories often serve the same purpose, but with the emphasis shifted toward the theoretical end of the scientific spectrum.

The development of technology has a strong bearing on its situation today and may be traced briefly. In the Middle Ages, there was natural philosophy and craftsmanship. Such science as existed was in the hands of the natural philosophers, and such technology as existed was in the hands of the craftsmen. There was precious little of either, for the exploration of the natural world was conducted by speculative philosophers, while the practical tasks were carried out by handicrafts employing comparatively simple tools, although there were exceptions: the windmill, for instance. There was little commerce between them, however, for their aims were quite different, and the effort to understand the existence of God took precedence over lesser pursuits.

Gradually, however, natural philosophy was replaced by experimental science, and handicraft by the power tool. The separation continued to be maintained, and for the same reasons; and this situation did not change until the end of the eighteenth century. At that time the foundations of technology shifted from craft to science. Technology and applied science ran together into the same powerful channels at the same time that the applications of pure science became more abundant. A craft is learned by the apprentice method; a science must be learned from the study of principles as well as from the practices of the laboratory, and while the practice may come from applied science the principles are those of pure science.

There is now only the smallest distinction between applied science, the application of the principles of pure science, and technology. The methods peculiar to technology: trial and error, invention aided by intuition, have merged with those of applied science: adopting the findings of pure science to the purposes of obtaining desirable practical consequences. Special training is required, as well as some understanding of applied and even of pure science. In general, industries are based on manufacturing processes which merely reproduce on a large scale effects first learned and practiced in a scientific laboratory. The manufacture of gasoline, penicillin, electricity, oxygen were never developed from technological procedures, but depended upon work first done by pure scientists. Science played a predominant role in such physical industries as steel, aluminum, and petroleum; in such chemical industries as pharmaceuticals and potash; in such biological industries as medicine and husbandry.

A concomitant development, in which the triumph of pure science over technology shows clear, is in the design and manufacture of instruments. The goniometer, for the determination of the refractive index of fluids (used in the chemical industries); the sugar refractometer, for the reading of the percentages by weight of sugar (used in sugar manufacture); the pyrometer, for the measurement of high temperatures (used in the making of electric light bulbs and of gold and silver utensils); the polarimeter, for ascertaining the amount of sugar in urine (used by the medical profession); these and many others—such as for instance the focometer for studies in the length of objectives, the anomaloscope for color blindness, and the spectroscope for the measurement of wave lengths—are precision instruments embodying principles not available to the technologist working unaided by a knowledge of pure science.

From Practice to Theory

In the course of pursuing practical ends abstract principles of science hitherto unsuspected are often discovered. The mathematical theory of probability was developed because some professional gamblers wished to know the odds in games of chance. Electromagnetics stimulated the development of differential equations, and hydrodynamics function theory. Carnot founded the pure science of thermodynamics as a result of the effort to improve the efficiency of steam and other heat engines. Aerodynamics and atomic physics were certainly advanced more swiftly because of the requirements of war. Air pollution, which accompanies big city "smog," has led a number of physical chemists to investigate the properties of extreme dilution. Hence it is not surprising that many advances in pure science have been made in industrial laboratories: from the Bell Telephone laboratories alone have come the discoveries by Davisson and Germer of the diffraction of electrons, by Jansky of radio astronomy, and by Shannon of information theory.

Technology has long been an aid and has furnished an impetus to experimental science. The development of the delicate mechanisms requisite for the carrying out of certain experiments calls on all the professional abilities of the instrument maker. Such a relation is not a new one; it has long existed. The skill of the Venetian glass blowers made possible many of Torricelli's experiments on gases. Indeed, glass can be followed through a single chain of development for several centuries, from the early microscopes to interferometers. The study of electromagnetics was responsible for the later commerce in electric power and the vast industries founded on it. But, contrariwise, thermodynamics grew up as a result of the problems arising from the use of steam in industry. We cannot afford to neglect in our considerations the economic support as well as the social justification which industry has furnished, and continues to be prepared to furnish, to research. The extraordinary rise of pure chemistry in Germany was not unrelated to the industry constructed on the basis of the aniline dyes, as well as cosmetics and explosives, during the nineteenth century.

The harm to practice of neglecting the development of pure theoretic science will not be felt until the limits of installing industries by means of applied science and technology, and of spreading its results, have been reached. Science can to some extent continue to progress on its own momentum together with such aid as the accidental or adventitious discoveries of pure science in technological laboratories can furnish it. But there are limits to this sort of progress. Thus far the Communist countries of the East have taken every advantage of the scientific developments achieved in the capitalist countries of the West. But after all, the applied science which the West has been able to furnish has been the result of its own preoccupations with pure science and with theoretic considerations which lie outside the purview of any practice. Industrial laboratories may occasionally contribute to pure science, but that is not their chief aim; and it is apt to be forgotten that such industries would not exist were it not for the fact that some centuries ago a handful of scientists with no thought of personal gain or even of social benefit tried to satisfy their curiosity about the nature of things. The restless spirit of science, never content with findings, hardly concerned with the applications of findings, is always actively engaged in pursuing methods in terms of assumptions, and must have some corner of the culture in which it can hope to be protected in its isolation. A wise culture will always provide it elbow room, with the understanding that in the future some amortized inquiry is bound to pay dividends. The ivory tower can be, and sometimes is, the most productive building in the market place.

Of course, applied science and technology cannot be independent of pure science, nor can pure science be independent of applied science and technology. The two developments work together and are interwoven. Gilbert discovered that the freely suspended magnetic needle (i.e., the compass) could be a practical aid to navigation at the same time he proposed that perhaps the earth was a gigantic magnet.

Problems which arise in the midst of practical tasks often suggest lines of theoretical inquiry. But there is more. Pragmatic evidence has always been held by logicians to have little standing. A scientific hypothesis needs more support than can be obtained from the practical fact that "it works." For who knows how long it will work or how well? What works best today may not work best tomorrow. A kind of practice which supports one theory may be supplanted by a more efficient kind of practice which supports quite a different theory. Relativity mechanics gives more accurate measurements than Newtonian mechanics. That use does not determine theory can be easily shown. Despite the theoretical success of the Copernican theory as refined and advanced by Kepler, Galileo, and Newton, we have never ceased to use the Ptolemaic conception in guiding our ships or in regulating our clocks. However, if the practical success achieved by the application of certain theories in pure science cannot be construed as a proof of their truth, neither can it be evidence of the contradictory: workability is no evidence of falsity, either. Newton is still correct within limits. Practicality suggests truth and supports the evidence in its favor even if offering no final proof. The practical uses of atomic energy do not prove that matter is transformable into energy, but they offer powerful support. Hence, the use of a scientific law in the control of nature constitutes the check of prediction and control.

Cross-field Applications

We have treated all too briefly the relations of theory to practice and of practice to theory in a given science. We have also mentioned the productive nature of cross-field research. It remains now to discuss a last dimension of relations between science and practice, and this is what we may call cross-field applications: the employment of the practical effects of one science in those of another.

The applications of science have been greatly aided by the cross-fertilization of techniques. Radio astronomy, which has proved so useful in basic research, being already responsible for the discovery of "radio stars," and for adding to our knowledge of meteor streams and the solar corona, owes its inception to a borrowed instrument. Cross-field application has a long history, dating at least to the early half of the eighteenth century when distilleries in England brought together the results of techniques of producing gin acquired from both chemistry and theories of heat energy. Perhaps the most prominent instance of this is the way in which medicine has drawn upon the physical technologies. The use of the vacuum tube amplifier and the cathode ray oscillograph in determining the electrical potential accompanying events in the nervous system, the entire areas covered by encephalography and by roentgenology show the enormous benefits which have accrued to medical studies and procedures. Scintillation counters, developed and used in physical research, have been employed to measure the rate at which the thyroid gland in a given individual removes iodine from the blood stream; to measure the natural radioactivity of the body; to determine the extent of ingested radioactive compounds in the body; to assay the radioactive iron in blood samples. Chemistry has been an equally potent aid to pharmacology, which would have hardly existed in any important sense without it.

Other instances abound, and indeed multiply every day. In 1948 the Armour Research Foundation sponsored a Crystallographic Center, of interest to pharmaceutical corporations because of the crystalline nature of some of the vitamins. The invention of automatic sequence controlled calculators and other types of computing machines has seen their immediate application to atomic research and to military problems of a technical nature. Medical knowledge is being placed at the service of airplane designers, who must estimate just what strains their aircraft will demand of pilots. Perhaps the most graphic illustration of cross-field application is in scientific agriculture. Here hardly a single science can be omitted: physics, chemistry, biology—all contribute enormously to the joint knowledge which it is necessary to have if soils are to respond to management.

The cross-field applications of science usually work upward in fields corresponding to the integrative levels of the sciences. Applications found in physics will be employed in chemistry or biology, those found in chemistry will be employed in biology or psychology, those found in biology will be employed in psychology or sociology, and so on. The use of physics in biology has in fact brought into existence the science of biophysics, in which are studied the physics of biological systems, the biological effects of physical agents, and the application of physical methods to biological problems.

The cross-fertilization of applied science—the use of techniques, skills, devices, acquired in one science to achieve gains in another—has effects which tend to go beyond either. They add up to a considerable acceleration in the speed with which the applied sciences affect the culture as a whole. In the brief space of some several hundred years, western culture has been altered out of all recognition by the employment of applied science and technology to purposes of industry, health, government, and war. Much of the alteration has been accomplished by means of the cooperation between the sciences. We now know that the shortest route to an effective practice lies indirectly through the understanding of nature. If there exists a human purpose of a practical kind, then the quickest as well as the most efficient method of achieving it is to apply the relevant natural laws of science to it.

Frederick W. Taylor (1856-1915) is best known as the founder of scientific management. Despite his upper-class background, Taylor chose to study mechanics both in school and in the shop. He found that both machine shops and American industry generally operated chaotically and inefficiently. At first, he attempted to develop new tools to fit particular jobs, but eventually he expanded his efforts to include all aspects of the work place by insisting that management was a science.

THE PRINCIPLES OF SCIENTIFIC MANAGEMENT

PRESIDENT ROOSEVELT, in his address to the Governors at the White House, prophetically remarked that "The conservation of our national resources is only preliminary to the larger question of national efficiency."

The whole country at once recognized the importance of conserving our material resources and a large movement has been started which will be effective in accomplishing this object. As yet, however, we have but vaguely appreciated the importance of "the larger question of increasing our national efficiency."

We can see our forests vanishing, our water-powers going to waste, our soil being carried by floods into the sea; and the end of our coal and our iron is in sight. But our larger wastes of human effort, which go on every day through such of our acts as are blundering, ill-directed, or inefficient, and which Mr. Roosevelt refers to as a lack of "national efficiency," are less visible, less tangible, and are but vaguely appreciated.

We can see and feel the waste of material things. Awkward, inefficient, or ill-directed movements of men, however, leave nothing visible or tangible behind them. Their appreciation calls for an act of memory, an effort of the imagination. And for this reason, even though our daily loss from this source is greater than from our waste of material things, the one has stirred us deeply, while the other has moved us but little.

As yet there has been no public agitation for "greater national efficiency," no meetings have been called to consider how this is to be brought about. And still there are signs that the need for greater efficiency is widely felt.

The search for better, for more competent men, from the presidents of our great companies down to our household servants, was never more vigorous than it is now. And more than ever before is the demand for competent men in excess of the supply.

What we are all looking for, however, is the ready-made, competent man; the man whom some one else has trained. It is only when we fully realize that our duty, as well as our opportunity, lies in systematically cooperating to train and to make this competent man, instead of in hunting for a man whom some one else has trained, that we shall be on the road to national efficiency.

In the past the prevailing idea has been well expressed in the saying that "Captains of industry are born, not made"; and the theory has been that if one could get the right man, methods could be safely left to him. In the future it will be appreciated that our leaders must be trained right as well as born right, and that no great man can (with the old system of personal management) hope to compete with a number of ordinary men who have been properly organized so as efficiently to cooperate.

In the past the man has been first; in the future the system must be first. This in no sense, however, implies that great men are not needed. On the contrary, the first object of any good system must be that of developing first-class men; and under systematic management the best man rises to the top more certainly and more rapidly than ever before.

This paper has been written:

First. To point out, through a series of simple illustrations, the great loss which the whole country is suffering through inefficiency in almost all of our daily acts.

Second. To try to convince the reader that the remedy for this inefficiency lies in systematic management, rather than in searching for some unusual or extraordinary man.

Third. To prove that the best management is a true science, resting upon clearly defined laws, rules, and principles, as a foundation. And further to show that the fundamental principles of scientific management are applicable to all kinds of human activities, from our simplest individual acts to the work of our great corporations, which call for the most elaborate cooperation. And, briefly, through a series of illustrations, to convince the reader that whenever these principles are correctly applied, results must follow which are truly astounding.

This paper was originally prepared for presentation to The American Society of Mechanical Engineers. The illustrations chosen are such as, it is believed, will especially appeal to engineers and to managers of industrial and manufacturing establishments, and also quite as much to all of the men who are working in these establishments. It is hoped, however, that it will be clear to other readers that the same principles can be applied with equal force to all social activities: to the management of our homes; the management of our farms; the management of the business of our tradesmen, large and small; of our churches, our philanthropic institutions, our universities, and our governmental departments.

CHAPTER I

Fundamentals of Scientific Management

THE principal object of management should be to secure the maximum prosperity for the employer, coupled with the maximum prosperity for each employé.

The words "maximum prosperity" are used, in their broad sense, to mean not only large dividends for the company or owner, but the development of every branch of the business to its highest state of excellence, so that the prosperity may be permanent.

In the same way maximum prosperity for each employé means not only higher wages than are usually received by men of his class, but, of more importance still, it also means the development of each man to his state of maximum efficiency, so that he may be able to do, generally speaking, the highest grade of work for which his natural abilities fit him, and it further means giving him, when possible, this class of work to do.

It would seem to be so self-evident that maximum prosperity for the employer, coupled with maximum prosperity for the employé, ought to be the two leading objects of management, that even to state this fact should be unnecessary. And yet there is no question that, throughout the industrial world, a large part of the organization of employers, as well as employés, is for war rather than for peace, and that perhaps the majority on either side do not believe that it is possible so to arrange their mutual relations that their interests become identical.

The majority of these men believe that the fundamental interests of employés and employers are necessarily antagonistic. Scientific management, on the contrary, has for its very foundation the firm conviction that the true interests of the two are one and the same; that prosperity for the employer cannot exist through a long term of years unless it is accompanied by prosperity for the employé, and *vice versa;* and that it is possible to give the workman what he most wants — high wages — and the employer what he wants — a low labor cost — for his manufactures.

It is hoped that some at least of those who do not sympathize with each of these objects may be led to modify their views; that some employers, whose attitude toward their workmen has been that of trying to get the largest amount of work out of them for the smallest possible wages, may be led to see that a more liberal policy toward their men will pay them better; and that some of those workmen who begrudge a fair and even a large profit to their employers, and who feel that all of the fruits of their labor should belong to them, and that those for whom they work and the capital invested in the business are entitled to little or nothing, may be led to modify these views.

No one can be found who will deny that in the case of any single individual the greatest prosperity can exist only when that individual has reached his highest state of efficiency; that is, when he is turning out his largest daily output.

The truth of this fact is also perfectly clear in the case of two men working together. To illustrate: if you and your workman have become so skilful that you and he together are making two pairs of shoes in a day, while your competitor and his workman are making only one pair, it is clear that after selling your two pairs of shoes you can pay your workman much higher wages than your competitor who produces only one pair of shoes is able to pay his man, and that there will still be enough money left over for you to have a larger profit than your competitor.

In the case of a more complicated manufacturing establishment, it should also be perfectly clear that the greatest permanent prosperity for the workman, coupled with the greatest prosperity for the employer, can be brought about only when the work of the establishment is done with the smallest combined expenditure of human effort, plus nature's resources, plus the cost for the use of capital in the shape of machines, buildings, etc. Or, to state the same thing in a different way: that the greatest prosperity can exist only as the result of the greatest possible productivity of the men and machines of the establishment — that is, when each man and each machine are turning out the largest possible output; because unless your men and your machines are daily turning out more work than others around you, it is clear that competition will prevent your paying higher wages to your workmen than are paid to those of your competitor. And what is true as to the possibility of paying high wages in the case of two companies competing close beside one another is also true as to whole districts of the country and even as to nations which are in competition. In a word, that maximum prosperity can exist only as the result of maximum productivity. Later in this paper illustrations will be given of several companies which are earning large dividends and at the same time paying from 30 per cent. to 100 per cent. higher wages to their men than are paid to similar men immediately around them, and with whose employers they are in competition. These illustrations will cover different types of work, from the most elementary to the most complicated.

If the above reasoning is correct, it follows that the most important object of both the workmen and the management should be the training and development of each individual in the establishment, so that he can do (at his fastest pace and with the maximum of efficiency) the highest class of work for which his natural abilities fit him.

These principles appear to be so self-evident that many men may think it almost childish to state them. Let us, however, turn to the facts, as they actually exist in this country and in England. The English and American peoples are the greatest sportsmen in the world. Whenever an American workman plays baseball, or an English workman plays cricket, it is safe to say that he strains every nerve to secure victory for his side. He does his very best to make the largest possible number of runs. The universal sentiment is so strong that any man who fails to give out all there is in him in sport is branded as a "quitter," and treated with contempt by those who are around him.

When the same workman returns to work on the following day, instead of using every effort to turn out the largest possible amount of work, in a majority of the cases this man deliberately plans to do as little as he safely can — to turn out far less work than he is well able to do — in many instances to do not more than one-third to one-half of a proper day's work. And in fact if he were to do his best to turn out his largest possible day's work, he would be abused by his fellow-workers for so doing, even more than if he had proved himself a "quitter" in sport. Underworking, that is, deliberately working slowly so as to avoid doing a full day's work, "soldiering," as it is called in this country, "hanging it out," as it is called in England, "ca canae," as it is called in Scotland, is almost universal in industrial establishments, and prevails also to a large extent in the building trades; and the writer asserts without fear of contradiction that this constitutes the greatest evil with which the working-people of both England and America are now afflicted.

It will be shown later in this paper that doing away with slow working and "soldiering" in all its forms and so arranging the relations between employer and employé that each workman will work to his very best advantage and at his best speed, accompanied by the intimate cooperation with the management and the help (which the workman should receive) from the management, would result on the average in nearly doubling the output of each man and each machine. What other reforms, among those which are being discussed by these two nations, could do as much toward promoting prosperity, toward the diminution of poverty, and the alleviation of suffering? America and England have been recently agitated over such subjects as the tariff, the control of the large corporations on the one hand, and of hereditary power on the other hand, and over various more or less socialistic proposals for taxation, etc. On these subjects both peoples have been profoundly stirred, and yet hardly a voice has been raised to call attention to this vastly greater and more important subject of "soldiering," which directly and powerfully affects the wages, the prosperity, and the life of almost every working-man, and also quite as much the prosperity of every industrial establishment in the nation.

The elimination of "soldiering" and of the several causes of slow working would so lower the cost of production that both our home and foreign markets would be greatly enlarged, and we could compete on more than even terms with our rivals. It would remove one of the fundamental causes for dull times, for lack of employment, and for poverty, and therefore would have a more permanent and far-reaching effect upon these misfortunes than any of the curative remedies that are now being used to soften their consequences. It would insure higher wages and make shorter working hours and better working and home conditions possible.

Why is it, then, in the face of the self-evident fact that maximum prosperity can exist only as the result of the determined effort of each workman to turn out each day his largest possible day's work, that the great majority of our men are deliberately doing just the opposite, and that even when the men have the best of intentions their work is in most cases far from efficient?

There are three causes for this condition, which may be briefly summarized as:

First. The fallacy, which has from time immemorial been almost universal among workmen, that a material increase in the output of each man or each machine in the trade would result in the end in throwing a large number of men out of work.

Second. The defective systems of management which are in common use, and which make it necessary for each workman to soldier, or work slowly, in order that he may protect his own best interests.

Third. The inefficient rule-of-thumb methods, which are still almost universal in all trades, and in practising which our workmen waste a large part of their effort.

CHAPTER II

THE PRINCIPLES OF SCIENTIFIC MANAGEMENT

THE writer has found that there are three questions uppermost in the minds of men when they become interested in scientific management.

First. Wherein do the principles of scientific management differ essentially from those of ordinary management?

Second. Why are better results attained under scientific management than under the other types?

Third. Is not the most important problem that of getting the right man at the head of the company? And if you have the right man cannot the choice of the type of management be safely left to him?

One of the principal objects of the following pages will be to give a satisfactory answer to these questions.

THE FINEST TYPE OF ORDINARY MANAGEMENT

Before starting to illustrate the principles of scientific management, or "task management" as it is briefly called, it seems desirable to outline what the writer believes will be recognized as the best type of management which is in common use. This is done so that the great difference between the best of the ordinary management and scientific management may be fully appreciated.

In an industrial establishment which employs say from 500 to 1000 workmen, there will be found in many cases at least twenty to thirty different trades. The workmen in each of these trades have had their knowledge handed down to them by word of mouth, through the many years in which their trade has been developed from the primitive condition, in which our far-distant ancestors each one practised the rudiments of many different trades, to the present state of great and growing subdivision of labor, in which each man specializes upon some comparatively small class of work.

The ingenuity of each generation has developed quicker and better methods for doing every element of the work in every trade. Thus the methods which are now in use may in a broad sense be said to be an evolution representing the survival of the fittest and best of the ideas which have been developed since the starting of each trade. However, while this is true in a broad sense, only those who are intimately acquainted with each of these trades are fully aware of the fact that in hardly any element of any trade is there uniformity in the methods which are used. Instead of having only one way which is generally accepted as a standard, there are in daily use, say, fifty or a hundred different ways of doing each element of the work. And a little thought will make it clear that this must inevitably be the case, since our methods have been handed down from man to man by word of mouth, or have, in most cases, been almost unconsciously learned through personal observation. Practically in no instances have they been codified or systematically analyzed or described. The ingenuity and experience of each generation — of each decade, even, have without doubt handed over better methods to the next. This mass of rule-of-thumb or traditional knowledge may be said to be the principal asset or possession of every tradesman. Now, in the best of the ordinary types of management, the managers recognize frankly the fact that the 500 or 1000 workmen, included in the twenty to thirty trades, who are under them, possess this mass of traditional knowledge, a large part of which is not in the possession of the management. The management, of course, includes foremen and superintendents, who themselves have been in most cases first-class workers at their trades. And yet these foremen and superintendents know, better than any one else, that their own knowledge and personal skill falls far short of the combined knowledge and dexterity of all the workmen under them. The most experienced managers therefore frankly place before their workmen the problem of doing the work in the best and most economical way. They recognize the task before them as that of inducing each workman to use his best endeavors, his hardest work, all his traditional knowledge, his skill, his ingenuity, and his good-will — in a word, his "initiative," so as to yield the largest possible return to his employer. The problem before the management, then, may be briefly said to be that of obtaining the best *initiative* of every workman. And the writer uses the word "initiative" in its broadest sense, to cover all of the good qualities sought for from the men.

On the other hand, no intelligent manager would hope to obtain in any full measure the initiative of his workmen unless he felt that he was giving them something more than they usually receive from their employers. Only those among the readers of this paper who have been managers or who have worked themselves at a trade realize how far the average workman falls short of giving his employer his full initiative. It is well within the mark to state that in nineteen out of twenty industrial establishments the workmen believe it to be directly against their interests to give their employers their best initiative, and that instead of working hard to do the largest possible amount of work and the best quality of work for their employers, they deliberately work as slowly as they dare while they at the same time try to make those over them believe that they are working fast.[1]

The writer repeats, therefore, that in order to have any hope of obtaining the initiative of his workmen the manager must give some *special incentive* to his men beyond that which is given to the average of the trade. This incentive can be given in several different ways, as, for example,

[1] The writer has tried to make the reason for this unfortunate state of things clear in a paper entitled "Shop Management," read before the American Society of Mechanical Engineers."

the hope of rapid promotion or advancement; higher wages, either in the form of generous piece-work prices or of a premium or bonus of some kind for good and rapid work; shorter hours of labor; better surroundings and working conditions than are ordinarily given, etc., and, above all, this special incentive should be accompanied by that personal consideration for, and friendly contact with, his workmen which comes only from a genuine and kindly interest in the welfare of those under him. It is only by giving a special inducement or "incentive" of this kind that the employer can hope even approximately to get the "initiative" of his workmen. Under the ordinary type of management the necessity for offering the workman a special inducement has come to be so generally recognized that a large proportion of those most interested in the subject look upon the adoption of some one of the modern schemes for paying men (such as piece work, the premium plan, or the bonus plan, for instance) as practically the whole system of management. Under scientific management, however, the particular pay system which is adopted is merely one of the subordinate elements.

Broadly speaking, then, the best type of management in ordinary use may be defined as management in which the workmen give their best *initiative* and in return receive some *special incentive* from their employers. This type of management will be referred to as the management of "*initiative and incentive*" in contradistinction to scientific management, or task management, with which it is to be compared.

The writer hopes that the management of "initiative and incentive" will be recognized as representing the best type in ordinary use, and in fact he believes that it will be hard to persuade the average manager that anything better exists in the whole field than this type. The task which the writer has before him, then, is the difficult one of trying to prove in a thoroughly convincing way that there is another type of management which is not only better but overwhelmingly better than the management of "initiative and incentive."

The universal prejudice in favor of the management of "initiative and incentive" is so strong that no mere theoretical advantages which can be pointed out will be likely to convince the average manager that any other system is better. It will be upon a series of practical illustrations of the actual working of the two systems that the writer will depend in his efforts to prove that scientific management is so greatly superior to other types. Certain elementary principles, a certain philosophy, will however be recognized as the essence of that which is being illustrated in all of the practical examples which will be given. And the broad principles in which the scientific system differs from the ordinary or "rule-of-thumb" system are so simple in their nature that it seems desirable to describe them before starting with the illustrations.

Under the old type of management success depends

almost entirely upon getting the "initiative" of the workmen, and it is indeed a rare case in which this initiative is really attained. Under scientific management the "initiative" of the workmen (that is, their hard work, their good-will, and their ingenuity) is obtained with absolute uniformity and to a greater extent than is possible under the old system; and in addition to this improvement on the part of the men, the managers assume new burdens, new duties, and responsibilities never dreamed of in the past. The managers assume, for instance, the burden of gathering together all of the traditional knowledge which in the past has been possessed by the workmen and then of classifying, tabulating, and reducing this knowledge to rules, laws, and formulæ which are immensely helpful to the workmen in doing their daily work. In addition to developing a *science* in this way, the management take on three other types of duties which involve new and heavy burdens for themselves.

These new duties are grouped under four heads:

First. They develop a science for each element of a man's work, which replaces the old rule-of-thumb method.

Second. They scientifically select and then train, teach, and develop the workman, whereas in the past he chose his own work and trained himself as best he could.

Third. They heartily cooperate with the men so as to insure all of the work being done in accordance with the principles of the science which has been developed.

Fourth. There is an almost equal division of the work and the responsibility between the management and the workmen. The management take over all work for which they are better fitted than the workmen, while in the past almost all of the work and the greater part of the responsibility were thrown upon the men.

It is this combination of the initiative of the workmen, coupled with the new types of work done by the management, that makes scientific management so much more efficient than the old plan.

Three of these elements exist in many cases, under the management of "initiative and incentive," in a small and rudimentary way, but they are, under this management, of minor importance, whereas under scientific management they form the very essence of the whole system.

The fourth of these elements, "an almost equal division of the responsibility between the management and the workmen," requires further explanation. The philosophy of the management of "initiative and incentive" makes it necessary for each workman to bear almost the entire responsibility for the general plan as well as for each detail of his work, and in many cases for his implements as well. In addition to this he must do all of the actual physical labor. The development of a science, on the other hand, involves the establishment of many rules, laws, and formulæ which replace the judgment of the individual workman and which can be effectively used only after having been systematically recorded, indexed, etc. The practical use of scientific data also calls for a room in which to keep the books, records,[1] etc., and a desk for the planner to work at. Thus all of the planning which under the old system was done by the workman, as a result of his personal experience, must of necessity under the new system be done by the management in accordance with the laws of the science; because even if the workman was well suited to the development and use of scientific data, it would be physically impossible for him to work at his machine and at a desk at the same time. It is also clear that in most cases one type of man is needed to plan ahead and an entirely different type to execute the work.

The man in the planning room, whose specialty under scientific management is planning ahead, invariably finds that the work can be done better and more economically by a subdivision of the labor; each act of each mechanic, for example, should be preceded by various preparatory acts done by other men. And all of this involves, as we have said, "an almost equal division of the responsibility and the work between the management and the workman."

To summarize: Under the management of "initiative and incentive" practically the whole problem is "up to the workman," while under scientific management fully one-half of the problem is "up to the management."

[1] For example, the records containing the data used under scientific management in an ordinary machine-shop fill thousands of pages.

Perhaps the most prominent single element in modern scientific management is the task idea. The work of every workman is fully planned out by the management at least one day in advance, and each man receives in most cases complete written instructions, describing in detail the task which he is to accomplish, as well as the means to be used in doing the work. And the work planned in advance in this way constitutes a task which is to be solved, as explained above, not by the workman alone, but in almost all cases by the joint effort of the workman and the management. This task specifies not only what is to be done but how it is to be done and the exact time allowed for doing it. And whenever the workman succeeds in doing his task right, and within the time limit specified, he receives an addition of from 30 per cent. to 100 per cent. to his ordinary wages. These tasks are carefully planned, so that both good and careful work are called for in their performance, but it should be distinctly understood that in no case is the workman called upon to work at a pace which would be injurious to his health. The task is always so regulated that the man who is well suited to his job will thrive while working at this rate during a long term of years and grow happier and more prosperous, instead of being overworked. Scientific management consists very largely in preparing for and carrying out these tasks.

The writer is fully aware that to perhaps most of the readers of this paper the four elements which differentiate the new management from the old will at first appear to be merely high-sounding phrases; and he would again repeat that he has no idea of convincing the reader of their value merely through announcing their existence. His hope of carrying conviction rests upon demonstrating the tremendous force and effect of these four elements through a series of practical illustrations. It will be shown, first, that they can be applied absolutely to all classes of work, from the most elementary to the most intricate; and second, that when they are applied, the results must of necessity be overwhelmingly greater than those which it is possible to attain under the management of initiative and incentive.

The first illustration is that of handling pig iron, and this work is chosen because it is typical of perhaps the crudest and most elementary form of labor which is performed by man. This work is done by men with no other implements than their hands. The pig-iron handler stoops down, picks up a pig weighing about 92 pounds, walks for a few feet or yards and then drops it on to the ground or upon a pile. This work is so crude and elementary in its nature that the writer firmly believes that it would be possible to train an intelligent gorilla so as to become a more efficient pig-iron handler than any man can be. Yet it will be shown that the science of handling pig iron is so great and amounts to so much that it is impossible for the man who is best suited to this type of work to understand the principles of this science, or even to work in accordance with these principles without the aid of a man better educated than he is. And the further illustrations to be given will make it clear that in almost all of the mechanic arts the science which underlies each workman's act is so great and amounts to so much that the workman who is best suited actually to do the work is incapable (either through lack of education or through insufficient mental capacity) of understanding this science. This is announced as a general principle, the truth of which will become apparent as one illustration after another is given. After showing these four elements in the handling of pig iron, several illustrations will be given of their application to different kinds of work in the field of the mechanic arts, at intervals in a rising scale, beginning with the simplest and ending with the more intricate forms of labor.

One of the first pieces of work undertaken by us, when the writer started to introduce scientific management into the Bethlehem Steel Company, was to handle pig iron on task work. The opening of the Spanish War found some 80,000 tons of pig iron placed in small piles in an open field adjoining the works. Prices for pig iron had been so low that it could not be sold at a profit, and it therefore had been stored. With the opening of the Spanish War the price of pig iron rose, and this large accumulation of iron was sold. This gave us a good opportunity to show the workmen, as well as the owners and managers of the works, on a fairly large scale the advantages of task work over the old-fashioned day work and piece work, in doing a very elementary class of work.

The Bethlehem Steel Company had five blast furnaces, the product of which had been handled by a pig-iron gang for many years. This gang, at this time, consisted of about 75 men. They were good, average pig-iron handlers, were under an excellent foreman who himself had been a pig-iron handler, and the work was done, on the whole, about as fast and as cheaply as it was anywhere else at that time.

A railroad switch was run out into the field, right along the edge of the piles of pig iron. An inclined plank was placed against the side of a car, and each man picked up from his pile a pig of iron weighing about 92 pounds, walked up the inclined plank and dropped it on the end of the car.

We found that this gang were loading on the average about $12\frac{1}{2}$ long tons per man per day. We were surprised to find, after studying the matter, that a first-class pig-iron handler ought to handle between 47 and 48 long tons per day, instead of $12\frac{1}{2}$ tons. This task seemed to us so very large that we were obliged to go over our work several times before we were absolutely sure that we were right. Once we were sure, however, that 47 tons was a proper day's work for a first-class pig-iron handler, the task which faced us as managers under the modern scientific plan was clearly before us. It was our duty to see that the 80,000 tons of pig iron was loaded on to the cars at the rate of 47 tons per man per day, in place of $12\frac{1}{2}$ tons, at which rate the work was then being done. And it was further our duty to see that this work was done without bringing on a strike among the men, without any quarrel with the men, and to see that the men were happier and better contented when loading at the new rate of 47 tons than they were when loading at the old rate of $12\frac{1}{2}$ tons.

Our first step was the scientific selection of the workman. In dealing with workmen under this type of management, it is an inflexible rule to talk to and deal with only one man at a time, since each workman has his own special abilities and limitations, and since we are not dealing with men in masses, but are trying to develop each individual man to his highest state of efficiency and prosperity. Our first step was to find the proper workman to begin with. We therefore carefully watched and studied these 75 men for three or four days, at the end of which time we had picked out four men who appeared to be physically able to handle pig iron at the rate of 47 tons per day. A careful study was then made of each of these men. We looked up their history as far back as practicable and thorough inquiries were made as to the character, habits, and the ambition of each of them. Finally we selected one from among the four as the most likely man to start with. He was a little Pennsylvania Dutchman who had been observed to trot back home for a mile or so after his work in the evening.

about as fresh as he was when he came trotting down to work in the morning. We found that upon wages of $1.15 a day he had succeeded in buying a small plot of ground, and that he was engaged in putting up the walls of a little house for himself in the morning before starting to work and at night after leaving. He also had the reputation of being exceedingly "close," that is, of placing a very high value on a dollar. As one man whom we talked to about him said, "A penny looks about the size of a cart-wheel to him." This man we will call Schmidt.

The task before us, then, narrowed itself down to getting Schmidt to handle 47 tons of pig iron per day and making him glad to do it. This was done as follows. Schmidt was called out from among the gang of pig-iron handlers and talked to somewhat in this way:

"Schmidt, are you a high-priced man?"

"Vell, I don't know vat you mean."

"Oh yes, you do. What I want to know is whether you are a high-priced man or not."

"Vell, I don't know vat you mean."

"Oh, come now, you answer my questions. What I want to find out is whether you are a high-priced man or one of these cheap fellows here. What I want to find out is whether you want to earn $1.85 a day or whether you are satisfied with $1.15, just the same as all those cheap fellows are getting."

"Did I vant $1.85 a day? Vas dot a high-priced man? Vell, yes, I vas a high-priced man."

"Oh, you're aggravating me. Of course you want

$1.85 a day — every one wants it! You know perfectly well that that has very little to do with your being a high-priced man. For goodness' sake answer my questions, and don't waste any more of my time. Now come over here. You see that pile of pig iron?"

"Yes."

"You see that car?"

"Yes."

"Well, if you are a high-priced man, you will load that pig iron on that car to-morrow for $1.85. Now do wake up and answer my question. Tell me whether you are a high-priced man or not."

"Vell — did I got $1.85 for loading dot pig iron on dot car to-morrow?"

"Yes, of course you do, and you get $1.85 for loading a pile like that every day right through the year. That is what a high-priced man does, and you know it just as well as I do."

"Vell, dot's all right. I could load dot pig iron on the car to-morrow for $1.85, and I get it every day, don't I?"

"Certainly you do — certainly you do."

"Vell, den, I vas a high-priced man."

"Now, hold on, hold on. You know just as well as I do that a high-priced man has to do exactly as he's told from morning till night. You have seen this man here before, haven't you?"

"No, I never saw him."

"Well, if you are a high-priced man, you will do exactly as this man tells you to-morrow, from morning till night. When he tells you to pick up a pig and walk, you pick it up and you walk, and when he tells you to sit down and rest, you sit down. You do that right straight through the day. And what's more, no back talk. Now a high-priced man does just what he's told to do, and no back talk. Do you understand that? When this man tells you to walk, you walk; when he tells you to sit down, you sit down, and you don't talk back at him. Now you come on to work here to-morrow morning and I'll know before night whether you are really a high-priced man or not."

This seems to be rather rough talk. And indeed it would be if applied to an educated mechanic, or even an intelligent laborer. With a man of the mentally sluggish type of Schmidt it is appropriate and not unkind, since it is effective in fixing his attention on the high wages which he wants and away from what, if it were called to his attention, he probably would consider impossibly hard work.

What would Schmidt's answer be if he were talked to in a manner which is usual under the management of "initiative and incentive"? say, as follows:

"Now, Schmidt, you are a first-class pig-iron handler and know your business well. You have been handling at the rate of 12½ tons per day. I have given considerable study to handling pig iron, and feel sure that you could do a much larger day's work than you have been doing. Now don't you think that if you really tried you could handle 47 tons of pig iron per day, instead of 12½ tons?"

What do you think Schmidt's answer would be to this?

Schmidt started to work, and all day long, and at regular intervals, was told by the man who stood over him with a watch, "Now pick up a pig and walk. Now sit down and rest. Now walk — now rest," etc. He worked when he was told to work, and rested when he was told to rest, and at half-past five in the afternoon had his 47½ tons loaded on the car. And he practically never failed to work at this pace and do the task that was set him during the three years that the writer was at Bethlehem. And throughout this time he averaged a little more than $1.85 per day, whereas before he had never received over $1.15 per day, which was the ruling rate of wages at that time in Bethlehem. That is, he received 60 per cent. higher wages than were paid to other men who were not working on task work. One man after another was picked out and trained to handle pig iron at the rate of 47½ tons per day until all of the pig iron was handled at this rate, and the men were receiving 60 per cent. more wages than other workmen around them.

The science which exists in most of the mechanic arts is, however, far simpler than the science of cutting metals. In almost all cases, in fact, the laws or rules which are developed are so simple that the average man would hardly dignify them with the name of a science. In most trades, the science is developed through a comparatively simple analysis and time study of the movements required by the workmen to do some small part of his work, and this study is usually made by a man equipped merely with a stop-watch and a properly ruled notebook. Hundreds of these "time-study men" are now engaged in developing elementary scientific knowledge where before existed only rule of thumb. Even the motion study of Mr. Gilbreth in bricklaying (described on pages 77 to 84) involves a much more elaborate investigation than that which occurs in most cases. The general steps to be taken in developing a simple law of this class are as follows:

First. Find, say, 10 or 15 different men (preferably in as many separate establishments and different parts of the country) who are especially skilful in doing the particular work to be analyzed.

Second. Study the exact series of elementary operations or motions which each of these men uses in doing the work which is being investigated, as well as the implements each man uses.

Third. Study with a stop-watch the time required to make each of these elementary movements and then select the quickest way of doing each element of the work.

Fourth. Eliminate all false movements, slow movements, and useless movements.

Fifth. After doing away with all unnecessary movements, collect into one series the quickest and best movements as well as the best implements.

This one new method, involving that series of motions which can be made quickest and best, is then substituted in place of the ten or fifteen inferior series which were formerly in use. This best method becomes standard, and remains standard, to be taught first to the teachers (or functional foremen) and by them to every workman in the establishment until it is superseded by a quicker and better series of movements. In this simple way one element after another of the science is developed.

In the same way each type of implement used in a trade is studied. Under the philosophy of the management of "initiative and incentive" each workman is called upon to use his own best judgment, so as to do the work in the quickest time, and from this results in all cases a large variety in the shapes and types of implements which are used for any specific purpose. Scientific management requires, first, a careful investigation of each of the many modifications of the same implement, developed under rule of thumb; and second, after a time study has been made of the speed attainable with each of these implements, that the good points of several of them shall be united in a single standard implement, which will enable the workman to work faster and with greater ease than he could before. This one implement, then, is adopted as standard in place of the many different kinds before in use, and it remains standard for all workmen to use until superseded by an implement which has been shown, through motion and time study, to be still better.

With this explanation it will be seen that the development of a science to replace rule of thumb is in most cases by no means a formidable undertaking, and that it can be accomplished by ordinary, every-day men without any elaborate scientific training; but that, on the other hand, the successful use of even the simplest improvement of this kind calls for records, system, and cooperation where in the past existed only individual effort.

4

Herbert J. Muller, THE CHILDREN OF FRANKENSTEIN

Lewis Mumford, AUTHORITARIAN AND DEMOCRATIC TECHNICS

1. What are the good and bad points of the Industrial Revolution according to Muller?

2. What are Muller's criticisms of modern technology? Do you agree with him?

3. How can we control technology according to Muller? Do you agree?

4. To what extent do Muller's attitudes toward technology resemble or differ from other writers in this theme?

5. Differentiate between what Mumford calls authoritarian and democratic technics. Give examples of each. Which is preferable and why? Does the preference have universal application?

6. How can the preferred technics be maximized?

7. According to Mumford, what caused the development of authoritarian technics? What may be the final consequences of their spread?

8. How does Mumford see technology as affecting the individual?

9. Does Mumford's view on technology differ from Muller's? From Vannevar's? From Bush's?

Herbert J. Muller (1905-1980), author of *Children of Frankenstein: A Primer on Modern Technology and Human Values*, was professor of English and government at Indiana University. He is well-known for his writings on the impact of science and technology on humanity.

THE CHILDREN OF FRANKENSTEIN

1

Definitions, Premises, and Basic Issues

According to my dictionary, technology is "the science or study of the practical or industrial arts." In modern usage emphasis has been on industrial techniques, based on the machine. I am using the term in a broader sense to cover as well the distinctive practices that have been generated by the rise of industrialism. These include large-scale organization throughout our society, professionalism in all activities, and the ways of thinking and doing indicated by such typically modern terms as "system," "systematic" methods, and "methodology." Modern technology may be broadly defined as the elaborate development of standardized, efficient means to practical ends. A comparable definition is Kenneth Galbraith's, "the systematic application of scientific or other organized knowledge to practical tasks." Jacques Ellul prefers to call all this simply "technique." In any case it is well to keep in mind that the term is not so precise or "scientific" as it may sound.

Practical arts or skills have of course always been vitally important as a means of providing for the material necessities of life. At the dawn of man's history we find him chipping flints, inaugurating his career ever after as a tool-maker—a career in which he would have far more success than in his quest of wisdom and virtue. A historical survey may make clearer the "naturalness" of technology, the continuities through the long ages, and the unquestionable progress that man has made in this fundamental endeavor. Not until our century, however, did men begin to write the history of technology, and the obvious reason why historians have so belatedly recognized its importance is the novelty of the supreme importance of modern technology. This is not only vastly superior in efficiency to the technology of pre-industrial societies but essentially different in kind, and has had many unprecedented consequences.

The fantastic pace of change, due initially to the increasingly systematic "invention of invention" as the Industrial Revolution gathered momentum, has in our time been steadily accelerated by the many billions spent on "R & D," research and development. The most conspicuous result has been the immense power over nature, now including the conquest of outer space, that man has achieved by what are being called "megatechnics." But most important for my purposes is the range of influence of modern technology. It is a commonplace that our society is dominated by science and technology, so much so that some writers (such as Jacques Barzun) have attacked C. P. Snow's thesis of the "two cultures" by maintaining that we really have only one culture. I think we have at least two, as the very complaints about science and technology suggest; but there is no question of their dominion. Hence many writers are echoing Ellul's insistence that our whole way of life is "unnatural." The extraordinary progress of technology is regarded as a problem, and though most men still celebrate the progress, it has certainly created problems such as past societies never had to worry over. As some scientist remarked, it has taught us how to become gods before we have learned to be men.

My primary concern is the systematic neglect or abuse of what I consider essential human values. (It is significant that values are regularly called "human" these days even though the term is strictly redundant—only human beings can have conscious values.) To be sure, anything that people want or think is good may be called a human value. Like Spinoza, I assume that people do not desire something because it is good, but that it is considered good because people desire it. Although money is often called a "false" value, it is plainly a real one. As plainly, however, it is only a means to some end. So too is our marvelous technology. The matrix of our problems, especially in America, is the common assumption in effect that it is an end in itself—an assumption fortified by the immense energy that goes into it, the worship of efficiency as the sovereign ideal, the boasts about our material wealth and power, and the national goal of steady economic growth. True, we cannot absolutely separate means and ends, since the achievement of necessary means may be satisfying in and for itself. As the end for many Americans has been not so much a pile of money as the game of money-making, so the mastery of technology may be its own reward, affording the satisfactions of a problem solved, a truth found, an adventure consummated, a craftsman's aim achieved—all amounting to a kind of esthetic satisfaction, in keeping with the definition of technology as a study of the practical "arts." Yet I assume that we still can and must roughly distinguish means and ends, both for our personal purposes and for purposes of sizing up our society. The distinction is all the more necessary because of our fabulous wealth of efficient means. And I am committed to the simple, old-fashioned assumption that the proper end for man is the good life.

Once upon a time men believed, or at least were constantly told, that their main business on earth was the salvation of their immortal soul, through the service of God. In America today the popular word for the good life is "happiness." This seems to me a quite normal goal, but the question remains what people mean by the term. The American pursuit of happiness can often look like a compulsive, joyless effort to escape boredom, and in any case a people blessed with far more material advantages than any other society has ever enjoyed is not clearly the happiest people on earth. One plain reason is a paltry conception of the good life, or what I have called the highest standard of low living in all history. But this only forces the basic question. What, then, is the good life?

The whole history of thought and culture may be summed up as an endless disagreement over this question. I take it that men never can or will agree on just what is the good life, unless the human race reaches the state of a universal anthill that Ellul and others predict. Neither do I wish for anything like complete agreement, in view of the rich, diverse potentialities of human nature and the manifold personal differences in interest and capacity. Yet the history of thought and culture also suggests that we can arrive at a general agreement on certain basic goods. They begin with the elementary goods of physical well-being, through health and the satisfaction of hunger and other physiological needs. They include the basic goods of social life, beginning with family life and broadening into simple comradeliness, fellow-feeling, and love. They include as well cultural goods—the satisfactions of natural curiosity, of the esthetic sense, and of the related craftsman's or creative impulse. These may be suspect to the man on the street for the same reason they are often called "higher," that they involve pleasures of the mind, but the powers of mind are precisely what distinguish man from other animals. All men normally enjoy using their heads, as children conspicuously do, and all known societies have made some provision for the cultivation of these goods. To some extent they are always known to the man on the street too. No Americans are as purely materialistic, or as narrowly practical, as they are often reputed to be. Though always inclined to ask where does it get you, they do all kinds of things that get them nowhere—except into a pleasurable state of mind.

Since most of us are conscientious relativists these days, and I have myself always shied away from high talk about the eternal verities, I venture some more elementary ideas. I am now operating on the premise that there *are* permanent, absolute values, at least so long as we assume that life is worthwhile. We cannot absolutely prove that it is, inasmuch as many men have decided otherwise, even to killing themselves; but this is the assumption that the rest of us perforce operate on, and these values are what warrant that assumption. They may be given some scientific warrant too in the common assumptions about the basic needs and drives of man, and lately the attention to the universals of culture, long neglected by anthropologists because of their stress on the more conspicuous diversity of cultures. I prefer, however, to emphasize their basis in common experience. They are absolute goods in that they are good for their own sake, they do not require either religious or scientific sanction, and they do not have to be demonstrated because men conscious of them simply know they are good. If they are not good enough to make life worthwhile for all men, the plainest reason is the deprivation of some basic good, such as health, or the frustration of some basic need. And especially on these needs I have not sought the authority of psychologists and social scientists. Usually concentrating on the obvious physiological drives, such as hunger, sex, and self-preservation, they have paid much less attention to natural curiosity, the esthetic sense, creative activity, and the desire for self-realization through such normal interests, if only because these are harder to deal with by any respectable methodology. I am assuming that the "higher" goods are also sprung from vital needs, and that life is seriously impoverished when people, however well-fed and well-sexed, have limited capacities or opportunities for enjoying them.

In insisting on the reality and importance of what may also be called "spiritual" values, I am expressing not merely my personal opinion as an English professor but something like the judgment of the human race, apparent in all known societies. It remains apparent as we approach the more complicated question of "civilized" values, another of my major concerns. The rise of civilization, with its seemingly artificial life, may nevertheless be regarded as a natural outgrowth of the absolute goods, a more conscious, elaborate, refined realization of human potentialities, or an extension and enrichment of consciousness, the seat of all human values. Similarly its "high" culture was a more deliberate pursuit of truth, beauty, and goodness, sometimes holiness. The historic outcome was the growth of learning, the flowering of literature and the fine arts, in time the rise of philosophy, the higher religions, and pure science itself—the astonishing belief in the sovereign value of a disinterested pursuit of truth, an absolute value that scientists may overlook because many of them taboo value-judgments on principle. The fuller consciousness that men had grown to amplified as well such simple goods as fellow-feeling, which flowered into the no less astonishing ideal of universal brotherhood. In this fuller consciousness disagreements over the good life also became more marked, but first we should note the broad agreement on what were the great works of art and thought—an agreement the more significant because today we know far more about the history of civilization, and the values of all other civilizations, than men ever did or could in the great societies of the past.

With distinctively civilized values, however, we can no longer talk easily of permanence or absoluteness. They came very late in the history of man, and we have learned that civilization is a precarious adventure. Most of its high culture, moreover, was never made available to most people, in particular the illiterate peasant masses, who made do with their prehistoric kind of folk culture. At that anthropologists insist that we have no right to regard civilized societies as superior to the primitive ones they study, and some writers offer variations on the perennial theme of primitivism, the belief harking back to antiquity that the values of civilization are not worth its costs. To me they are worth it; but I would argue only that the great works of art and thought provide a richer experience than primitive or folk art (which I enjoy too), and that almost all people who have learned to appreciate them value this experience, as both a component of the good life and a possible guide to better conceptions of it.

More complex issues are raised by the distinctive values of our own civilization, above all the ideals of personal freedom, and the rights, opportunities, and incentives it has extended to the individual. In my efforts to write the history of freedom I was soon forced to question the congenial belief embedded in American tradition, that men have not only a natural right to freedom but a natural passion for it. On the historic record the great majority of men through the long ages manifested no such passion, they put up with what now looks like oppressive servitude, and they hardly dreamed of the rights and opportunities that most Americans take for granted as their birthright. Here I am arguing only that on the record man nevertheless has some potentialities for freedom, that once people have known and enjoyed more freedom they generally recognize it as a precious good, and that they have sound reason for this value judgment. Similarly with the ideals of individuality and self-realization that have come to be considered essential for the good life. They became conscious ideals with only a few peoples in the past, notably the ancient Greeks, and only in modern times have they been held up to the common people; there was no worry in past societies over the pressures to conformism that alarm contemporaries because the leaders of these societies wanted nothing more than conformity or unquestioning obedience in the common people. I assume only that these ideals have a biological basis in the fact that man is the most highly individualized of animals, as well as a historical basis in the wider range of choice and self-expression developed by civilization, and that on the record of Western civilization, and the basis of our own experience, there is again good reason for the common belief that they are precious ideals.

The excuse for all these elementary observations, at any rate, is that they may help as we begin to size up the social and cultural consequences of modern technology, the themes to be developed in later chapters. Most obviously it is by far the most efficient means yet devised by man to provide for the absolute goods of physical well-being, beginning with the bread man needs before he can realize that he cannot live on bread alone. Because these are "material" goods it has become all too easy to disparage them, and hard to realize that before our times untold millions of people starved to death, many more millions all through history had to get along without enough bread—as in the non-Western world they still do.[1] Almost all of us enjoy many other of our abundant material goods. Readers may forgive the repetition of the commonplace that it is quite possible to lead the good life in a comfortable home with plumbing, central heating, electrical appliances, even a garage. Very few critics of modern civilization actually spurn all these comforts and conveniences, and I know of none who are putting on hair shirts. Today the drive of the whole non-Western world to modernize confirms the judgment of the human race that the basic material goods are absolute goods. Those who think the human race is wrong—as no doubt it can be—might look harder at their own experience.

---

1. The many who tend to idealize the more wholesome life of the past might consult a UNESCO study of the typical life of a seventeenth-century Frenchman. He could expect to lose one of his parents by the time he was fifteen years old, two or three of his brothers and sisters before he was married, and if he reached the age of fifty he would have lived through two or three famines and three or four near famines, as well as two or three major epidemics. Americans should note, however, that there is still much downright hunger in their own most affluent society—a recent "discovery" that is all the more disgraceful because theirs has been called the first great nation in which more people die of eating too much than eating too little.

Almost as plain, but also often overlooked by critics, are the cultural goods that have flowed out of modern technology. It provided the wealth that made possible the effort of nations to educate their entire populace, as no pre-industrial society could do even if its rulers had dreamed of such a thing. It also provided the incentive because of the needs of industry; a predominantly agricultural society can get along with an illiterate peasantry, as all did in the past, but an industrial society has to have not merely laborers but many kinds of literate workers. Likewise modern technology has made much more widely available a wealth of cultural goods in paperbacks, musical albums, photographs and reproductions of paintings, the better offerings of the mass media, etc. To the extent of their interest and capacity, ordinary people may now share in the values of civilization as the vast majority never did in the past. And because of the ample provision for both material needs and the uses of the mind they enjoy more effective freedom, freedom not only from want but to want, with more opportunities for self-realization, a much wider range of choices and a greater power of choice.

What they commonly choose, however, brings up the depressing products of modern technology—the superabundance of trivial goods that have been made into needs, the flood of trash in the mass media, the incessant, high-powered advertising of a vulgar conception of the good life, and so on *ad nauseam*. I shall be obliged to return to such too familiar complaints about the uses of our wealth and power, and of literacy. At this point I return to my premise of absolute goods. Because men can agree on them they can also agree on some absolute evils. No one would deny, for example, that the pollution of our rivers and the air we breathe is a bad thing. No one would argue either that dropping nuclear bombs on people is good for them. A nuclear war would be bad for everybody, except maybe the "victors"—and people are beginning to realize tardily that in an all-out war there would be no victors, only some wretched survivors. Then the question arises: Why all the pollution? Why has the richest, most powerful nation in all history only begun to do something about it? Why must it go on making bigger and better nuclear bombs when it already has many more than enough to blast the whole earth? I am brought to the compulsions of modern technology, the reasons why it so often neglects or even outrages essential human values, and why it has also drastically limited human freedom.

From the beginning the Industrial Revolution introduced new compulsions. It would condemn millions of people to the routines of factory work, force them day in and day out to go through the same mechanical operations. In a real sense slaves to their machines when at work, they also had to live where machines were at home and so lived in dreary places like Gary, Indiana, once described as a city inhabited by four blast furnaces and a hundred thousand people. In various ways people have been subjected to the compulsions of the new technology in the interests of efficiency and economy, conceived in terms of money values, too often without regard to human costs. I shall have to elaborate the tiresome theme of how our technology has tended to standardize and regiment people, mechanize and dehumanize life, generate massive pressures against the individual it had helped to liberate.

More paradoxical, at once subtler and more tyrannical, are the compulsions resulting from the astoundingly rapid progress of technology since World War II. Together with a wealth of new wonders, from television to space ships, this has created the "affluent society," one so much wealthier than any in the past that it is different in kind. As has often been said, the primary function of Americans today is to be consumers. They *must* keep on consuming faithfully, arduously, to the end of their days. If most of them appear to be doing their duty gladly enough, it nevertheless illustrates the odd kind of tyranny of our economy; if they didn't keep on buying the latest model, and all kinds of superfluous goods or gadgets, our economy would collapse. To assist them we have built up an immense advertising industry, which spends many billions on strictly unproductive but now essential purposes. One of the forms of propaganda required by an advanced technological society, it is a reminder that the immense power man has achieved is increasingly a power over not only nature but people, through ways of manipulating people. These may be for their own good, but thereby they raise again the question of what is good for people, and what people are good for.

4

The Industrial Revolution

In 1835, when a "railway mania" was exciting Britain, the editor of the journal *John Bull* was much alarmed by the latest invention. "Railroads, if they succeed," he warned, "will give an unnatural impetus to society, destroy all the relations which exist between man and man, overthrow all mercantile regulations, and create, at the peril of life, all sorts of confusion and distress." Needless to say, railroads did succeed. A generation later Britishers rejoiced in the thousands of miles of them that crisscrossed their land. Although trains had been invented chiefly to transport freight, they had soon attracted droves of passengers, and by the end of the century would be carrying well over a billion riders a year. Europeans on the Continent had been as pleased to imperil life by building railways, and Americans were simply thrilled when their first transcontinental railroad was completed in 1869. Everywhere the locomotive, whistling through the countryside at sixty miles an hour, was the most popular symbol of the new powers man was acquiring in an industrial age. The editor, long since forgotten, would have seemed a ludicrous old fogey.

Yet he was quite right. It is now hard to realize that before the railway era the great majority of people the world over spent their entire life in the region where they were born, never leaving the village except perhaps to go to the nearest market town, always remaining set in their traditional ways. The railroad accordingly did represent an "unnatural impetus," and a profoundly disruptive one. It signaled the end of the old social and political order as it broke down both geographical and social barriers by carrying ever more millions of passengers, drawn from all classes. It became the popular symbol of a technological revolution that was in fact destroying the traditional relations between men and creating "all sorts of confusion and distress." Even its literal peril to life was dramatized upon the inauguration of the first railroad: a director was run over by the train, prophetically named the Rocket. Today we can better understand the disruption of a traditional society by industrialism, for this is the drama being enacted all over the non-Western world as it seeks to "modernize."

More to the point, the editor of *John Bull* still spoke for many respectable people of his day, and indeed for the vast majority of mankind throughout history. Men had always tended to resist fundamental change, any radical innovation. Though the Industrial Revolution had started at least fifty years before the editor wrote, it had proceeded slowly enough to obscure the deepest change taking place, also symbolized by the railroad. This was again the attitude toward change itself, now in daily life as well as government. It was the growing disposition to accept innovation, even to welcome it. Not only thinkers but ordinary people were calling change "progress." The once novel faith in progress was on its way to becoming common sense. But then we must add that that stodgy editor is still not really a stranger in our revolutionary world. Although change has long since become routine, most people welcome only superficial novelty—the latest models, gadgets, thrills. They still resent and resist any call for fundamental change in their ways of thinking; as Bertrand Russell once observed, most people would sooner die than think, and in fact do so. Especially in America they do not at all welcome radically new ideas; despite our boasts about our greatness, Americans seem more afraid of radicals or revolutionaries in our midst than are any other people in the world. And none more so than the business leaders who keep on revolutionizing the economy, accelerating the drive of our technology—the most influential radicals of our day.

This brings up a deeper paradox. The Industrial Revolution was the work of many inventive, enterprising, daring individuals. At first largely confined to a few districts in England, it was a natural, understandable development of what had started in the Elizabethan Age, but it was by no means inevitable, automatic, or predetermined by any known iron laws of history. Although broadly anticipated by Francis Bacon, its actual course was foretold by no thinker of the time, including Adam Smith, the most acute analyst in the days when machines were growing up. Yet by the same token the Industrial Revolution as a whole was quite unplanned. Its pioneers did not get together and say, Let us try for a change a society based on the machine. Like the inventor of the railroad, and the medieval burghers before them, they were unconscious revolutionaries who hardly foresaw, much less intended, the profound changes their innovations would bring throughout the whole society. The revolution illustrated the "vast, impersonal forces" of history that today we hear so much about. Slowly as it moved in the first fifty years, it came to seem impersonal and automatic as one invention called out another, and the impetus given it by the railroad made it well nigh as irresistible as irreversible. Basically the whole process was by no means so "rational" as historians of technology now make it appear. And all along society was even slower to realize the changes that were coming over it, to deal adequately with the problems they created. It illustrated the difficulties men always have in understanding and catching up with the history they have somehow made—difficulties that we now call cultural lag, but that in spite of our knowingness are as great as ever because of the terrific drive of our technology.

With these ironies in mind, let us review a few of the inventions that started revolutionizing life in a society that had known and valued machines since its beginning. James Watt's steam engine deserves its textbook fame as the key invention, no less because he only improved an engine long used to pump water out of mines. A pupil of a maker of "philosophical-instruments," Watt did not build on new scientific knowledge, but like the technicians of the past merely experimented in a practical spirit. Although he went broke on his early experiments with engines, he persevered, in 1769 patented one good for pumping, and then spent twenty years adding improvements that made it capable of driving all kinds of machinery, in factories as well as mines. Powerful, regular, and itself movable, it had a great advantage over water power in its independence of geography and season; henceforth industry could grow up in any region. Watt helped to sell his engine by establishing a standard measure called "horsepower," though without realizing that the animal would be replaced by the "iron horse"; as an old man he rejected the new-fangled idea of a steam carriage on railways. By that time the steam engine had nevertheless become the prime-mover of a new industrial age, now called the Age of Steam.

One reason why it took half a century to become the main source of industrial power was that the early engines had to be made by hand and could not be made to precise standards. The Age of Steam got really under way with the development of the machine tool industry, which signaled a fantastic development by turning out "machines that make machines," or breeding monsters like Frankenstein's; by 1850 all the basic ones were invented or perfected. The challenge of engines also led to the rise of a new profession, engineering, destined to take its place among the major professions.

Another early key to the Industrial Revolution that deserves its textbook reputation was the textile industry. A series of inventions known to most people (including me) only by name if at all —the water frame, the jenny, the mule—was climaxed by the power loom of Edmund Cartwright, a clergyman turned inventor to meet the increasing need of inventions. This made the industry the first major one to be revolutionized; machinery now did the work, making nonsense of the original meaning of "manufacture" —to make by hand. Likewise it accentuated the essential difference between modern machines and tools or neolithic machines. Whereas the latter required skilled craftsmen, most of the modern ones were automatic, requiring only operators to keep them going. Cotton manufacturers could accordingly begin producing for mass consumption, in foreign markets too. They did most to make Britain, already the greatest trading nation, the "workshop of the world." A grateful House of Commons awarded Cartwright in his old age 10,000 pounds for the benefits he had conferred on his country.

A title was enough for Richard Arkwright, the most enterprising of the early textile manufacturers, who had started out as a barber's apprentice. Described by Carlyle as "that bag-cheeked, pot-bellied, much enduring, much inventing barber," he had to endure the loss of his patent for the water frame, an invention he had stolen from other men; but he earned both a fortune and the title Sir Richard by developing the major social invention required by the new technology—the factory system. Unlike the spinning jenny, but like other of the new machines, his water frame was much too big and heavy to be installed in a worker's cottage, or to be powered by hand or foot; so he and his partners built cotton mills run by water power, in which they employed hundreds of workers. Factories had indeed long been known in Europe, in particular arms factories for its professional armies, but now many more were needed because they were the natural home of the machine. In his mills Arkwright mastered the new art of managing large establishments with machine processes, teaching workers (as he put it) "to conform to the regular celerity of the machine." He employed mostly children because they were not only cheap labor but quicker at learning this strange habit; men with skills were apt to make poor machine-tenders. Arkwright's title, still a rare honor for a businessman, perhaps did not entitle him to be considered one of the "real noblemen" envisaged by Saint-Simon; but in any case it foreshadowed the increasing prestige businessmen would win in the new society.

With his partners Arkwright heralded as well a new era of capitalism, primarily industrial. Such partnerships were common in the early days of the Industrial Revolution, inasmuch as few men had at once the capital to set up a mill and the technical and commercial skills needed to run it successfully. With the growth of industry small firms grew into joint-stock companies as other men were induced to invest their savings, and from mid-century on such companies became the standard form of business organization, producing the bulk of manufactured goods. The "organizational revolution" was under way, a major consequence of the new technology that would culminate in giant corporations, and with them big organizations all through the society. Capitalism was also equipped with a philosophy, such as it had lacked in the heyday of the Fuggers. In *The Wealth of Nations* Adam Smith had advanced his novel theory of laissez faire or free private enterprise, based on what is technically called the principle of the self-regulating market. The theory was that unregulated competition in self-interest best produced the goods people wanted, and that a free market would regulate itself, as if by an "unseen hand," because the incompetent would be weeded out, only those would succeed who knew how to produce the goods cheaply; granting a free rein to self-interest thus benefited the whole society. The new creed, which rationalized common business practice, was admirably suited to the interests of a rising business class that in England had to contend with an often hostile landed aristocracy in control of the government. In America, where business met much less opposition, the new theory became a gospel long before the end of the century. Here economic freedom would be hailed as the most fundamental freedom, the heart of the American Way of Life.

Early in the century America had contributed its pioneer in the factory system, Eli Whitney. Best known to schoolboys for his cotton gin, he also ran a small arms factory in which he introduced the manufacture of interchangeable parts, the rudiments of the assembly line.[1] Harking back to the printing press, his ingenious "American system" speeded up the mass production more characteristic of the new age. And among the new industries to adopt it was harvesting machinery, dating from Cyrus McCormick's invention of the mechanical reaper in 1834, which started the revolutionizing of agriculture too by the machine. Immediately it was another reminder that it took a lot of bold persons to create the "vast, impersonal forces" that made the Industrial Revolution. Twelve years after McCormick's invention, an observer in America—the celebrated land of opportunity and enterprise—marveled at a factory that was daring enough to manufacture a hundred reapers. "It was difficult indeed," he reported, "to find parties with sufficient boldness or pluck and energy to undertake the hazardous enterprise of building reapers, and quite as difficult to

1. Schoolboys may not be taught either that Whitney's cotton gin revived the dying institution of slavery by making it immensely profitable. Improved technology therefore helped to bring on the Civil War.

prevail upon farmers to take their chances of cutting their grain with them, or to look favorably upon such innovation."

Because of such attitudes the "unnatural impetus" of railroads takes on more importance, warranting another new name—the Railway Age. While in England they soon became a "mania" as a profitable source of investment, elsewhere they brought countries to what is now called the "take-off" stage in modernizing. They not only constituted a big new industry, a booming market for iron and coal, but stimulated the growth of all other industries. Whereas transport of goods had been slow and expensive, except for towns located on rivers or the sea, railroads could swiftly deliver all raw materials and finished products, giving manufacturers a much wider choice both in sites for their factories and in markets for their goods. And outside of Britain they gave still more impetus to industrialism by bringing in the aid of government. In America the government lavished huge grants of public lands on the transcontinental railroad builders, as well as generous loans for every mile of track they finished; such public enterprise enabled these entrepreneurs to make fortunes with little personal investment, at as little personal risk. On the Continent the railroad systems were generally planned when not built by the state.

Hence the revolution that had started in Britain now spread to all of Europe. Some countries took to industrialism much more slowly than others, remaining "backward," but everywhere it introduced the same tendencies—more use of mechanical power, more material goods, more standardization of products, and more business. By the end of the century a machine civilization was well on its way to becoming the first civilization in history in which most people would make their living through business instead of agriculture. Likewise it was becoming the first predominantly urban civilization, in which most people in the advanced industrial countries would live in or about the city instead of the village.

We may now pause over an early climax of the Industrial Revolution, the Great Exposition of London in 1851. Held in the Crystal Palace, a monumental building of iron and glass designed especially for the occasion (though not by an architect), it was still more dazzling as the first international exhibition of the wonders being produced by industry. The London *Economist* described the Palace as a temple erected to the honor of "the mightiest empire of the globe—the empire in which industry is most successfully cultivated, and in which its triumphs have been greatest." This was simple fact, no mere boast, but Victorians might be more edified by the deeper significance the *Economist* pointed out in the exhibits in this temple. They testified to the "honour" in which "humble industry" was now held, the "moral improvement" already apparent, the "devotion to peace," the promise of a still more peaceful future—all in all, the "irresistible assurances that a yet higher destiny awaits our successors even on earth."

So we are brought to the social consequences of the Industrial Revolution. Since it has become all too easy to resist the complacent assurances of the *Economist,* I should note first of all that its editor had some good reasons for his complacence. Men had obviously achieved much more of the power over nature that the human race had always cherished, and they could rightly expect to acquire ever more power, and with it more wealth. In his *Communist Manifesto* Karl Marx himself had already been pleased to grant that the bourgeois economy was by far the most productive in history. Unlike others it was producing an abundance of cheap goods; its main market was ordinary people, not the rich. Although economists are still debating to what extent, if any, real wages went up in the early decades of industrialism, by mid-century the trend was clearly upward; workers were on their way to the highest standard of living in history. C. P. Snow has reminded the literary world of a simple truth too often overlooked, that for all its evils the Industrial Revolution was the only hope of the poor—the great majority of mankind who had lived in poverty ever since the rise of civilization. With the prospects of increasing plenty it was not unreasonable to hope for increasing peace too. Marx could dream of his classless society, in which industrialism would provide an abundance for all.

In view of such real gains we may discount somewhat the admitted evils of early industrialism—the many women and children working twelve or fourteen hours a day in factories and mines, the many men slaving at starvation wages, the foul living conditions to which all were condemned in the industrial towns or city slums. Herbert Heaton, an economic historian, has observed that these were ancient forms of evil, taken for granted in the past, and that the revolutionary development was the realization that they were evils. It was the growing indignation over them, the belief that they were remediable, and then the positive resolve to do something about them. We know much more about them than we do about the lot of workers in the past because of a series of full, detailed reports by Royal Commissions and Committees of Inquiry; these were the main source not only of Marx's documentation of the evils of capitalism but of the indignation of many bourgeois, and of government enterprise in a series of laws to check the worst abuses.[2] And the spreading faith in progress that may seem ludicrous in the *Economist*'s hymn to the Great Exposition had much to do with the concerted efforts at reform, which in the past had been rare and sporadic. It was not merely a complacent belief in an automatic material progress, but an active will to improve life on earth. Among the many inspired by the faith in progress had been Robert Owen, an early industrialist who by his own success demonstrated that the current evils of the factory system were economically unnecessary, and who preached that "the new powers which men are about to acquire" made any poverty unnecessary.

Workers themselves were stirred by the new spirit. The violent uprisings that had broken out as early as the Middle Ages could be called the beginning of class war against the bourgeois, but the still unbaptized "proletariat" had then had no political program. In the 1830's the great Chartist movement swept over England, leading to enough riots by workers to look like a revolution, but also gathering more than a million signatures to a petition for a "People's Charter" of democratic government. If this owed most to the democratic ideals proclaimed by the American and French Revolutions, it also owed much to the Industrial Revolution. The industrial proletariat was the fastest growing class in England, it was concentrated in cities, and as Marx would

---

2. We might also note that the human costs of industrial development were as heavy in the Soviet, established in the name of Marx. The enforced collectivization of agriculture cost millions of lives.

say, it was being "armed" by the bourgeois. It was growing more literate and class-conscious, learning from the successful efforts of the bourgeois to wrest political rights from the ruling aristocracy. So it began its protracted struggle for the right to organize in labor unions, the right to vote, and the right to free public education. By the end of the century it had largely won these rights in the advanced industrial countries. With their rising standard of living, workers thereby had the means to more effective freedom than they had enjoyed in any previous society.

Yet they always had to struggle, in America especially against the almost solid opposition of the ruling business class. Although the most shocking abuses of the early years have been eliminated, there remain plenty of evils. We are still struggling with the basic problems created by the triumph of industrialism. I think we still need to dwell on the neglect of elementary human values that was so glaring at the outset.

To begin with, most of the workers long remained wretchedly poor. This was a repetition of an old story, beginning with the rise of civilization, but it was more demoralizing in a society that was acquiring far more wealth and power than any before it. Another common kind of victim of industrial progress was the hand-weaver, the many men left with useless skills, fighting a doomed battle against machines, in their fury now and then smashing them. Comfortable economists, always at home with abstractions, would point out that the new technology caused only "temporary dislocations," inasmuch as its victims would in time be absorbed by new industries, but the dislocations were always painful for human beings, who usually lacked the means of promptly moving into new jobs. Later on economists also grew aware of the "business cycle," the periodic panics of depressions that came with the new economy, throwing men out of work for no fault of their own; but they regarded this cycle coolly as a kind of natural law of business and proposed no means of preventing depressions, nor of taking care of their victims. And despite the accelerating industrial progress, and the gradual rise of real wages, there always remained millions of very poor people. Americans have only begun to realize belatedly how many millions of them there are in our most affluent society. An increasing abundance of material goods made more glaring an elementary failure of industrialism: all along it failed to provide a great many workers with the minimum necessities of a decent life—an adequate diet, adequate medical care, decent homes, pleasant surroundings.

Living conditions were most appalling in the new industrial towns, the heralds of the first predominantly urban civilization in history. In the same year when the editor of *John Bull* was alarmed by the prospect of railroads, Alexis de Tocqueville reported on the city of Manchester, the foremost center of industrialism: "From this foul drain the greatest stream of human industry flows out to fertilize the whole world. From this filthy sewer pure gold flows. Here humanity attains its most complete development and its most brutish, here civilization works its miracles and civilized man is turned almost into a savage." The smoke-filled towns were the drabbest, grimiest, and ugliest in all history. As they went about "making a new Heaven and a new Earth—both black," they made little or no provision for sanitation and recreation, parks and playgrounds, or any open space where people could gather and relax. The industrial slums that started growing up in the old cities too made drabness, filthiness, and ugliness seem still more like a natural, normal condition of industrial progress. Respectable people were not shocked by the foul surroundings in which men were learning to live, and could hope to live with any contentment only by virtue of deadened senses, with the help of gin.

One might excuse such failings by saying that it would naturally take men some time to learn how to live with machines and factories. Thus municipal government in Britain was quite unprepared for the new conditions, and lacked anyway the authority to deal with them; in time it would acquire the authority, make some provision for the elementary needs of a decent civic life. Yet the slums would remain, above all in wealthy America, and with them other root evils that would grow worse. Industry soon began polluting rivers and the atmosphere; not until American cities were smog-bound did people begin to realize, too late, how intolerable the pollution had become. Likewise people grew so inured to ugliness that they seemed quite unaware of it—as most Americans still seem unaware of the commercial blight on their landscape, once known as God's own country. Progressive industrial man has never shown the respect for the natural environment that his benighted ancestors did in creating and tending the landscape of Europe.

Hence the obvious question: Why all such neglect or even contempt of elementary human values? The immediate answer seems to me as obvious: it was due to the vaunted free private enterprise that created industrialism, for the sake of private profit. So in fairness I would again first acknowledge the very real enterprise of the pioneers in the Industrial Revolution, and the technical ingenuity and resourcefulness of their followers in building up industry, producing the goods that people did indeed want. No other class at the time was so enterprising. In America, where businessmen were given most freedom and opportunity, they rewarded the nation by conquering a continent. Having a particular need of machinery for this purpose, they responded by developing a genius for mechanizing, organizing, and standardizing their far-flung operations. In but a generation after the Civil War they made America the greatest industrial power on earth. Allan Nevins has called them "the heroes of our material growth"—the growth that Americans have always been proud of.

Yet the heroes also distinguished themselves by exploitation, plunder, and fraud, on a colossal scale unknown in Britain. They made this generation the most flagrant in the nation's history for routine corruption, in both government and business; they thoroughly earned another name for themselves, the "robber barons." With the help of the Republican Party, which ruled the country in their interests to the end of the century, they also made America the most backward of the industrial countries in social legislation to protect workers against the abuses of private enterprise and the hazards of industry—just as the interests of business had helped to make it the last "Christian" nation to abolish slavery. The economic freedom they prized, no less because the authors of the Constitution had neglected to mention it, was a freedom only for themselves, the men on top; they fought bitterly the enterprise of workers who sought to achieve more freedom for themselves by organizing in labor unions. They forced the basic issue that is still with us, in an era of much bigger business.

5

The "Neotechnic" Phase

Toward the end of the nineteenth century the Industrial Revolution entered a new phase, which in his pioneering *Technics and Civilization* Lewis Mumford called "neotechnic" to distinguish it from the long initial phase he called "paleotechnic" and regarded as on the whole a "disastrous interlude." Men now began mastering their new technology. They not only developed much more rapidly all kinds of practical possibilities but realized more of the ideal possibilities of the machine, feeling more at home with it. They exercised more social control over their technology, in the interests of long neglected human values. Mumford added, however, that barbarous paleotechnic ideals still largely dominated industry and politics, and in a review of his book written twenty-five years later he played down the successes of the neotechnic phase. In my somewhat broader treatment of modern technology I should stress still more both the persistence of old problems and the emergence of new ones. On all counts we may get a perspective on the further gains made in our own time, and also on the much worse problems confronting us.

6

"Post-Industrial" Society

Having at last reached our own time, the generation since World War II, I feel obliged to begin by recording more history. Much of it will already seem like ancient history to young people under twenty-five, who now make up almost half the population of America. They must find it hard to realize what occurred to me only a few years ago, when I noted that in this one generation we have lived through more radical, sweeping, startling changes than past civilizations had experienced in a thousand years. These changes incidentally included a host of new technical marvels—radar, television, wonder drugs, transistors, computers, jets, sputniks, intercontinental missiles, etc.—most of which did not startle me when I first saw or heard of them; I suppose young people too soon learned to take them as a matter of course. We have all grown used to the momentous events on the world stage, such as the dawn of the Atomic Age, the emergence of the Soviet as a world power rivaling America, the Cold War, the conquest of outer space, and the revolt and rise of the whole non-Western world, the dozens of new nations in the United Nations. It is already unnecessary to use quotation marks with once sensational phrases like the population explosion and the knowledge explosion. In our society exploding appears to be a normal mode of expansion.

All these changes were due to technology. In alliance with science, it is now unmistakably the basic determinant of our history.

The upshot was the now notorious "military-industrial complex," another new power in the land. This oligarchy was due simply to the requirements of military technology—not to any theory of either capitalism or socialism. Even so President Eisenhower warned against it as a possible menace to democracy in his farewell address to the nation.[3] Although conservatives have been most fearful of bureaucratic power, they have otherwise expressed little alarm over this extreme example of it, the immense power exercised by a small inner group not directly responsible to the public, the possibly life-or-death decisions made often in secrecy without public debate. In view of the obsession with the Cold War most Americans do not complain either because their government has been directing our fabulous technology primarily to military purposes.

With some absent-minded exceptions, businessmen have been outgrowing their suspicion of another government enterprise, described by Daniel Moynihan as the "most powerful development" of this generation—a more wholesome effort to avert depressions and promote steady economic growth by new fiscal policies based on forecasts of economic advisers. Although growth was at first retarded by periodic "recessions," which were hard on unemployed workers even when they were given the still gentler name of "rolling readjustments," post-industrial society has so far been able to prevent the severe depressions that for a century were regular occurrences. Businessmen and Congressmen are much less horrified by "government planning," which they had regarded as something that only the Russians did, hence *ipso facto* an un-American activity bad for business. (They never objected, of course, to extensive planning by big corporations.) Similarly they are less frightened by deviations from archaic notions of "sound fiscal policy," the fetish of the balanced budget that hobbled the Eisenhower administration. Altogether, it appears that the economy is under far better control than it ever was in the

3. Malcolm Moos wrote the passage he recited, which is now considered by far the most important and enduring of his messages. Eisenhower had not previously indicated such misgivings, nor did he afterward to my knowledge. He tried to hold down the extravagant demands of the military because of his passion for economy, but his awe of big business precluded serious worries about its influence.

palmy days of free private enterprise; though laymen might wonder how it would fare without the big war business.

Ordinary Americans have mostly welcomed a quite different reason for the expansion of government—the growth of the Welfare State, dating from the New Deal. This has been much slower and spottier than the growth of the Warfare State. Congressmen who voted billions for military defense expenditures without batting an eye were still disposed to regard expenditures on public welfare services as reckless extravagance; so for years the nation spent on these services a smaller proportion of its growing national income than it had been spending, while a supposedly alarming increase in the federal debt was actually a decrease in relation to this income. Public demand for them grew, however, if only because an affluent society could plainly afford them. The necessity of keeping up with the Russians finally persuaded Congress of the need of federal aid to education, at first to turn out more scientists and engineers. More tardily the government began to respond to the crying public needs of decent housing and adequate medical care, which private enterprise was either unable or indisposed to provide except at a cost beyond the means of many Americans. At length the government launched a piddling but unprecedented "war" on poverty—the poverty that had persisted all through the triumphant progress of industrialism, and that in a land of unparalleled wealth and abundance had at last come to seem both unnecessary and unjust. President Johnson announced that the national goal was the Great Society.

But perhaps the most significant change in post-industrial society has come about through the vast sums that the government has poured into research and development, especially in the universities. (The last time I looked, the sums had reached $17 billion a year.) As I noted at the outset, the "knowledge industry" has become the biggest business in America. Daniel Bell predicts that the university will become the primary institution of the new society, more important even than the big corporation, and that all the major institutions will be intellectual, not business institutions. If businessmen may long retain more prestige as well as wealth, a new professional elite of brainworkers will make the major decisions about policy.

So all of us in the universities might rejoice in our new

eminence. All, that is, except maybe those engaged in the old-fashioned business of liberal education, concentrated on human values. When Clark Kerr writes, in the kind of language now popular, that the "production, distribution, and consumption" of knowledge accounts for 29 per cent of the gross national product, and is growing at about twice the rate of the economy, teachers of the humanities may be reminded that their product represents an insignificant fraction of the GNP. The billions are going chiefly to science and technology. Bell emphasizes "the new centrality of *theoretical* knowledge, the primacy of theory over empiricism," or one might add over the common sense of businessmen; but what about *philosophical* knowledge? It appears that political scientists now distinguish between political theory and political philosophy, which in my innocence I had thought were virtually synonymous. Theory is what goes with the study of political behavior, the current vogue; it eschews value judgments and has outmoded philosophy. I remain more concerned with philosophy, the issues of the good state and its service of the good life.

In government the host of technicians and experts, who have given the name of "technocratic" to our society, raise a cluster of questions. By definition a technocrat is a technician who has been endowed with power, enough to influence when not to make important political decisions, and he is rarely elected to office. Some thinkers argue that technocracy dooms democracy, makes it already a mere appearance, or let us say an incidental nuisance to experts. Others might rejoice that the status of the new elite is based upon knowledge and skill, not success in pursuing economic or political self-interest; but what are their ruling values? Are their ends primarily the interests of people, or of system? Specifically, Secretary of Defense Robert McNamara developed an influential cost-effectiveness analysis for making key decisions, or what is called "value-engineering"; and does this technique take into account values other than economy? Can any such technique promote wisdom—an imprecise term, but a quality especially needed in a technological society with tremendous power at its disposal? At the moment all such questions are obscured by the Cold War. Daniel Bell has remarked the irony that whereas thinkers from Saint-Simon on who hailed industrialism assumed that the "new men" it would make leaders of would be hostile to the wasteful military spirit, it is primarily war that has elevated the technocrats. But if or when the obsession with the Cold War lifts, what then will come out of all the system?

Since I shall return to such questions, I shall say here only that I do not think the answers are simple, clear, or foreordained. As for the cultural values of the good life, government need not serve as their principal custodian or promoter; enough if it maintains the conditions that make it possible for people to cultivate these values. Immediately I am more concerned with the issues forced by postwar society regarded simply as an affluent society, in which most people have as never before the means to pursue the good life.

Among many other things, the affluent society has produced the historically novel type of the teenager, with whole industries catering to his tastes, and tending to make him not an endearing type. It also produced the education explosion that has sent millions of teenagers to college, most of them with the primary goal of becoming "well-adjusted" to such a society. With the coming of automation it holds out the prospect of increasing leisure in which to enjoy the abundance. This apparent boon is regarded, rather oddly, as a "problem"; but the problem seems less odd when one considers a favorite means of passing the time—staring at TV. A society in which the vast majority own a television set has made it the most popular and influential of the mass media, and so raises the issue of mass culture and the notions of the good life it inculcates. Meanwhile the big networks suggest some reservations about the prediction that the university will replace the corporation as the primary institution, and an intellectual class will make the major decisions affecting public policy. As the networks keep making their own decisions with little interference from government, so the big corporations have been holding their own. They too have been directing our fabulous technology, in the interests of private profit. Hence they force another basic question: How well is big business serving the public interest?

For the time being, all these issues may be illustrated by the mammoth automobile industry. Although invented in Europe, its product became distinctively American as the most popular status symbol, transformed American life more than has any other single product, and indeed has had a deeper, more lasting influence on people than had any historic conqueror. After Henry Ford achieved the mass production of the standardized Model T, a sturdy, economical car once beloved, he had to meet the competion of another wizard, Alfred Sloan, who built up General Motors by producing a range of fancier cars at higher prices, featuring annual new models. Since the war GM has dominated the industry, producing more than half the cars. In the early days of the industry Ford and other entrepreneurs had had a real passion for making automobiles, but Sloan had a more effective strategy. As he explained in *My Years with General Motors*: "The primary object of the corporation . . . was to make money, not just to make motor cars." He knew it could make more money by turning out cars with excess power and chrome, not just sturdy, well-designed, efficient cars. His kind of commercial rather than mechanical efficiency also included careful planning to assure a 20 per cent return on capital, with such success that GM actually averaged a higher return, almost double the national average, once hitting a record 50 per cent; though this did not lead to lower prices for consumers.

To these record profits government contributed by spending lavishly on services for the automobile—one kind of public service that conservatives whole-heartedly favored. Congress under frugal President Eisenhower approved a federal highway program costing many billions, while states spent more billions to speed the traffic on superhighways and freeways, and cities ate their heart out to provide more parking space. By now more than a hundred billion dollars a year goes into automobiles and their servicing—gas, tires, repairs, insurance and finance, maintenance of highways, etc. Americans own enough cars for every man, woman, and child in the country to ride in the front seat, since some ten million families have at least two of them.

The transformation of the country began when the automobile brought modern life to Main Street, the tractor to the farm: except in some backwoods districts, the villager and the farmer were no longer "hayseeds" or "yokels." The many shiny highways everywhere with all their accessories—gas stations, roadside stands, lunchrooms, billboards, etc.—have changed the face of the countryside. The automobile stimulated the mass migration to the suburbs, which have been spreading fast all over the land, changing cities into undefined "metropolitan areas." As it created such other novelties as supermarkets, shopping centers, and drive-ins, suburbanites grew absolutely dependent on it. It took fathers to work, mothers to the store, youngsters to school. Teenagers used it to make love in or to go on joy-rides. And from the beginning it had brought joy to the hearts of Americans, an exhilarating sense of freedom and power as they drove, and of possible social mobility too.

Some by-products of their pleasure, however, illustrate a neglect of human values that has become more conspicuous in the affluent society. The most publicized is the slaughter on the highways, realizing the "peril to life" that the editor of *John Bull* feared in the railroad: many more Americans have been killed by automobiles than were killed in both world wars. The automobile industry has contributed to the slaughter by its primary concern with what made money, including excessive power rather than safety. Highway engineers have been more responsible after their fashion but in the interest only of efficiency and economy, bulldozing their way through landscape and cityscape without regard for natural beauty or civic concerns. Business interests have shown less respect for the landscape as they defiled it with garish billboards, designed to attract attention by not blending into it. The automobile itself ends up in the junkyards that provide more eyesores, speeding the transformation of the country into "God's own junkyard."

The migration to the suburbs it promoted was more rational —except that one reason for it is another basic irrationality of a technological society, the blight of the American city. In the most urban of civilizations, and the richest nation on earth, one might reasonably expect to find beautiful cities, spacious, airy, comfortable; but in fact most of the big ones have notoriously been growing dirtier and shoddier, more congested and polluted, more unfit to live in. The automobile has been chiefly responsible for the congestion and pollution; ever more highways are built to enable more drivers to get into the heart of the city. Hence the people who have fled the city do not really escape it. They help to enact one of the most absurd daily spectacles in affluent America: millions of people driving to and from work, one to an automobile, bumper to bumper, often through smog, often tense and irritable—unable even to absorb the message of all the billboards and neon signs. Lewis Mumford summed up the whole story in a protest that this "Sacred Cow of the American Way of Life is overfed and bloated; that the milk she supplies is poisonous; that the pasturage this species requires wastes acres of land that could be used for more significant human purposes; and that the vast herd of sacred cows, allowed to roam everywhere, like their Hindu counterparts, are trampling down the vegetation, depleting the wild life, and turning both urban and rural areas into a single smudgy wasteland, whose fancy sociological name is Megalopolis."

Still, the Sacred Cow evidently remains a primary need of Americans in their pursuit of happiness. Granted that the industry spends hundreds of millions to create their needs by advertising, there is no question that most Americans like their cars big, powerful, and ostentatious. As certainly they want all the services that government lavishes on the automobile; as taxpayers they are more willing to pay for highway programs than for better schools. They seldom protest against all the bulldozing, the erosion of the heart of cities by parking areas. Once behind the wheel, they do not mind all the ugliness along the highways. Ultimately, in short, the American people must be blamed for the neglect of human values. They may complain about this or that, chiefly about the taxes they have to pay for having it so good; but apparently they would rather have life this way than support a national effort to make America fit for civilized living. Their primary duty remains consuming, buying instead of merely making a living to the end of their days, suitably marked by a costly funeral. Theirs not to question why—theirs only to go and buy.

So regarded, Americans scarcely look like a great people—the kind of people needed to create a Great Society, or to appreciate one. They support rather the forecast of Aldous Huxley, that technology is preparing them for life in something like Brave New World—a benevolent totalitarian state in which almost everybody will be kept happy and hollow. Automation will provide the machinery for running such a world. The technocratic elite is learning how to govern it. Big business is able and willing to cooperate by producing the needed superfluity of material goods. Admen will keep the people consuming and conforming, television will keep them entertained. Experts in motivational research—another new kind of technician—will condition them more perfectly to a state of mindless happiness, while wonder drugs will keep them tranquil, remove the uneasiness of such mind as they have left. In *Post-Historic Man* Roderick Seidenberg foresees a society in which all individuality and spontaneity will be ironed out and consciousness itself will finally disappear.

Again I do not think that the trend toward such a state is irresistible, for reasons I will go into at the end; there is much more to be said about the American people. Meanwhile the clearest reason is frightening, quite apart from the constant possibility of nuclear war. To the peoples of the non-Western world—the great majority of mankind—worries over the possibility of a Brave New World are simply academic. Stirred by the rising expectations of the fruits of modern technology, they will have to struggle indefinitely with the problems of scarcity, in India even with the threat of mass starvation. The affluence of America and western Europe is steadily widening the gulf between the few haves and the many more have-not nations, while the population explosion aggravates the difficulties and kindles other explosive possibilities. Americans may still dream of going their own way toward ever more affluence, ease, and comfort, their notion of happiness; but they can never hope for the security and stability of a Brave New World so long as the aroused non-Western world suffers from acute want. If consciousness is doomed to disappear, it can be trusted to remain painful enough for the foreseeable future. Today the problem in America is to alert it to the world-wide consequences of modern technology.

So let us consider C. P. Snow's celebrated, much abused lecture "The Two Cultures," the sciences and the humanities. His thesis was that these cultures, alike essential to an adequate education today, are unfortunately separated by mutual incomprehension, to some extent even by hostility. Critics were quick to point out that there are more than two cultures—one could easily make out two dozen—and that there is as little communication within them as between them. Sir Charles was more vulnerable because of his complacence over science and technology. He took for granted the greater importance and value of science, dwelling chiefly on the faults of literary people; he said very little about the values of literature, the reasons why scientists ought to be better acquainted with it. But after such qualifications his basic thesis seems to me clearly valid. Most specialists in science do not have an adequate grounding in the humanities; most specialists in the humanities do not know enough about the fundamentals of science and technology. And the current stress on research and development is widening and deepening the gulf between them.

In a following lecture Snow recognized the social sciences as a "third culture," a potential bridge between the other two. Ideally they could serve as a mediator because their subject is man and society. Actually, however, they are generally taught and practiced as sciences, not as humanistic studies, and if anything they have been tending to widen the gulf, arouse more suspicion or hostility among literary people, for reasons I suggested previously. Similarly with the study of engineering. Eric Ashby has said that it could unite science and humanism because it is inseparable from men and communities, concerned with the application of science to their needs. In fact most students of technology acquire very little humanistic culture, and they tend to be more indifferent to it than are students of the natural sciences.[5] And technical or vocational studies have been attracting a larger proportion of the students flooding our colleges than have either the humanities or the natural sciences. They might be grouped in another separate culture, distinctive in its incomprehension of both the other two. Too many of their graduates are cultural illiterates.

The best excuse for such narrowness, all the practical training that students need to fit them for their profession, may now look antiquated. As long ago as 1897 John Dewey pointed out a peculiar difficulty of education today:

> The only possible adjustment which we can give to the child under existing conditions is that which arises through putting him in complete possession of all his powers. With the advent of democracy and modern industrial conditions, it is impossible to foretell definitely just what civilization will be twenty years from now. Hence it is impossible to prepare the child for any precise set of conditions.

Since Dewey wrote, the much faster pace of technological change has made his comment still more pertinent, above all for college students. For years engineers and other technical students who went into business have never used much of the supposedly practical training they got, but today much of it is strictly obsolescent, in process of being outmoded by the latest innovations. Many technical students are being prepared thoroughly, meticulously, for a "set of conditions" that is ceasing to exist. They may find themselves in not only a rut but the wrong rut. In this view, as Robert Hutchins argues, a liberal education may be considered the most practical kind. The humanities in particular are supposed to put students in a fuller possession of all their powers.

But we therefore need to look more closely into their claims, which traditionalists may too easily take for granted. Students in the humanities also concentrate on specialized subject matter, and while it will not be simply superseded as scientific theories and engineering techniques often are, one may ask what purpose is served by all the knowledge or "book learning" they acquire. It is clearly useful for students who are going on to become teachers or scholars in their subject, but what of the many more who are going out into the world? Specifically, the guardians of the humanities might ponder some pointed questions asked by a responsible business leader who wanted to know more about their value. Just why are they considered so important? "Speaking quite practically, what can the humanities do for me, for my family, for my business, for my community? . . . Do they make people better? Do they make people happier? Do they make people more capable? How do you know?"

He could read many resounding answers. To cite a few

random examples, the humanities preserve our rich cultural heritage; they are "the very soul of our culture," the heart of the "civilizing process"; they elevate the spirit by giving us "a sense of man's innate worth and of his infinite capacities;" they further our understanding "of such enduring values as justice, freedom, virtue, beauty, and truth"; they are a "storehouse of wisdom" that enables us to make better judgments; they endow us with purpose and "concern for man's ultimate destiny"; they "help men to live more fully and creatively and to expand their dignity, self-direction, and freedom"; and scientists themselves are wont to say that without a knowledge of the humanities "science would lack vision, inspiration, and purpose." Or the businessman might ponder Cardinal Newman's classic statement of the ideal values of a liberal education:

> It aims at raising the intellectual tone of society, at cultivating the public mind, at purifying the national taste, at supplying true principles to popular enthusiasm and fixed aims to popular aspiration, at giving enlargement and sobriety to the ideas of the age, at facilitating the exercise of political power, and refining the intercourse of private life. It is the education which gives a man a clear conscious view of his own opinions and judgments, a truth in developing them, an eloquence in expressing them, and a force in urging them. It teaches him to see things as they are, to go right to the point, to disentangle a skein of thought, to detect what is sophistical, and to discard what is irrelevant. It prepares him to fill any post with credit, and to master any subject with facility. It shows him how to accommodate himself to others, how to throw himself into their state of mind, how to bring before them his own, how to influence them, how to come to an understanding with them, how to bear with them. He is at home in any society, he has common ground with every class; he knows when to speak and when to be silent; he is able to converse, he is able to listen; he can ask a question pertinently, and gain a lesson seasonably, when he has nothing to impart himself; he is ever ready, yet never in the way; he is a pleasant companion, and a comrade you can depend upon; he knows when to be serious and when to trifle, and he has a sure tact which enables him to trifle

> with gracefulness and to be serious with effect. He has the repose of a mind which lives in itself, while it lives in the world, and which has resources for its happiness at home when it cannot go abroad. He has a gift which serves him in public, and supports him in retirement, without which good fortune is but vulgar, and with which failure and disappointment have a charm. The art which tends to make a man all this is in the object which it pursues as useful as the art of wealth or the art of health, though it is less susceptible of method, and less tangible, less certain, less complete in its result.

Still, Newman paid this tribute at a time when Oxford and Cambridge were offering a routine classical education that now looks pretty sterile. The important question remains unanswered: Do the humanities today actually achieve these high goals, or come anywhere near them? As the businessman asked, "*How do you know?*"

The honest answer is that we don't know. I doubt that we can know, since the lofty objectives ascribed to the humanities do not lend themselves to measurement. As it is, we go on giving examinations that test chiefly a student's knowledge of a particular subject, scarcely his wisdom, vision, creativity, humanity, soulfulness, or what not. We do not know either what kind of subject matter would help most in attaining the lofty objectives, but go on teaching all the traditional subjects in more or less traditional ways. As an English teacher I am disposed to believe that a proper command of language and literature could contribute most to self-realization, or broadly to a fuller realization of both our common humanity and one's personal identity, but the history of literary studies offers no convincing evidence that they make men more humane, and recent history gives more reason for doubt; Dr. Goebbels was only one of the many German doctors of letters who welcomed Hitler. I am sure only that many students specializing in the humanities achieve at most a mastery of a particular subject, or rather of some subdivision of it. Their studies have made them more capable in some respects, and we must hope on the whole better people, but we must wonder how much they can contribute.

Yet we of course can and must talk about human nature, as we do about "man," "mankind," the "human condition," and our "common humanity." Biologically man is a distinct species, with the same structure and needs that he had fifty thousand years ago. "It is absolutely certain," writes Rene Dubos, "that all of his physiological needs and drives, his potentialities and limitations, his responses to environmental stimuli, are still determined by the same twenty thousand pairs of genes that controlled his life when he was a palaeolithic hunter or a neolithic farmer." Dubos explains that we know very little about the pathological effects on man of a polluted, crowded environment and the stresses of modern life because the problems they create appear chiefly late in life, past the reproductive age that could call into play the mechanisms of natural selection; but he is certain that the new environment and ways of life have had deleterious effects. Likewise biologists and sociologists know hardly anything about man's adaptive potentialities, which would require studies of long-range consequences; but again it is certain that man cannot adapt himself successfully to any kind of life in any kind of environment. Meanwhile technology keeps plunging on in ignorance, in ways largely determined by economic expediency, with little regard for the basic nature and needs of man. So the accelerating pace of change may make one wonder: Just how much change can man stand? Can he adapt himself successfully to a continuously revolutionary world?

As a humanistic scientist who believes that "quiet, privacy, independence, initiative, and open space" are "real biological necessities," Dubos recalls me to my own premises. At the outset I stated my belief that despite the conspicuous diversity of cultures and relativity of values we *can* talk of absolute values, permanent as long as man exists; or in other words, these derive from "human nature." The psychologist William Maslow maintains that even the "higher" human needs, as for friendship, love, dignity, self-respect, and self-fulfillment, are biological in origin, and human nature has been sold short because they have not been so considered by most scientists; but in any case they have long been vital needs. As for our own society, he attributes the growing discontent and rebelliousness of young people to the frustration of the idealism natural to them. Academic psychology, economics, and most social science offer them a limited view of human nature, with emphasis on the lower or material needs. The dominant positivistic and behavioristic theories reduce the highest values and virtues to mere appearances or illusions. The whole society operates on a low conception of human nature. "How could young people not be disappointed and disillusioned?" Maslow concludes. "What else could be the result of *getting* all the material and animal gratifications and then *not being happy*, as they were led to expect, not only by the theorists, but also by the conventional wisdom of parents and teachers, and the insistent grey lies of the advertisers?"

The youth also best exemplify the peculiar complication of human nature today. A person's genes determine not primarily his traits but his potentialities; man is distinguished by the remarkable range of his potentialities, most of which the individual never realizes or develops. Our technological society has carried much further the ambiguous tendencies that arose with civilization, which at once widened the range of opportunities for self-realization and for most people narrowed it by a division of labor as well as inequality of opportunity. Young people now have an extraordinary range of choice in profession, habitat, and way of life, while most of them also feel impelled to become specialists and conform to the requirements of a technological society, in an urban environment becoming increasingly uniform. And collective man, always at once creator and creature of his culture, has become both to a more marked extent. While developing immense power to change the environment that shapes his nature, and now more power to change his nature directly, technological man—the most thoroughly conditioned creature of his culture—feels impelled to go on acquiring still more power, however dangerous, and to use it for purposes that many young people find revolting. They have been growing more aware of rich potentialities unrealized in the American way of life. Change is now so rapid that it has taken only a few years to create a wide gap between the generations.

In this view of human nature "alienation" can become more meaningful as a neglect or frustration of higher needs. First I would repeat that it is not a plight peculiar to modern man. By virtue of his plastic nature and his powers of choice, man has always been capable of stupid, perverted, self-defeating, possibly fatal behavior, as his whole history abundantly proves. Yet alienation in modern America is in some respects historically new. It may spring from mechanical routines to which man is not clearly adapted, biologically or psychologically. The many symptoms of it in ordinary people, including the feelings of emptiness and impotence, are quite different from the fatalistic resignation to which the peasant masses were always prone. And the very achievements of America, in creating opportunities and incentives for self-fulfillment, have alienated many people because they have higher expectations, make higher demands on their society than people did in the past.

Hence the question now is, are Americans likely to get more alienated in the years ahead? Much will depend on what the nation does about all its serious problems, including the man-made scene. Whether in particular the many millions of poor, above all the blacks, remain alienated or get more so depends wholly on whether it recognizes its debt to them. Their prospects are dimmer because the problem in my present terms is to improve not merely their material but their spiritual condition, by building up their self-respect, enabling them to have more feeling of dignity, giving them more opportunities for self-fulfillment, bringing them into the society as full-fledged members; and this calls for more patience, humility, tact, and concern for the higher needs than are displayed by either most technologists or most prosperous Americans. While not at all confident that the nation will commit itself to a serious effort to eliminate poverty, discrimination, and social injustice, I take it that at best the problem will not be solved by the year 2000. When dealing with such problems one may realize, as William James did in his later years, that for the purposes of a lifetime human nature isn't so damn plastic after all.

As for the subtler "alienation" of prosperous Americans, I would not venture a prediction. Increasing wealth might fortify their complacence and their hedonistic tendencies, or it might intensify their underlying discontents and anxieties. If the latter, it might produce more violent outbursts of irrationality and inhumanity, or more concern for basic human values and higher human needs. Most likely the affluence will have all these effects on people, and then some, but in proportions impossible to predict with assurance in a society changing so rapidly. All I should say is that our hope lies in the growing awareness of these conflicting tendencies, the persistent criticism of our society, and the characteristic theme of "challenges." Having announced that alienation is "almost total," Erich Fromm nevertheless writes *The Revolution of Hope: Toward a Humanized Technology*. Maslow believes that a "Humanist Revolution" is under way in all fields of thought, with a centering of concern on human needs. Ralph Siu, a fervent technologist, likewise believes that we are on the threshold of an "Age of Holistic Humanism." If so, people may in time grow better adapted to affluence and leisure, as over the ages they have to toil.

At least the symptoms of alienation might lessen the fear that we are headed toward Brave New World. They belie the assumption of Huxley and other critics that people are growing ever more content with their affluence, their mindless role as consumers, or the satisfaction of merely material needs. In this respect "human nature" does appear to be resisting the trend toward Brave New World. But here we run into all the other technological threats to humanistic ideals. In particular they threaten the ideals of individuality and personal freedom, the most obvious basis for resistance to the whole trend.

As I have said, these ideals cannot on the historic record be considered an absolute need of man, or permanent values, but are nevertheless clearly basic potentialities of human nature. All societies have enabled the individual to achieve some measure of self-fulfillment through the development of skills. Since our own civilization has developed the fullest consciousness of selfhood, the value of self-realization, and then of personal freedom as a necessary means to this end, these possibilities of human life can for us be regarded as among the higher needs of man. "The excellent becomes the permanent," Aristotle observed, "—once seen it is never completely lost." Or in historical terms, human nature is a product of a development that has included some values the human race has hung on to. Hence the widespread alarm over the trend to uniformity and the pressures to conformity is quite "natural" even though it is historically novel. No less natural are the fears of increasing organization. As Roderick Seidenberg sees it, organization breeds organization, it accomplishes what used to be the work of creative individuals, it steadily shrinks the autonomy of the individual, and it moves inexorably toward uniformity and universality, the ideal of frictionless mechanism or perfect efficiency. Lewis Mumford agrees that already the foundations have been laid for a society that could produce Seidenberg's "post-historic man," lacking all individuality:

> If the goal of human history is a uniform type of man, reproducing at a uniform rate, in a uniform environment, kept at constant temperature, pressure, and humidity, living a uniformly lifeless existence, with his uniform physical needs satisfied by uniform goods, all inner waywardness brought into conformity by hypnotics and sedatives, or by surgical extirpation, a creature under constant mechanical pressure from incubator to incinerator, most of the problems of human development would disappear. Only one problem would remain: Why should anyone, even a machine, bother to keep this kind of creature alive?

There is no immediate prospect, however, of any such uniformity in society or human nature. Granted all the plain pressures against the individual, the signs of resistance to them have become plainer too, especially among the many critical or rebellious young people. If this may be a passing fashion, the individual remains very much alive in ordinary capacities. Even in systems analysis he is still considered real and important as a means of "feedback." Administrators of public welfare programs soon learn (as August Heckscher did in the Parks Department of New York) that they are not dealing with atomized mass-men, for they run into plenty of positive opinions from spokesmen of local groups or communities. And they make plainest that the technological threats to individuality are not all simply threats.

A planned society need not be totalitarian, and if the experience of the Soviets means anything, it cannot be totally or permanently planned. Karl Mannheim's idea of "planning for freedom" may be questionable, but at least planning is not necessarily fatal to freedom and individuality, any more than would be a program for eliminating poverty; some planners want to get rid of much unnecessary interference with people. Neither, to repeat, is organization necessarily fatal. This is obviously necessary not only for accomplishing the massive action called for by some problems, but for combatting the organizations of vested interests, as the young radicals have discovered. While feeling most strongly that the "Establishment" is hostile to the ideal of self-determination, these radicals are also learning that the old-time uncompromising individualist, or the loner like Thoreau, cannot hope to be effective today, that in organized movements they have to provide for cooperativeness and tolerance of diverse opinion, and that in these movements they can nevertheless enjoy more sense of personal freedom and self-fulfillment. (No class, incidentally, is more addicted to "groupism" than the hippies.) On a national scale the extraordinary number of diverse organizations—public, private, professional—likewise afford the individual opportunities for self-fulfillment, and suggest that the real problem is to develop a different kind of individualist. Nor is rationality or intelligence so blind or ultimately stupid as Seidenberg pictures it. Intelligence can of course be sensitive to its own abuses, just as what men used to call "reason" can be aware of its limitations and its excesses. It is reason or intelligence that warns us against the dangers of organization, the common excesses of technical rationality, the trends toward Brave New World. Human nature as it has developed has some capacity for reasonableness, which is the necessary basis for what hope we can muster.

In any case Seidenberg is looking ahead thousands of years, to a future we need not worry about at the moment. Thus he sees society becoming ever more unified, moving toward a state of complete and permanent stability; but we have to deal with the realities of a divided people and dynamic change, a highly uncertain future. He likewise argues that freedom must disappear for the sake of perfect justice, since injustice would make impossible the frictionless mechanism to which organization aspires; but we still have plenty of social injustice to cope with, for millions of Americans it remains the chief threat to their effective freedom, and we can count indefinitely on ample friction. For such reasons too there is little or no prospect of Brave New World in the near future. Meanwhile it is too easy to blame everything on technology. Our most serious problems are political and moral, strictly up to us. With these problems we could do with more utopian thinking, appealing to the best in the American people, whose tradition has it that human nature can be improved by democratic institutions, or appealing to the many who still are concerned with the higher human needs.

But so far I have been writing with an eye chiefly to the nature of Americans in their man-made scene. For the sake of perspective I wish finally to take a brief look at the rest of mankind, specifically the non-Western world, which remains the most obvious reason why we will not have to worry about Brave New World for a long time to come. It also throws some light on the nature of man, who has grown capable of much more rapid adaptation to change, but at enough biological and psychological cost to make one wonder how much change he can stand. Today the disorders all over the world, including the student revolts, suggest that the whole human race may be facing a crisis.

Herman Kahn and Anthony Wiener forecast that by the year 2000 Japan will have joined the United States and the Western democracies as post-industrial societies; so presumably it will know all the blessings and the problems of affluence. A few countries, including the Soviet Union and several of its satellites, will have reached the early post-industrial stage, and more the mass-consumption stage. But such large countries as China, India, Pakistan, Indonesia, Brazil, and Nigeria, with an estimated total population then of over 3 billion, will still be only partially industrialized and will have a per capita income ranging from only $200 to $600. Behind them will still be most of Africa, two-thirds of the Arab world, and some scattered small countries, with a total population of 750 million, greater than that of the post-industrial countries. Hence an island of wealth will be surrounded by "misery." Although the poor countries will have a higher standard of living, they will be much farther behind the affluent ones than they now are.

This prospect, which will give Americans more trouble than they realize even though they have begun to learn that we have fewer friends in the world than ever, is still not bringing out the best in them. The $30 billion we have spent on aid to pre-industrial countries represent much less than one per cent of the national income, and of late the foreign aid program has been steadily whittled as nominally Christian politicians and voters complain of "give-aways." Apparently they do not know that American business has profited from the program, inasmuch as about 80 per cent of the aid is spent here. Now more of it is being tied to exports, while seldom giving the poor countries better opportunities to export their products. Although the disappointing results of the aid program have been largely due to the common inefficiency and corruption of government in the poor nations, and the selfishness of their wealthy class, they have also been due to politics at home. Backed by Congress, our aid missions have usually favored industrialization projects more than projects for improving agriculture; these might lead to competition with U.S. farm products. In short, Uncle Sam has been neither the Santa Claus nor the sucker portrayed in cartoons.[1] Not to mention that a people very proud of their own Revolution remain about the least capable of understanding or tolerating revolutions in the rest of the world.

---

1. In fairness, the Soviet Union has been no more disinterested. With increasing affluence both it and the European democracies could easily afford more generous aid, but my guess is that the poor countries had better not count on it. Meanwhile the U.S. Arms Control and Disarmament Agency has reported that the soaring cost of the global arms race has reached more than $182 billion a year, averaging $53 for every man, woman, and child on earth—more than many of them have to live on; and it estimates that if the increasing cost continues, the world will spend $4 trillion in the next ten years. Such is the cost of a "security" that no nation enjoys, including America, the leading spender.

I shall not go into the staggering economic and political problems of the new nations, except to say that they cannot be solved by any magic of free private enterprise, or merely economic rationality. My concern here is some questions raised by their recognition of the goods of modern technology, their confirmation of its apparently irresistible drive, and the consequences of their efforts to modernize, in the name of "progress." They have already made enough progress to produce a growing uniformity, immediately recognizable in their airports and in the new buildings and boulevards of their capitals. Industrial societies are bound to be much more alike than agricultural, handicraft societies. Railroads, automobiles, cement factories, steel mills, power plants, and all kinds of machinery and machine products are everywhere basically the same, alike under Communism or capitalism. People grown used to them might be expected to be more alike.

Yet industrialism by no means requires the total elimination of traditional cultures. When Japan, for example, developed a factory system, it was quite different from the American system. A worker employed by a company assumed he would work for it the rest of his life and would not quit to take a higher paid job in another factory, while the company felt as obligated to him, only under extreme circumstances discharging him or even laying him off temporarily. Still by far the most successful of the non-Western countries in modernizing, Japan accordingly refutes the assumption of Jacques Ellul that technology is wholly autonomous and automatically imposes its requirements. In India, where Nehru gave up Gandhi's opposition to industrialism, leaders are still trying to adapt it to their traditional culture; some resent the term "underdeveloped" because it implies that economic growth and technological advance are the supreme goals for a nation. Although leaders are typically eager to modernize, few of the non-Western peoples set such store by economic values as Americans do, or assume that economic motives are the most powerful.[2] They are

2. Americans might be bewildered by the attitude of peasants in Java. When the price of coconuts goes up, they are likely to bring fewer instead of more coconuts to market, for the sensible reason that

not simply awed either by American know-how or modes of efficiency, which are usually not suited to their culture and environment.

For such reasons they are likely to be handicapped in industrializing, especially because their traditional ways often conflict with the regular habits and technical rationality it requires. Nevertheless their leaders also have a potential advantage in being aware of the evils that can come with it. Although most of them now tend to be arrogant and short-sighted in the exercise of their new powers, the "developing" nations (to use the politer term) may hope to escape the inhumanity of early or "palaeolithic" industrialism, and to retain more respect for esthetic and other human values than most Americans appear to have. Kahn and Wiener, who remarked that the "Sensate" culture of the West has been spreading over the whole world, consider it an open question whether this trend will continue in the next generation. The tendencies to "Late Sensate" are certainly much less conspicuous in the non-Western world.

In the very long run, if or when all countries reach the post-industrial stage, I suppose it is conceivable that there might be an almost complete uniformity, and "human nature" will everywhere be more literally the same. Even by the year 2000 there may be a considerable degree of drab uniformity, especially if television "unifies" peoples. In my own travels abroad I have been mostly depressed by the changes that have taken place in just thirty years, because of both technology and tourism. Yet it seems to me unlikely that the world will ever have but one culture, any more than that all people will speak the same language and worship the same God. Countries with advanced technology can still retain much of their traditional culture and have different styles of living, as European countries do; my generalizations about the nature of Americans and their man-made scene would not all apply in Europe. In the non-Western world an intense nationalism —a sentiment that Americans think of as just human nature, but that was actually acquired under Western domination—creates a

the higher price makes it unnecessary for them to work as hard to satisfy their needs.

more conscious pride in traditional values or native variations, and strengthens a resolve to be not just like Americans.

Unfortunately it also heightens the dangers of war—the all too real possibility that there may be no long run for modern man. The non-Western world throws into relief the most tragic aspect of the failures of America in its professed role as leader of the "free world." As an Italian historian remarked to Walter Lippmann, we have come down off the pedestal and entered history, joined all the other nations that have insoluble problems and suffer defeats; yet the tragedy remains that there is no nation clearly fit to take our place or carry the torch for the non-Western world. Despite de Gaulle's dreams of *gloire,* France is no more ready or able than Great Britain, West Germany, or Japan. The Soviet Union still adheres to a narrow conception of its own national interests; it antagonized most of the world by its ruthless suppression of the Czechs. The future may belong to teeming China, but who would now look to it for world leadership? Or to India, Pakistan, Indonesia, Egypt, Nigeria? By the year 2000 there will no doubt be surprising changes in the status and the alignment of nations, but the critical period is the coming generation, during which the danger of a nuclear explosion will be most acute; and as we face it, the fact is that no nation on earth has both the material means and the will to provide the enlightened aid needed by the poor countries.

Otherwise the plainest lesson for Americans in the turbulent drama of the non-Western world is the need of not only more sympathetic understanding of the different ways of other peoples, but more self-understanding, a more critical attitude toward their own accepted values and notions of the good life, or simply more awareness that their own nature is not just human nature, their ways are too often a perversion or defiance of the developed higher needs of man. Else there is also a real possibility that America—the pace-setter in technology, and the country in which above all it has run wild—may go down as the greatest failure in history, or that it will be spared this ignominy only because history will have come to an end.

Lewis Mumford (1895-1970), an outstanding social critic, philosopher, journalist, and historian, has written a four volume philosophy of civilization. The first volume, *Technics and Civilization,* is considered a classic. Other outstanding works of his include *The City in History* (1961), *Technics and Human Development* (1967) and *The Pentagon of Power* (1970).

This paper, first presented at a convocation of the Fund for the Republic, was published in *Technology and Culture (Vol. V, No. 1).*

Authoritarian and Democratic Technics

"Democracy" is a term now confused and sophisticated by indiscriminate use, and often treated with patronizing contempt. Can we agree, no matter how far we might diverge at a later point, that the spinal principle of democracy is to place what is common to all men above that which any organization, institution, or group may claim for itself? This is not to deny the claims of superior natural endowment, specialized knowledge, technical skill, or institutional organization: all these may, by democratic permission, play a useful role in the human economy. But democracy consists in giving final authority to the whole, rather than the part; and only living human beings, as such, are an authentic expression of the whole, whether acting alone or with the help of others.

Around this central principle clusters a group of related ideas and practices with a long foreground in history, though they are not always present, or present in equal amounts, in all societies. Among these items are communal self-government, free communication as between equals, unimpeded access to the common store of knowledge, protection against arbitrary external controls, and a sense of individual moral responsibility for behavior that affects the whole community. All living organisms are in some degree autonomous, in that they follow a life-pattern of their own; but in man this autonomy is an essential condition for his further development. We surrender some of our autonomy when ill or crippled: but to surrender it every day on every occasion would be to turn life itself into a chronic illness. The best life possible—and here I am consciously treading on contested ground—is one that calls for an ever greater degree of self-direction, self-expression, and self-realization. In this sense, personality, once the exclusive attribute of kings, belongs on democratic theory to every man. Life itself in its fullness and wholeness cannot be delegated.

In framing this provisional definition I trust that I have not, for the sake of agreement, left out anything important. Democracy, in the primal sense I shall use the term, is necessarily most visible in relatively small communities and groups, whose members meet frequently face to face, interact freely, and are known to each other as persons. As soon as large numbers are involved, democratic association must be supplemented by a more abstract, depersonalized form. Historical experience shows that it is much easier to wipe out democracy by an institutional arrangement that gives authority only to those at the apex of the social hierarchy than it is to incorporate democratic practices into a well-organized system under centralized direction, which achieves the highest degree of mechanical efficiency when those who work it have no mind or purpose of their own.

The tension between small-scale association and large-scale organization, between personal autonomy and institutional regulation, between remote control and diffused local intervention, has now created a critical situation. If our eyes had been open, we might long ago have discovered this conflict deeply embedded in technology itself.

I wish it were possible to characterize technics with as much hope of getting assent, with whatever quizzical reserves you may still have, as in this description of democracy. But the very title of this paper is, I confess, a controversial one; and I cannot go far in my analysis without drawing on interpretations that have not yet been adequately published, still less widely discussed or rigorously criticized and evaluated. My thesis, to put it bluntly, is that from late neolithic times in the Near East, right down to our own day, two technologies have recurrently existed side by side: one authoritarian, the other democratic, the first system-centered, immensely powerful, but inherently unstable, the other man-centered, relatively weak, but resourceful and durable. If I am right, we are now rapidly approaching a point at which, unless we radically alter our present course, our surviving democratic technics will be completely suppressed or supplanted, so that every residual autonomy will be wiped out, or will be permitted only as a playful device of government, like national balloting for already chosen leaders in totalitarian countries.

The data on which this thesis is based are familiar; but their significance has, I believe, been overlooked. What I would call democratic technics is the small-scale method of production, resting mainly on human skill and animal energy but always, even when employing machines, remaining under the active direction of the craftsman or the farmer, each group developing its own gifts, through appropriate arts and social ceremonies, as well as making discreet use of the gifts of nature. This technology had limited horizons of achievement, but, just because of its wide diffusion and its modest demands, it had great powers of adaptation and recuperation. This democratic technics has underpinned and firmly supported every historical culture until our own day, and redeemed the constant tendency of authoritarian technics to misapply its powers. Even when paying tribute to the most oppressive authoritarian regimes, there yet remained within the workshop or the farmyard some degree of autonomy, selectivity, creativity. No royal mace, no slave-driver's whip, no bureaucratic directive left its imprint on the textiles of Damascus or the pottery of fifth-century Athens.

If this democratic technics goes back to the earliest use of tools, authoritarian technics is a much more recent achievement: it begins around the fourth millennium B.C. in a new configuration of technical invention, scientific observation, and centralized political control that gave rise to the peculiar mode of life we may now identify, without eulogy, as civilization. Under the new institution of kingship, activities that had been scattered, diversified, cut to the human measure, were united on a monumental scale into an entirely new kind of theological-technological mass organization. In the person of an absolute ruler, whose word was law, cosmic powers came down to earth, mobilizing and unifying the efforts of thousands of men, hitherto all-too-autonomous and too decentralized to act voluntarily in unison for purposes that lay beyond the village horizon.

The new authoritarian technology was not limited by village custom or human sentiment: its herculean feats of mechanical organization rested on ruthless physical coercion, forced labor and slavery, which brought into existence machines that were capable of exerting thousands of horsepower centuries before horses were harnessed or wheels invented. This centralized technics drew on inventions and scientific discoveries of a high order: the written record, mathematics and astronomy, irrigation and canalization: above all, it created complex human machines composed of specialized, standardized, replaceable, interdependent parts—the work army, the military army, the bureaucracy. These work armies and military armies raised the ceiling of human achievement: the first in mass construction, the second in mass destruction, both on a scale hitherto inconceivable. Despite its constant drive to destruction, this totalitarian technics was tolerated, perhaps even welcomed, in home territory, for it created the first economy of controlled abundance: notably, immense food crops that not merely supported a big urban population but released a large trained minority for purely religious, scientific, bureaucratic, or military activity. But the efficiency of the system was impaired by weaknesses that were never overcome until our own day.

To begin with, the democratic economy of the agricultural village resisted incorporation into the new authoritarian system. So even the Roman Empire found it expedient, once resistance was broken and taxes were collected, to consent to a large degree of local autonomy in religion and government. Moreover, as long as agriculture absorbed the labor of some 90 per cent of the population, mass technics were confined largely to the populous urban centers. Since authoritarian technics first took form in an age when metals were scarce and human raw material, captured in war, was easily convertible into machines, its directors never bothered to invent inorganic mechanical substitutes. But there were even greater weaknesses: the system had no inner coherence: a break in communication, a missing link in the chain of command, and the great human machines fell apart. Finally, the myths upon which the whole system was based—particularly the essential myth of kingship—were irrational, with their paranoid suspicions and animosities and their paranoid claims to unconditional obedience and absolute power. For all its redoubtable constructive achievements, authoritarian technics expressed a deep hostility to life.

By now you doubtless see the point of this brief historical excursus. That authoritarian technics has come back today in an immensely magnified and adroitly perfected form. Up to now, following the optimistic premises of nineteenth-century thinkers like Auguste Comte and Herbert Spencer, we have regarded the spread of experimental science and mechanical invention as the soundest guarantee of a peaceful, productive, above all democratic, industrial society. Many have even comfortably supposed that the revolt against arbitrary political power in the seventeenth century was causally connected with the industrial revolution that accompanied it. But what we have interpreted as the new freedom now turns out to be a much more sophisticated version of the old slavery: for the rise of political democracy during the last few centuries has been increasingly nullified by the successful resurrection of a centralized authoritarian technics—a technics that had in fact for long lapsed in many parts of the world.

Let us fool ourselves no longer. At the very moment Western nations threw off the ancient regime of absolute government, operating under a once-divine king, they were restoring this same system in a far more effective form in their technology, reintroducing coercions of a military character no less strict in the organization of a factory than in that of the new drilled, uniformed, and regimented army. During the transitional stages of the last two centuries, the ultimate tendency of this system might be in doubt, for in many areas there were strong democratic reactions; but with the knitting together of a scientific ideology, itself liberated from theological restrictions or humanistic purposes, authoritarian technics found an instrument at hand that has now given it absolute command of physical energies of cosmic dimensions. The inventors of nuclear bombs, space rockets, and computers are the pyramid builders of our own age: psychologically inflated by a similar myth of unqualified power, boasting through their science of their increasing omnipotence, if not omniscience, moved by obsessions and compulsions no less irrational than those of earlier absolute systems: particularly the notion that the system itself must be expanded, at whatever eventual cost to life.

Through mechanization, automation, cybernetic direction, this authoritarian technics has at last successfully overcome its most serious weakness: its original dependence upon resistant, sometimes actively disobedient servomechanisms, still human enough to harbor purposes that do not always coincide with those of the system.

Like the earliest form of authoritarian technics, this new technology is marvelously dynamic and productive: its power in every form tends to increase without limits, in quantities that defy assimilation and defeat control, whether we are thinking of the output of scientific knowledge or of industrial assembly lines. To maximize energy, speed, or automation, without reference to the complex conditions that sustain organic life, have become ends in themselves. As with the earliest forms of authoritarian technics, the weight of effort, if one is to judge by national budgets, is toward absolute instruments of destruction, designed for absolutely irrational purposes whose chief by-product would be the mutilation or extermination of the human race. Even Ashurbanipal and Genghis Khan performed their gory operations under normal human limits.

The center of authority in this new system is no longer a visible personality, an all-powerful king: even in totalitarian dictatorships the center now lies in the system itself, invisible but omnipresent: all its human components, even the technical and managerial elite, even the sacred priesthood of science, who alone have access to the secret knowledge by means of which total control is now swiftly being effected, are themselves trapped by the very perfection of the organization they have invented. Like the pharaohs of the Pyramid Age, these servants of the system identify its goods with their own kind of well-being: as with the divine king, their praise of the system is an act of self-worship;

and again like the king, they are in the grip of an irrational compulsion to extend their means of control and expand the scope of their authority. In this new systems-centered collective, this Pentagon of power, there is no visible presence who issues commands: unlike Job's God, the new deities cannot be confronted, still less defied. Under the pretext of saving labor, the ultimate end of this technics is to displace life, or rather, to transfer the attributes of life to the machine and the mechanical collective, allowing only so much of the organism to remain as may be controlled and manipulated.

Do not misunderstand this analysis. The danger to democracy does not spring from any specific scientific discoveries or electronic inventions. The human compulsions that dominate the authoritarian technics of our own day date back to a period before even the wheel had been invented. The danger springs from the fact that, since Francis Bacon and Galileo defined the new methods and objectives of technics, our great physical transformations have been effected by a system that deliberately eliminates the whole human personality, ignores the historical process, overplays the role of the abstract intelligence, and makes control over physical nature, ultimately control over man himself, the chief purpose of existence. This system has made its way so insidiously into Western society that my analysis of its derivation and its intentions may well seem more questionable—indeed more shocking—than the facts themselves.

Why has our age surrendered so easily to the controllers, the manipulators, the conditioners of an authoritarian technics? The answer to this question is both paradoxical and ironic. Present-day technics differs from that of the overtly brutal, half-baked authoritarian systems of the past in one highly favorable particular: it has accepted the basic principle of democracy, that every member of society should have a share in its goods. By progressively fulfilling this part of the democratic promise, our system has achieved a hold over the whole community that threatens to wipe out every other vestige of democracy.

The bargain we are being asked to ratify takes the form of a magnificent bribe. Under the democratic-authoritarian social contract, each member of the community may claim every material advantage, every intellectual and emotional stimulus he may desire, in quantities hardly available hitherto even for a restricted minority: food, housing, swift transportation, instantaneous communication, medical care, entertainment, education. But on one condition: that one must not merely ask for nothing that the system does not provide, but likewise agree to take everything offered, duly processed and fabricated, homogenized and equalized, in the precise quantities that the system, rather than the person, requires. Once one opts for the system no further choice remains. In a word, if one surrenders one's life at source, authoritarian technics will give back as much of it as can be mechanically graded, quantitatively multiplied, collectively manipulated and magnified.

"Is this not a fair bargain?" those who speak for the system will ask. "Are not the goods authoritarian technics promises real goods? Is this not the horn of plenty that mankind has long dreamed of, and that every ruling class has tried to secure, at whatever cost of brutality and injustice, for itself?" I would not belittle, still less deny, the many admirable products this technology has brought forth, products that a self-regulating economy would make good use of. I would only suggest that it is time to reckon up the human disadvantages and costs, to say nothing of the dangers, of our unqualified acceptance of the system itself. Even the immediate price is heavy; for the system is so far from being under effective human direction that it may poison us wholesale to provide us with food or exterminate us to provide national security, before we can enjoy its promised goods. Is it really humanly profitable to give up the possibility of living a few years at Walden Pond, so to say, for the privilege of spending a lifetime in *Walden Two?* Once our authoritarian technics consolidates its powers, with the aid of its new forms of mass control, its panoply of tranquilizers and sedatives and aphrodisiacs, could democracy in any form survive? That question is absurd: life itself will not survive, except what is funneled through the mechanical collective. The spread of a sterilized scientific intelligence over the planet would not, as Teilhard de Chardin so innocently imagined, be the happy consummation of divine purpose: it would rather ensure the final arrest of any further human development.

Again: do not mistake my meaning. This is not a prediction of what *will* happen, but a warning against what *may* happen.

What means must be taken to escape this fate? In characterizing the authoritarian technics that has begun to dominate us, I have not forgotten the great lesson of history: prepare for the unexpected! Nor do I overlook the immense reserves of vitality and creativity that a more humane democratic tradition still offers us. What I wish to do is to persuade those who are concerned with maintaining democratic institutions to see that their constructive efforts must include technology itself. There, too, we must return to the human center. We must challenge this authoritarian system that has given to an underdimensioned ideology and technology the authority that belongs to the human personality. I repeat: life cannot be delegated.

Curiously, the first words in support of this thesis came forth, with exquisite symbolic aptness, from a willing agent —but very nearly a classic victim!—of the new authoritarian technics. They came from the astronaut, John Glenn, whose life was endangered by the malfunctioning of his automatic controls, operated from a remote center. After he barely saved his life by personal intervention, he emerged from his space capsule with these ringing words: "Now let man take over!"

That command is easier to utter than obey. But if we are not to be driven to even more drastic measures than Samuel Butler suggested in *Erewhon,* we had better map out a more positive course: namely, the reconstitution of both our science and our technics in such a fashion as to insert the rejected parts of the human personality at every stage in the process. This means gladly sacrificing mere quantity in order to restore qualitative choice, shifting the seat of authority from the mechanical collective to the human personality and the autonomous group, favoring variety and ecological complexity, instead of stressing undue uniformity and standardization, above all, reducing the insensate drive to extend the system itself, instead of containing it within definite human limits and thus releasing man himself for other purposes. We must ask, not what is good for science or technology, still less what is good for General Motors or Union Carbide or IBM or the Pentagon, but what is good for man: not machine-conditioned, system-regulated, mass-man, but man in person, moving freely over every area of life.

There are large areas of technology that can be redeemed by the democratic process, once we have overcome the infantile compulsions and automatisms that now threaten to cancel out our real gains. The very leisure that the machine now gives in advanced countries can be profitably used, not for further commitment to still other kinds of machine, furnishing automatic recreation, but by doing significant forms of work, unprofitable or technically impossible under mass production: work dependent upon special skill, knowledge, aesthetic sense. The do-it-yourself movement prematurely got bogged down in an attempt to sell still more machines; but its slogan pointed in the right direction, provided we still have a self to do it with. The glut of motor cars that is now destroying our cities can be coped with only if we redesign our cities to make fuller use of a more efficient human agent: the walker. Even in childbirth, the emphasis is already happily shifting from an officious, often lethal, authoritarian procedure, centered in hospital routine, to a more human mode, which restores initiative to the mother and to the body's natural rhythms.

The replenishment of democratic technics is plainly too big a subject to be handled in a final sentence or two: but I trust I have made it clear that the genuine advantages our scientifically based technics has brought can be preserved only if we cut the whole system back to a point at which it will permit human alternatives, human interventions, and human destinations for entirely different purposes from those of the system itself. At the present juncture, if democracy did not exist, we would have to invent it, in order to save and recultivate the spirit of man.

5

Jacques Ellul, THE TECHNOLOGICAL ORDER

Emmanuel G. Mesthene, HOW TECHNOLOGY WILL SHAPE THE FUTURE

1. What negative characteristics does technology have according to Ellul? Consider these characteristics carefully; they are not easy to understand. What examples of each can you identify in modern society?

2. From Ellul's point of view, why does technology destroy human values?

3. What does Ellul say can be done to avoid the negative consequences of technology? Do you agree? If not, what do you think can be done to avoid the negative aspects of technology that Ellul identifies?

4. Can technological progress ever be good according to Ellul? Do you agree?

5. How would you describe Mesthene's view of technology compared to Ellul's?

6. What type of impact does technology have on values according to Mesthene?

7. Why does technology create change? What type of change does technology create?

8. How does technology affect politics according to Mesthene?

Jacques Ellul (1912-) is Professor of History and Contemporary Sociology at the University of Bourdeaux in France. He is also a prominent Protestant layman who has been active in the ecumenical movement. In his writings, he has produced a powerful critique of technology which each of us needs to consider because it says a great deal about our freedom in a technological society.

The Technological Order

Jacques Ellul

From *The Technological Order*, ed. Carl F. Stover (Detroit: Wayne State University Press, 1963).

I. I refer the reader to my book *The Technological Society* for an account of my general theses on this subject. I shall confine myself here to recapitulating the points which seem to me to be essential to a sociological study of the problem:

1. Technique has become the new and specific milieu in which man is required to exist, one which has supplanted the old milieu, namely, that of nature.

2. This new technical milieu has the following characteristics:

a. It is artificial;
b. It is autonomous with respect to values, ideas, and the state;
c. It is self-determining in a closed circle. Like nature, it is a closed organization which permits it to be self-determinative independently of all human intervention;
d. It grows according to a process which is causal but not directed to ends;
e. It is formed by an accumulation of means which have established primacy over ends;
f. All its parts are mutually implicated to such a degree that it is impossible to separate them or to settle any technical problem in isolation.

3. The development of the individual techniques is an "ambivalent" phenomenon.[1]

4. Since Technique has become the new milieu, all social phenomena are situated in it. It is incorrect to say that economics, politics, and the sphere of the cultural are influenced or modified *by* Technique; they are rather situated *in* it, a novel situation modifying all traditional social concepts. Politics, for example, is not modified by Technique as one factor among others which operate upon it; the political world is today *defined* through its relation to the technological society. Traditionally, politics formed a part of a larger social whole; at the present the converse is the case.

5. Technique comprises organizational and psychosociological techniques. It is useless to hope that the use of techniques of organization will succeed in compensating for the effects of techniques in general; or that the use of psychosociological techniques will assure mankind ascendancy over the technical phenomenon. In the former case, we will doubtless succeed in averting certain technically induced crises, disorders, and serious social disequilibrations; but this will but confirm the fact that Technique constitutes a closed circle. In the latter case, we will secure human psychic equilibrium in the technological milieu by avoiding the psychobiologic pathology resulting from the individual techniques taken singly and thereby attain a certain happiness. But these results will come about through the *adaptation of human beings to the technical milieu.* Psychosociological techniques result in the *modification* of men in order to render them happily subordinate to their new environment, and by no means imply any kind of human domination over Technique.

6. The ideas, judgments, beliefs, and myths of the man of today have already been essentially modified by his technical milieu. It is no longer possible to reflect that, on the one hand, there are techniques which may or may not have an effect on the human being; and, on the other, there is the human being himself who is to attempt to invent means to master his techniques and subordinate them to his own ends by *making a choice* among them. Choices and ends are both based on beliefs, sociological presuppositions, and myths which are a function of the technological society. Modern man's state of mind is completely dominated by technical values, and his goals are represented only by such progress and happiness as is to be

achieved through techniques. Modern man in choosing is already incorporated within the technical process and modified in his nature by it. He is no longer in his traditional state of freedom with respect to judgment and choice.

II. To understand the problem posed to us, it is first of all requisite to disembarrass ourselves of certain fake problems.

1. We make too much of the disagreeable features of technical development—for example, urban overcrowding, nervous tension, air pollution, and so forth. I am convinced that all such inconveniences will be done away with by the ongoing evolution of Technique itself, and indeed, that it is only by means of such evolution that this can happen. The inconveniences we emphasize are always dependent on technical solutions, and it is only by means of techniques that they can be solved. This fact leads to the following two considerations:

a. Every solution to some technical inconvenience is able only to reinforce the system of techniques *in their ensemble*;
b. Enmeshed in a process of technical development like our own, the possibilities of human survival are better served by more technique than less, a fact which contributes nothing, however, to the resolution of the basic problem.

2. We hear too often that morals are being threatened by the growth of our techniques. For example, we hear of greater moral decadence in those environments most directly affected technically—say, in working class or urbanized milieux. We hear, too, of familial disintegration as a function of techniques. The falseness of this problem consists in contrasting the technological environment with the moral values inculcated by society itself.[2] The presumed opposition between ethical problematics and technological systematics probably at the present is, and certainly in the long run will be, false. The traditional ethical milieu and the traditional moral values are admittedly in process of disappearing, and we are witnessing the creation of a *new* technological ethics with its own values. We are witnessing the evolution of a morally consistent system of imperatives and virtues, which tends to replace the traditional system. But man is not necessarily left thereby on a morally inferior level, although a moral relativism is indeed implied—an attitude according to which everything is well, *provided* that the individual obeys some ethic or other. We *could* contest the value of this development *if* we had a clear and adequate concept of what good-in-itself is. But such judgments are impossible on the basis of our general morality. On *that* level, what we are getting is merely a substitution of a new technological morality for a traditional one which Technique has rendered obsolete.

3. We dread the "sterilization" of art through technique. We hear of the artist's lack of freedom, calm, and the impossibility of meditation in the technological society. This problem is no more real than the two preceding. On the contrary, the best artistic production of the present is a result of a close connection between art and Technique. Naturally, new artistic form, expression, and ethic are implied, but this fact does not make art less art than what we traditionally called such. What assuredly is *not* art is a fixation in congealed forms, and a rejection of technical evolution as exemplified, say, in the neoclassicism of the nineteenth century or in present-day "socialist realism." The modern cinema furnishes an artistic response comparable to the Greek theater at its best; and modern music, painting, and poetry express, not a canker, but an authentic esthetic expression of mankind plunged into a new technical milieu.

4. One last example of a false problem is our fear that the technological society is completely *eliminating* instinctive human values and powers. It is held that systematization, organization, "rationalized" conditions of labor, overly hygienic living conditions, and the like have a tendency to repress the forces of instinct. For some people the phenomenon of "beatniks," "*blousons noirs*,"[3] and "hooligans" is explained by youth's violent reaction and the protestation of youth's vital force to a society which is overorganized, overordered, overregulated, in short, technicized.[4] But here too, even if the facts are established beyond question, it is

very likely that a superior conception of the technological society will result in the integration of these instinctive, creative, and vital forces. Compensatory mechanisms are already coming into play; the increasing appreciation of the aesthetic eroticism of authors like Henry Miller and the rehabilitation of the Marquis de Sade are good examples. The same holds for music like the new jazz forms which are "escapist" and exaltative of instinct; *item*, the latest dances. All these things represent a process of "*defoulement*"[5] which is finding its place in the technological society. In the same way, we are beginning to understand that it is impossible indefinitely to repress or expel religious tendencies and to bring the human race to a perfect rationality. Our fears for our instincts *are* justified to the degree that Technique, instead of provoking conflict, tends rather to *absorb* it, and to *integrate* instinctive and religious forces by giving them a place within its structure, whether it be by an adaptation of Christianity[6] or by the creation of new religious expressions like myths and mystiques which are in full compatibility with the technological society.[7] The Russians have gone farthest in creating a "religion" compatible with Technique by means of their transformation of communism into a religion.

III. What, then, is the real problem posed to men by the development of the technological society? It comprises two parts: 1. Is man able to remain master[8] in a world of means? 2. Can a new civilization appear inclusive of Technique?

1. The answer to the first question, and the one most often encountered, seems obvious: Man, who exploits the ensemble of means, *is* the master of them. Unfortunately, this manner of viewing matters is purely theoretical and superficial. We must remember the autonomous character of Technique. We must likewise not lose sight of the fact that the human individual himself is to an ever greater degree the *object* of certain techniques and their procedures. He is the object of pedagogical techniques, psychotechniques, vocational guidance testing, personality and intelligence testing, industrial and group aptitude testing, and so on. In these cases (and in countless others) most men are treated as a collection of objects. But, it might be objected, these techniques are exploited by other men, and the exploiters at least remain masters. In a certain sense this is true; the exploiters *are* masters of the particular techniques they exploit. But they, too, are subjected to the action of yet other techniques—as, for example, propaganda. Above all, they are spiritually taken over by the technological society; they believe in what they do; they are the most fervent adepts of that society. They themselves have been profoundly technicized. They never in any way affect to despise Technique, which to them is a thing good in itself. They never pretend to assign values to Technique, which to them is in itself an entity working out its own ends. They never claim to subordinate it to any value because for them Technique *is* value.

It may be objected that these individual techniques have as their end the best adaptation of the individual, the best utilization of his abilities, and, in the long run, his happiness. This, in effect, is the objective and the justification of all techniques. (One ought not, of course, to confound man's "happiness" with capacity for mastery with, say, freedom.) If the first of all values is happiness, it is likely that man, thanks to his techniques, will be in a position to attain to a certain state of this good. But happiness does not contain everything it is thought to contain, and the *absolute disparity between happiness and freedom* remains an ever real theme for our reflections. To say that man should remain *subject* rather than *object* in the technological society means two things—namely, that he be capable of giving direction and orientation to Technique, and that, to this end, he be able to master it.

Up to the present he has been able to do neither. As to the first, he is content passively to participate in technical progress, to accept whatever direction it takes automatically, and to admit its autonomous meaning. In the circumstances he can either proclaim that life is an absurdity without meaning or value; *or* he can predicate a number of indefinitely sophisticated values. But neither attitude accords with the fact of the technical phenomenon any more than it does with the other. Modern declarations of the absurdity of life are not based on modern technological efflorescence, which none (least of all the existentialists) think an absurdity. And the predication of values is a purely theoretical matter, since these values are not equipped with any means for putting them into practice. It is easy to reach agreement on what they are, but it is quite another matter to make them have any effect whatever on the technological society, or to cause them to be accepted in such a way that techniques must evolve in order to realize them. The values spoken of in the technological society are simply there to justify what is; *or* they are generalities without consequence; *or* technical progress realizes them automatically as a matter of course. Put otherwise, neither of the above alternatives is to be taken seriously.

The second condition *that man be subject rather than object*—that is, the imperative that he exercise mastery over technical development—is facilely accepted by everyone. But factually it simply does not hold. Even more embarrassing than the question How? is the question Who? We must ask ourselves realistically and concretely just who is in a position to choose the values which give Technique its justification and to exert mastery over it. If such a person or persons are to be found, it must be in the Western world (inclusive of Russia). They certainly are not to be discovered in the bulk of the world's population which inhabits Africa and Asia, who are, as yet, scarcely confronted by technical problems, and who, in any case, are even less aware of the questions involved than we are.

Is the arbiter we seek to be found among the *philosophers*, those thinking specialists? We well know the small influence these gentry exert upon our society, and how the technicians of every order distrust them and rightly refuse to take their reveries seriously. Even if the philosopher could make his voice heard, he would still have to contrive means of mass education so as to communicate an effective message to the masses.

Can the *technician* himself assume mastery over Technique? The trouble here is that the technician is *always* a specialist and cannot make the slightest claim to have mastered any technique but his own. Those for whom Technique bears its meaning in itself will scarcely discover the values which lend meaning to what they are doing. They will not even look for them. The only thing they can do is to apply their technical specialty and assist in its refinement. They cannot *in principle* dominate the totality of the technical problem or envisage it in its global dimensions. *Ergo*, they are completely incapable of mastering it.

Can the *scientist* do it? There, if anywhere, is the great hope. Does not the scientist dominate our techniques? Is he not an intellectual inclined and fit to put basic questions? Unfortunately, we are obliged to re-examine our hopes here when we look at things as they are. We see quickly enough that the scientist is as specialized as the technician, as incapable of general ideas, and as much out of commission as the philosopher. Think of the scientists who, on one tack or another, have addressed themselves to the technical phenomenon: Einstein, Oppenheimer, Carrel. It is only too clear that the ideas these gentlemen have advanced in the sphere of the philosophic or the spiritual are vague, superficial, and contradictory *in extremis*. They really ought to stick to warnings and proclamations for, as soon as they assay anything else, the other scientists and the technicians rightly refuse to take them seriously, and they even run the risk of losing their reputations as scientists.

Can the *politician* bring it off? In the democracies the politicians are subject to the wishes of their constituents who are primarily concerned with the happiness and well-being which they think Technique assures them. Moreover, the further we get on, the more a conflict shapes up between the politicians and the technicians. We cannot here go into the matter which is just beginning to be the object of serious study.[9] But it would appear that the power of the politician is being (and will continue to be) outclassed by the power of the technician in modern states. Only dictatorships can impose their will on technical evolution. But, on the one hand, human freedom would gain nothing thereby, and, on the other, a dictatorship thirsty for power has no recourse at all but to push toward an excessive development of various techniques at its disposal.

Any of us? An individual can doubtless seek the soundest attitude to dominate the techniques at his disposal. He can inquire after the values to impose on techniques in his use of them, and search out the way to follow in order to remain a man in the fullest sense of the word within a technological society. All this is extremely difficult, but it is far from being useless, since it is apparently the only solution presently possible. But the individual's efforts are powerless to resolve in any way the technical problem in its universality; to accomplish this would mean that *all* men adopt the same values and the same behavior.

2. The second real problem posed by the technological society is whether or not a new civilization can appear which is inclusive of Technique. The elements of this question are as difficult as those of the first. It would obviously be vain to deny all the things that can contribute something useful to a new civilization: security, ease of living, social solidarity, shortening of the work week, social security, and so forth. But a civilization in the strictest sense of the term is not brought into being by all these things.[10]

A threefold contradiction resides between civilization and Technique of which we must be aware if we are to approach the problem correctly:

a. The technical world is the world of material things; it is put together out of material things and with respect to them. When Technique displays any interest in man, it does so by converting him into a material object. The supreme and final authority in the technological society is fact, at once ground and evidence. And when we think on man as he exists in this society it can only be as a being immersed in a universe of objects, machines, and innumerable material things. Technique indeed guarantees him such material happiness as material objects can. But, the technical society is not, and cannot be, a genuinely humanist society since it puts in first place not man but material things. It can only act on man by lessening him and putting him in the way of the quantitative. The radical contradiction referred to exists between technical perfection and human development because such perfection is only to be achieved through quantitative development and necessarily aims exclusively at what is measurable. Human excellence, on the contrary, is of the domain of the qualitative and aims at what is not measurable. Space is lacking here to argue the point that spiritual values cannot evolve as a function of material improvement. The transition from the technically quantitative to the humanly qualitative is an impossible one. In our times, technical growth monopolizes all human forces, passions, intelligences, and virtues in such a way that it is in practice nigh impossible to seek and find anywhere any distinctively human excellence. And if this search is impossible, there cannot be any civilization in the proper sense of the term.

b. Technical growth leads to a growth of power in the sense of technical means incomparably more effective than anything ever before invented, power which has as its object only power, in the widest sense of the word. The possibility of action becomes limitless and absolute. For example, we are confronted for the first time with the possibility of the annihilation of all life on earth, since we have the means to accomplish it. In every sphere of action we are faced with just such absolute possibilities. Again, by way of example, governmental techniques, which amalgamate organizational, psychological, and police techniques, tend to lend to government absolute powers. And here I must emphasize a great law which I believe to be essential to the comprehension of the world in which we live—namely, that when power becomes absolute, values disappear. When man is able to accomplish anything at all, there is no value which can be proposed to him; when the means of action are absolute, no goal of action is imaginable. Power eliminates in proportion to its growth, the boundary between good and evil, between the just and the unjust. We are familiar enough with this phenomenon in totalitarian societies. The distinction between good and evil disappears beginning with the moment that the ground of action (for example, the *raison d'état*, or the instinct of the proletariat) claims to have absolute power and thus to incorporate *ipso facto* all value. Thus it is that the growth of technical means tending to absolutism forbids the appearance of values, and condemns to sterility our search for the ethical and the spiritual. Again, where Technique has place, there is the implication of the impossibility of the evolution of civilization.

c. The third and final contradiction is that Technique can never engender freedom. Of course, Technique frees mankind from a whole collection of ancient constraints. It is evident, for example, that it liberates him from the limits imposed on him by time and space; that man, through its agency, is free (or at least tending to become free) from famine, excessive heat and cold, the rhythms of the seasons, and from the gloom of night; that the race is freed from certain social constraints through its commerce with the universe, and from its intellectual limitations through its accumulation of information. But is this what it means really to be free? Other constraints as oppressive and rigorous as the traditional ones are imposed on the human being in today's technological society through the agency of Technique. New limits and technical oppressions have taken the place of the older, natural constraints, and we certainly cannot aver that much has been gained. The problem is deeper—the operation of Technique is the contrary of freedom, an operation of determinism and necessity. Technique is an ensemble of rational and efficient practices; a collection of orders, schemas, and mechanisms. All of this expresses very well a necessary order and a determinate process, but one into which freedom, unorthodoxy, and the sphere of the gratuitous and spontaneous cannot penetrate. All that these last could possibly introduce is discord and disorder. The more technical actions increase in society, the more human autonomy and initiative diminish. The more the human being comes to exist in a world of ever increasing demands (fortified with technical apparatus possessing its own laws to meet these demands), the more he loses any possibility of free choice and indi-vidiuality in action. This loss is greatly magnified by Technique's character of self-determination, which makes its appearance among us as a kind of fatality and as a species of perpetually exaggerated necessity. But where freedom is excluded in this way, an authentic civilization has little chance. Confronted in this way by the problem, it is clear to us that no solution can exist, in spite of the writings of all the authors who have concerned themselves with it. They all make an unacceptable premise, namely, rejection of Technique and return to a pretechnical society. One may well regret that some value or other of the past, some social or moral form, has disappeared; but, when one attacks the problem of the technical society, one can scarcely make the serious claim to be able to revive the past, a procedure which, in any case, scarcely seems to have been, globally speaking, much of an improvement over the human situation of today. All we know with certainty is that it was different, that the human being confronted other dangers, errors, difficulties, and temptations. Our duty is to occupy ourselves with the dangers, errors, difficulties, and temptations of modern man in the modern world. All regret for the past is vain; every desire to revert to a former social stage is unreal. There is no possibility of turning back, of annulling, or even of arresting technical progress. What is done is done. It is our duty to find our place in our present situation and in no other. Nostalgia has no survival value in the modern world and can only be considered a flight into dreamland.

We shall insist no further on this point. Beyond it, we can divide into two great categories the authors who search for a solution to the problem posed by Technique: the first class is that of those who hold that the problem will solve itself; the second, of those who hold that the problem demands a great effort or even a great modification of the whole man. We shall indicate a number of examples drawn from each class and beg to be excused for choosing to cite principally French authors.

Politicians, scientists, and technicians are to be found in the first class. In general, they consider the problem in a very concrete and practical way. Their general notion seems to be that technical progress resolves all difficulties *pari passu* with their appearance, and that it contains within itself the solution to everything. The sufficient condition for them, therefore, is that technical progress be not arrested; everything which plagues us today will disappear tomorrow.

The primary example of these people is furnished by the Marxists, for whom technical progress is the solution to the plight of the proletariat and all its miseries, and to the problem posed by the exploitation of man by man in the capitalistic world. Technical progress, which is for Marx the motive force of history, *necessarily* increases the forces of production, and simultaneously produces a progressive conflict between forward-moving factors and stationary social factors like the state, law, ideology, and morality, a conflict occasioning the periodic disappearance of the outmoded factors. Specifically, in the world of the present, conflict necessitates the disappearance of the structures of capitalism, which are so constituted as to be completely unable to absorb the economic results of technical progress, and are hence obliged to vanish. When they do vanish, they of necessity make room for a socialist structure of society corresponding perfectly to the sound and normal utilization of Technique. The Marxist solution to the technical problems is therefore an automatic one since the transition to socialism is *in itself* the solution. Everything is *ex hypothesi* resolved in the socialist society, and humankind finds therein its maturation. Technique, integrated into the socialist society, "changes sign": from being destructive it becomes constructive; from being a means of human exploitation it becomes humane; the contradiction between the infrastructures and the suprastructures disappears. In other words, all the admittedly difficult problems raised in the modern world belong to the structure of capitalism and not to that of Technique. On the one hand, it *suffices* that social structures become socialist for social problems to disappear; and, on the other, society *must necessarily* become socialist by the very movement of Technique. Technique, therefore, carries in itself the response to all the difficulties it raises.

A second example of this kind of solution is given by a certain number of technicians—for example, Frisch. All difficulties, according to Frisch, will inevitably be resolved by the technical growth which will bring the technicians to power. Technique admittedly raises certain conflicts and problems, but their cause is that the human race remains attached to certain political ideologies and moralities and loyal to certain outmoded and antiquated humanists whose sole visible function is to provoke discord of heart and head, thereby preventing men from adapting themselves and from entering resolutely into the path of technical progress. *Ergo*, men are subject to distortions of life and consciousness which have their origin, *not* in Technique, but in the conflict between Technique and the false values to which men remain attached. These fake values, decrepit sentiments, and outmoded notions must inevitably be eliminated by the invincible progress of Technique. In particular, in the political domain, the majority of crises arise from the fact that men are still wedded to certain antique political forms and ideas—for example, democracy. All problems will be resolved if power is delivered into the hands of the technicians who alone are capable of directing Technique in its entirety and making of it a positive instrument for human service. This is all the more true in that, thanks to the so-called "human techniques" (for example, propaganda), they will be in a position to take account of the human factor in the technical context. The technocrats will be able to use the totality of Technique without destroying the human being, but rather by treating him as he should be treated so as to become simultaneously useful and happy. General power accorded to the technicians become technocrats is the only way out for Frisch, since they are the only ones possessing the necessary competence; and, in any case, they are being carried to power by the current of history, the fact which alone offers a quick enough solution to technical problems. It is impossible to rely on the general improvement of the human species, a process which would take too long and would be too chancy. For the generality of men, it is necessary to take into account that Technique establishes an inevitable discipline, which, on the one hand, they must accept, and, on the other, the technocrats will humanize.

The third and last example (it is possible that there are many more) is furnished by the economists, who, in very different ways, affirm the thesis of the automatic solution. Fourastié is a good example of such economists. For him, the first thing to do is to draw up a balance between that which Technique is able to deliver and that which it may destroy. In his eyes there is no real problem: what Technique can bring to man is incomparably superior to that which it threatens. Moreover, if difficulties *do* exist, they are only temporary ones which will be resolved beneficially, as was the case with the similar difficulties of the last century. Nothing decisive is at stake; man is in no mortal danger. The contrary is the case: Technique produces the foundation, infrastructure, and suprastructure which will enable man really to become man. What we have known up to now can only be called the *prehistory* of a human race so overwhelmed by material cares, famine, and danger that the truly human never had an opportunity to develop into a civilization worthy of the name. Human intellectual, spiritual, and moral life will, according to Fourastié, never mature except when life is able to start from a complete satisfaction of its material needs, complete security, including security from famine and disease. The growth of Technique, therefore, initiates the genuinely human history of the whole man. This new type of human being will clearly be different from what we have hitherto known; but this fact should occasion no complaint or fear. The new type cannot help being superior to the old in every way, *after* all the traditional (and exclusively material) obstacles to his development have vanished. Thus progress occurs automatically, and the inevitable role of Technique will be that of guaranteeing such material development as allows the intellectual and spiritual maturation of what has been up to now only potentially present in human nature.

The orientation of the other group of doctrines affirms, on the contrary, that man is dangerously imperiled by technical progress; and that human will, personality, and organization must be set again to rights if society is to be able to guard against the imminent danger. Unfortunately, these doctrines share with their opposites the quality of being too optimistic, in that they affirm that their thesis is even feasible and that man is really capable of the rectifications proposed. I will give three very different examples of this, noting that the attitude in question is generally due to philosophers and theologians.

The orientation of Einstein, and the closely related one of Jules Romains, are well known —namely, that the human being must get technical progress back again into his own hands, admitting that the situation is so complicated and the data so overwhelming that only some kind of "superstate" can possibly accomplish the task. A sort of spiritual power integrated into a world government in possession of indisputable moral authority might be able to master the progression of techniques and to direct human evolution. Einstein's suggestion is the convocation of certain philosopher-scientists, whereas Romain's idea is the establishment of a "Supreme Court of Humanity." Both of these bodies would be organs of meditation, of moral quest, before which temporal powers would be forced to bow. (One thinks, in this connection, of the role of the papacy in medieval Christianity vis-à-vis the temporal powers.)

A second example of this kind of orientation is given by Bergson, at the end of his work, *The Two Sources of Morality and Religion*. According to Bergson, initiative can only proceed from humanity, since in Technique there is no "*force des choses*." Technique has conferred disproportionate power on the human being, and a disproportionate extension to his organism. But "in this disproportionately magnified body, the soul remains what it was, i.e., too small to fill it and too feeble to direct it. Hence the void between the two." Bergson goes on to say that "this enlarged body awaits a supplement of soul, the mechanical demands the mystical," and "that Technique will never render service proportionate to its powers unless humanity, which has bent it earthwards, succeeds by its means in reforming itself and looking heavenwards." This means that humanity has a task to perform, and that man must grow proportionately to his techniques, but that he must *will* it and *force* himself to make the experiment. This experiment is, in Bergson's view, a possibility, and is even favored by that technical growth which allows more material resources to men than ever before. The required "supplement of soul" is, therefore, of the order of the possible and will suffice for humans to establish mastery over Technique. The same position, it may be added, has in great part been picked up by E. Mounier.

A third example is afforded by a whole group of theologians, most of them Roman Catholic. Man, in his actions in the domain of the technical, is but obeying the vocation assigned him by his Creator. Man, in continuing his work of technical creation, is pursuing the work of his Creator. Thanks to Technique, this man, who was originally created "insufficient," is becoming "adolescent." He is summoned to new responsibilities in this world which do not transcend his powers since they correspond exactly to what God expects of him. Moreover, it is God Himself who through man is the Creator of Technique, which is something not to be taken in itself but in its relation to its Creator. Under such conditions, it is clear that Technique is neither evil nor fraught with evil consequences. On the contrary, it is good and cannot be dangerous to men. It can only become evil to the extent that man turns from God; it is a danger only if its true nature is misapprehended. All the errors and problems visible in today's world result uniquely from the fact that man no longer recognizes his vocation as God's collaborator. If man ceases to adore the "creature" (i.e., Technique) in order to adore the true God; if he turns Technique to God and to His service, the problems must disappear. All of this is considered the more true in that the world transformed by technical activity *must* become the point of departure and the material support of the new creation which is to come at the end of time.

Finally, it is necessary to represent by itself a doctrine which holds at the present a place of some importance in the Western world—that is, that of Father Teilhard de Chardin, a man who was simultaneously a theologian and a scientist. His doctrine appears as an intermediate between the two tendencies already sketched. For Chardin, evolution in general, since the origin of the universe, has represented a constant progression. First of all, there was a motion toward a diversification of matter and of beings; then there supervened a motion toward Unity—that is, a higher Unity. In the biological world, every step forward has been effected when man has passed from a stage of "dispersion" to a stage of "concentration." At the present, technical human progress and the spontaneous movement of life are in agreement and in mutual continuity. They are evolving together toward a higher degree of organization, and this movement manifests the influence of Spirit. Matter, left to itself, is characterized by a necessary and continuous degradation. But, on the contrary, we note that progress, advancement, improvement do exist, and, hence, a power contradicting the spontaneous movement of matter, a power of creation and progress exists which is the opposite of matter—that is, it is Spirit. Spirit has contrived Technique as a means of organizing dispersed matter, in order simultaneously to express progress and to combat the degradation of matter. Technique is producing at the same time a prodigious demographic explosion—that is, a greater density of human population. By all these means it is bringing forth "communion" among men; and, likewise, creating from inanimate matter a higher and more organized form of matter which is taking part in the ascension of the cosmos toward God. Granting that it is true that every progression in the physical and biological order is brought about by a condensation of the elements of the preceding period, what we are witnessing today, according to Chardin, is a condensation, a concentration of the whole human species. Technique, in producing this, possesses a function of unification *inside* humanity, so that humanity becomes able thereby to have access to a sort of unity. Technical progress is, therefore, synonymous with "socialization," this latter being but the political and economic sign of communion among men, the temporary expression of the "condensation" of the human species into a whole. Technique is the irreversible agent of this condensation; it prepares the new step forward which humanity must make. When men cease to be individual and separate units, and all together form a total and indissoluble communion, then humanity will be a single body. This material concentration is always accompanied by a spiritual concentration—that is, a maturation of the spirit, the commencement of a new species of life. Thanks to Technique, there is socialization, the progressive concentration on a planetary scale of disseminated spiritual personalities into a suprapersonal unity. This mutation leads to another Man, spiritual and unique, and means that humanity in its ensemble and in its unity, has attained the supreme goal, that is, its fusion with that glorious Christ who must appear at the end of time. Thus Chardin holds that in technical progress man is "Christified," and that technical evolution tends inevitably to the "edification" of the cosmic Christ.

It is clear that in Chardin's grandiose perspective, the individual problems, difficulties, and mishaps of Technique are negligible. It is likewise clear how Chardin's doctrine lies midway between the two preceding ones: on the one hand, it affirms a natural and involuntary ascension of man, a process inclusive of biology, history, and the like, evolving as a kind of will of God in which Technique has its proper place; and, on the other, it affirms that the evolution in question implies consciousness, and an intense *involvement* on the part of man who is proceeding to socialization and thus *committing* himself to this mutation.

We shall not proceed to a critique of these different theories, but content ourselves with noting that all of them appear to repose on a too superficial view of the technical phenomenon; and that they are *practically* inapplicable because they presuppose a certain number of *necessary* conditions which are not given. None of these theories, therefore, can be deemed satisfactory.

IV. It does not seem that at the present we are in a position to give a satisfactory reply to the complex of technical problems. All that appears possible is to inquire into the above-mentioned *necessary* conditions for a possible solution.

In any case, it seems to me that we can set forth the following thesis: The further technical progress advances, the more the social problem of mastering this progress becomes one of an ethical and spiritual kind. In proportion to the degree that man extricates himself from the domain of the material, dominates it, and multiplies thereby the means of exploiting matter, the problem ceases to be one of human possibilities and limits and becomes one rather of knowing which man (or group of men) will exploit technical means, and what will be the enabling moral and spiritual qualities. (In this point I am not far from that, for example, of Bergson.) It is essential not to consider the problem resolved once this has been said; the current attitude is false according to which, once a matter has been pronounced a matter of morality, it is something simple and also automatically resolvable. On the contrary, the more decision depends on a man or a group of them, the more difficult it appears, *if* we take a realistic view of the matter and refuse to admit *a priori* that man is good, democratic, liberal, reasonable, and so on. The difficulty resides in the following points:

a. It is impossible to trust the spontaneous employment which men will make of the available technical means;

b. Man, as we have already indicated, is *integrated* into the technological process;

c. *If* we desire to preserve man's freedom, dignity, and responsibility, it is precluded to act upon him by technical means, like psychology, and so forth. To transform a man into a reasonable being and a good exploiter of techniques *through* certain psychological procedures is precisely to destroy him as a spiritual and ethical subject.

We are thus caught in a dilemma before the decisive question, the question which may well be the penultimate one.

With this preliminary, what are these necessary conditions? I shall note them as they appear to me at the present, starting from that which is more general and working toward that which is more particular.

1. The first thing needed is a correct diagnosis and an effort to achieve a genuine consciousness of the problem. It is necessary to see the situation clearly and to pose the problem correctly if it be desired to know just what is to be done and if adequate answers are to be forthcoming. Inexact formulation of the problem affords no hope of getting a solution. The diagnostic element, on which I do not insist, must be accompanied by a becoming conscious—by passing from the intellectual to the existential, which means that mankind must accept the fact that his existence is "engaged" and involved in this venture, and that his very freedom is at stake. It is necessary to become conscious of the fact that in every domain Technique has established stricter and stricter domination over the human being. But this consciousness must not be negative—no scientific determinism or divine fatalism before which man can only bow and confess himself unfree. On the contrary, it must be recog-

nized that man *qua* free is subject to constraints and determinations which his vocation to be free must make him combat and rise clear of. *But*, to the extent that man clings to the illusion of the present that he *is* free (and uses the vocabulary of freedom), conceiving liberty as inalienable, *or*, to the extent that he holds to the conviction that all will be well though he sees that the Technique actually diminishes the area of freedom, and dreams that possibilities of freedom still exist—in all these cases, his natural inertia is leading him to accept a condition of slavery and to pay for his technological happiness with his freedom. It is only by making men conscious to what degree they have become slaves in becoming "happy," that there is any hope of regaining liberty by asserting themselves, perhaps at the cost of much sacrifice, over the Technique which has come to dominate them. Short of attaining to such consciousness, there is no reason for any human being to lift a finger to secure mastery over his technology.

2. A second essential element consists in ruthlessly destroying the "myth" of Technique—that is, the whole ideological construction and the tendency to consider technology something possessing sacred character. Intellectuals attempt to insert the technical phenomenon into the framework of their respective intellectual or philosophical systems by attributing to it a quality of supreme excellence; for example, when they demonstrate that Technique is an instrument of freedom, or the means of ascent to historical destiny, or the execution of a divine vocation, and the like. All such constructions have the result of glorifying and sanctifying Technique and of putting the human being at the disposal of some indisputable historical law or other. A further aspect of this element is the *sacred*—that is, the human tendency spontaneously to attribute sacred value to what so manifestly possesses transcendent power. Technique, in this view, is not solely an ensemble of material elements, but that which gives meaning and value to life, allowing man not only to live but to live well. Technique is intangible and unattackable precisely because *everything* is subject and subordinate to it. Man unconsciously invests with a holy prestige that against which he is unable to prevail. It seems to me that the only means to mastery over Technique is by way of "desacralization" and "deideologization." This means that *all* men must be shown that Technique is nothing more than a complex of material objects, procedures, and combinations, which have as their sole result a modicum of comfort, hygiene, and ease; and that it possesses nothing worthy of the trouble of devoting one's whole life to it, or of commanding an excessive respect, or of reposing in it one's success and honor, or of massacring one's fellow men. Men must be convinced that technical progress is not humanity's supreme adventure, but a commonplace fabrication of certain objects which scarcely merit enthusiastic delirium even when they happen to be Sputniks. As long as man worships Technique, there is as good as no chance at all that he will ever succeed in mastering it.

3. A consequence of this is that, in practice, it is necessary to teach man in his employment of Techniques a certain detachment, an independence with respect to them—and humor. It is, naturally, very difficult to accomplish this; and, above all, to get him to give up his illusions, not pretending to be completely free with respect to automobiles, television sets, or jobs, when the plain fact is that he is totally enslaved to them. Man must be capable of questioning at every step his use of his technical goods, able to refuse them and to force them to submit to determining factors other than the technical—say, the spiritual. He must be able to exploit all these goods without becoming unduly attached to them and without becoming convinced that even his most imposing technical conquests are to be taken seriously. Such recommendations must, of course, appear scandalous to contemporary eyes. To affirm that these things have no importance at all in respect to truth and freedom, that it is a matter of no *real* importance whether man succeeds in reaching the moon, or curing disease with antibiotics, or upping steel production, is really a scandal. As long as man does not learn to use technical objects in the right way he must remain their slave.

What I am saying refers to Technique itself and *not* to the individual's use of individual techniques. These two problems are situated on different levels. But, *if* the *individual* cannot attain personal liberty with respect to technical objects, there is no chance that he will be able to respond to the general problem of Technique. Let us recall once more that what we are setting forth are certain necessary conditions for finding a solution to this general problem.

4. Everything we have said presupposes an effort at reflection which might be thought of as philosophic. If we admit that the technical adventure is a genuine novelty for the human race, that all that it has excogitated up to now can scarcely be of any use to it at the present; if we admit that it can only be by means of a fundamental and arduous search that we will be able to extricate ourselves from the mess we are in, a *truly* philosophic reflection will be necessary. But modern philosophic systems, like existentialism and phenomenology, have small utility because they limit themselves into desuetude with their assertions that philosophy *in principle* can have no purchase on Technique. How, in the nature of things, can a philosophy which is nothing more than a research into the meaning of words, get any grip on the technical phenomenon? Preoccupation with "semantics" is the reason why modern philosophy immures itself in a refusal to come to grips with Technique. As Ducassé has put it in his *Les Techniques et le philosophe*: "Between the refusal of the philosophers, who claim to open up existence to themselves while evading the technical nature of the existent, and the hypocritical humility of the technicians manifested by an ambition stronger than their discipline, some very peculiar enterprises get under way, which might be termed pseudophilosophics and pseudotechniques, respectively, and which usurp in man the place of philosophy's absent mediation."[11] Authentic philosophy of real meaning would bring us to precisely that possibility of mediation between man and the technical phenomenon without which any legitimate attitude is inconceivable. But for such a philosophy to exist would mean that philosophy would first have to cease to be a purely academic technique with a hermetically sealed vocabulary, to become again the property of *every* man who thinks while he is engaged in the business of being alive.

5. Finally, it is necessary to point out the importance of the relation between the technicians and those who try to pose the technical problem. None of the preceding is more difficult than this, since the technicians have become an authoritarian and closed world. They are armed with good consciences, but likewise with the conviction of their essential rightness and the persuasion that *all* discourse and reflection of a nontechnical nature are verbalisms of no importance. To get them to engage in the dialogue or to question their own creation is an almost superhuman task, the more so that he who will enter this dialogue must be completely aware of what he wants, just what the technician is driving at, and what the technician is able to grasp of the problem. But, as long as such interchange does not take place, nothing will happen, since influencing Technique necessarily means influencing the technicians. It seems to me that this dialogue can only come about by making contact which will represent a *permanent* and *basic* confrontation between Technique's pretensions to resolve all human problems and the human will to escape technical determinism.

Such, I think, are the five conditions necessary that an opening on the technical problem can even become a possibility.

Note on the Theme: Technical Progress Is Always Ambiguous

It cannot be maintained that technical progress is in itself either good or bad. In the evolution of Technique, contradictory elements are always indissolubly connected. Let us consider these elements under the following four rubrics:

1. All technical progress exacts a price;
2. Technique raises more problems than it solves;

3. Pernicious effects are inseparable from favorable effects; and
4. Every technique implies unforeseeable effects.

1. ALL TECHNICAL PROGRESS EXACTS A PRICE

What is meant here is not that technical progress exacts a price in money or in intellectual effort, but that, when technical progress adds something, on the one hand, it inevitably subtracts something on the other. It is always difficult to interpret satisfactorily the bald statement that "technical progress is an established fact," because some people cling to traditional social forms, tending to deny any value at all to such progress, and deeming that nothing can be called progress if it casts doubt on established social values. Other persons, on the contrary, hold that Technique produces extraordinary things of a prodigious novelty, bringing about the consequent disappearance of all sorts of valueless junk.

The fact is that, viewed objectively, technological progress produces values of unimpeachable merit, while simultaneously destroying values no less important. As a consequence, it cannot be maintained that there is absolute progress or absolute regress.

Let me give a few simple examples of this reciprocal action. In the first place, let us consider the fact that modern man, thanks to hygiene in particular and to technical progress in general, enjoys a greater life span than ever before. Life expectancy in France today is approximately sixty years, compared, say, to thirty-five years in 1890 and thirty years about 1800.[12] But, even with this indubitable extension of the average life span, all physicians are in agreement that, proportionate to this extension, life has become very much more precarious—that is, our general state of health has become very much more fragile. Human beings of the present have neither the same resistance as their ancestors to disease or to natural conditions nor the same endurance; they suffer from a certain nervous "fragility" and a loss of general vitality, sensitiveness of their senses, and so on. In the sixty years during which such studies have been carried out, regression in all these respects has been marked. Thus, though we live longer, we live a reduced life with nothing resembling the vital energy of our ancestors. It is clear that diminution, on the one hand, has been accompanied by augmentation, on the other.

In the sphere of labor, the technical progress of the present has effected a considerable economy of muscular effort; but at the same time this progress has come to demand a greater and greater nervous effort so that tension and wear and tear on our nerves have inversely increased. Here, again, a kind of equilibrium has asserted itself between savings and expense.

To take an instance from the sphere of economics, technical progress allows the creation of new industries. But a just view of the matter would compel us to take into consideration the accompanying destruction of resources. To take a French example, the so-called Lacq case is beginning to be well known. An industrial complex for the exploitation of sulphur and natural gas has been established at Lacq, a simple technical fact. But, from the economic point of view, this is far from being the case, since a serious agricultural problem has arisen because of the excessive destruction of farm products in the region. Up to now, the government has not seen fit to take the matter seriously, although it has been officially estimated in reports to the Chamber that, for 1960, agricultural losses have aggregated two billion francs. Now the vineyards of Jurançon are being attacked by the sulfurous gases and are disappearing, a not inconsiderable economic loss.

To calculate from the economist's point of view the profits of an industry of this kind, it would at the minimum be necessary to deduct the value of what has been destroyed —in this case two billion francs. It would likewise be necessary to deduct the very considerable expenses of all the necessary protective devices, hospitals (which, incidentally, have not yet been constructed), schools —in short, of the whole urban complex which has not yet been brought into being but which is nevertheless indispensable. We must have knowledge of how to calculate the *whole*. The Lacq enterprise, counting all the expenses of which we have been speaking, must be reckoned a "deficit" enterprise.

Our last example has to do with the problem of the intellectual culture of the masses. True, today's technical means permit a mass culture to exist. Television allows people who never visited a theater in their lives to see performances of the great classics. *Paris-Match*, through its articles, allows masses of people who would be in total ignorance without such articles to attain to a certain literary (and even to a certain aesthetic) culture. But, on the other side of the ledger, it must be recorded that this same technical progress leads to an ever increasing cultural superficiality. Technical progress absolutely forbids certain indispensable conditions of a genuine culture—namely, reflection and opportunity for assimilation. We are indeed witnessing the creation of knowledge, since we are in possession of the means of knowing what we could never have known before; but it is, nevertheless, a superficial development because it is one which is purely *quantitative*.

The intellectual no longer has any time to meditate on a book and must choose between two alternatives: *either* he reads through a whole collection of books rapidly, of which a little later but a few fragments survive—scattered bits of vague knowledge; *or* he takes a year to peruse a few books thoroughly. I should like to know who today has the time to take Pascal or Montaigne seriously. To do them justice would require months and months; but today's Technique forbids any such thing. Exactly the same holds for the problem of the "Musée Imaginaire," which Malraux has put so well. We can be in contact with the whole painting and sculpture of humanity; but this availability has no cultural value comparable to that enjoyed by Poussin, who, in his voyage to Rome, passed several years in studying, statue by statue, the ensemble of artistic works at his disposal. He clearly knew nothing of Polynesian or Chinese art, but what he did know had infinitely more educational value for him because it penetrated his personality slowly.

So, once again, we see that Technique

allows us to progress quantitatively to the level of culture spoken of, but at the same time interdicts us from making any progress in depth. In the circumstances, is it really possible to speak of "culture" at all? All technical progress exacts a price. We cannot believe that Technique brings us nothing; but we must not think that what it brings it brings free of charge.

2. THE PROBLEMS POSED BY TECHNICAL PROGRESS

The second aspect of the ambiguity of technical progress concerns the following point: When Technique evolves, it does so by solving a certain number of problems, and by raising others.

The further we advance into the technological society, the more convinced we become that, in any sphere whatever, there are nothing but technical problems. We conceive all problems in their technical aspect, and think that solutions to them can only appear by means of further perfecting techniques. In a certain sense, we are right; it is true that Technique permits us to solve the majority of the problems we encounter. But we are compelled to note (perhaps not often enough) that each technical evolution raises new problems, and that, as a consequence, there is never *one* technique which solves *one* problem. The technological movement is more complicated; one technique solves one problem, but at the same time creates others.

Let us take some simple examples of this fact. We are well acquainted with the details of the gravest sociological problem faced by the nineteenth century—that is, that of the proletariat, a problem which we are only now in process of solving (with difficulty). The phenomenon of the proletariat is not to be considered a simple one, and Marx himself did not describe it as "merely" the exploitation of the workers by certain wicked capitalists. His explanation of the "proletarian condition" was very much more profound; he demonstrated that the proletariat was a result of the division and the mechanization of labor. He expressly states that "it is necessary to pass through the stage represented by the proletariat." For Marx, therefore, the problem is not, say, a moral one, with "bad guys exploiting good guys." Marx never puts the problem in this way; he always poses it as lying outside good or bad moral qualities, external to value judgments, and on the level of fact. And the fact is the fact of the division of labor, and of the machine, giving rise to a society in which exploitation is inevitable—that is, drawing off surplus values. The phenomenon of the proletariat is, therefore, even in the Marxian analysis, the result of technical progress. The machine and the division of labor allowed, from the economic point of view, an extraordinary expansion, but, at the same time, and as a result of the same movement, posed the social problem which it has taken a whole century to resolve.

Let us consider in the same way the extension of the above problem as it appears in the questions which will eventually but certainly be posed by the so-called automation. Again, automation is not just another simple economic fact; indeed, we are gradually coming to realize that it will entail difficulties which, from our present point of view, can only be characterized as insurmountable. First of all, automation implies a production of goods in a relatively constant series of types. This means that when production has been automated, it is no longer possible to vary types, so that an unavoidable condition of immobilism with regard to production must ensue. An automated production line, considered in its full context of operation, is so expensive that amortization must occur over terms so long that the exclusive production of certain types of goods without any possibility of modification must be a consequence. *But*, up to the present, no commercial market of the capitalist world is suited to the absorption of the production of an unchanging line of goods. No presently existing Western economic organization, *on the commercial plane*, is prepared to find an answer to automated production.

Another difficulty of automation is the fact that it will result in a massive diminution of the necessary labor force. The simplistic reaction to this problem will clearly be to hold that the solution is easy. It is not necessary to cut down on the number of the workers, but only to diminish the number of daily working hours of each. This solution is quite clearly impossible for a very simple reason. Automation cannot be applied to any arbitrarily selected industry or production, and this for reasons which are basic and not due to the temporary exigencies of, say, the money market. Certain kinds of production can and will be automated; certain others cannot and will never be automated. Consequently, it is not possible to cut down working hours over the working class as a whole. There are industrial sectors in which the workers would conceivably work one hour per day, whereas in others the workers would have to continue working a normal day of eight hours. Hence, as a result of automation, there will be extended sectors of the economy emptied of manpower, while other sectors will continue on the normal standard.

Diebold estimates that in the single year 1955–1956, in the United States, automation reduced the total number of working hours by 7 percent. In the automated plants of the Ford Motor Company there was a reduction of personnel by 25 percent; and in 1957, in industrial branches in which automation gained most (in particular in the manufacture of electric bulbs and in the very highly automated chemical industry), it was possible to dispense with the services of eight hundred thousand workers. In other words, automation does not result in labor saving favorable to the workers, but is expressed through unemployment and employment disequilibration.

It might be alleged that the situation is true of capitalist countries but cannot be identical in socialist. This statement is not exact; in socialist countries the problem likewise is posed, primarily because of socialist egalitarianism. The problem is the same for the Soviet Union, for example, where automation is commencing, as for the United States. There will be specialized workers in some industries who will be freed from the necessity to work in one way or another, while in other branches of industry the eight-hour day will have to remain in force, a situation clearly unacceptable to the egalitarian theories of socialism.

A second problem is bound to arise in connection with the *retraining* of the "liberated" workers for jobs in new industrial sectors in which there is a shortage of manpower. But such retraining more often than not presents enormous difficulties, since the disemployed worker is generally semiskilled (or unskilled) and a completely new apprenticeship is implied of such a nature as to steer him toward other branches of industry.

A third difficulty occasioned by automation is the problem of *wages*. The wage problem produced by automation has, up till now, not been solved. How is it possible to fix a wage scale for automated industrial plants? It cannot be done on the piecework plan—machines do all the work. It cannot be done on the basis of time put in on the job. If it is desired to reduce unemployment by reducing the work day to, say, two or three hours, a given worker would only be employed for a very short period each day. Should such a worker, then, be paid according to a wage schedule which pays him for two hours of work at the equivalent of a worker who must work eight? The injustice of such a procedure is clear. How, then, should wages be calculated in an automated industry? One is forced to the admission that the relation between wages and productivity, on the one hand, and between wages and job time, on the other, *must* disappear. Wages will be calculated only as a function of the purchasing power given to the worker (with a view to maximum consumption) by dividing the total production value by the total number of workers. Such a method is really the only one feasible. Since 1950, in Russia, it has actually been tried twice. But the results were unsatisfactory, and it very soon became necessary to return to the system of hourly wages, since, in the present state of affairs, the necessary calculations prove unfeasible. But then the difficulties mentioned above (inherent in calculating either according to job time or according to production) return, and, at the moment, wage calculation in automated industries is completely shrouded in uncertainties.

Still another problem is presented by the fact that modern economic crises most often result from a "distortion" between the different economic sectors—more exactly, from unequal growth of the different sectors. Here automation must prove to be an economic factor much to be feared. There will not only be disparity of economic growth between the automated and the non-automated industrial sectors, but still more between industry and agriculture. *Either* capitalist countries must look forward to an increase of crises due to automation, *or* they must adopt planning aimed at rectifying the distortions (and planning by authoritarian measures, as in the Soviet Union). At the present time even the Soviet planners find that their planning is insufficient to meet the problems of automation, since it is not "flexible" enough, on the one hand, and not "extensive" enough to re-equilibrate the out-of-phase sectors, on the other.

Here, then, are a number of problems (and there are a great many others) with which we must expect to be confronted by the fact of automation, all of which furnish us with examples of our thesis that Technique raises, in proportion to its progress, problems of greater and greater difficulty.

Let me indicate one final example of this, that is, the problem of overpopulation, resulting from the application of medical and prophylactic health techniques, the final result of which is the suppression of infant mortality and the prolongation of human life. The phenomenon of overpopulation, in its turn, produces the tragic phenomenon of underconsumption. A century hence, all of us *without exception* will be menaced by a general underconsumption which will afflict the whole human race, *if* the expansion of the world's population increases. Here we are confronted by a problem clearly provoked by certain techniques, certain *positive techniques*.

The common factor of all these examples is that technical progress raises whole complexes of problems which we are in no position to solve. Examples of such problems are literally innumerable.

3. THE EVIL EFFECTS OF TECHNIQUE ARE INSEPARABLE FROM THE GOOD

An idea frequently to be encountered in superficial inquiries concerning Technique is the following: "At bottom, everything depends on the way Technique is employed; mankind has only to use Technique for the good and avoid using it for the bad." A common example of this notion is the usual recommendation to employ techniques for the beneficent purposes of peace and eschew them for the maleficent purposes of war. All then will go well.

Our thesis is that technical progress contains simultaneously the good *and* the bad. Consider automation, the problem which we have just been discussing. It is indisputable that technological unemployment is the result of mechanical progress. It cannot be otherwise. All mechanical progress necessarily entails a saving of labor and, consequently, a necessary technological unemployment. Here we have an ill-omened effect indissolubly connected with one which is in itself beneficial. The progress of mechanization necessarily entails unemployment. The technological unemployment so produced can be resolved by either of two means, which are the only two means economically or politically possible—namely, spreading it out either in *space* or in *time*.

A capitalist economist holds that the solution to unemployment is "that technological unemployment ultimately dies out of itself." This means that the workers who have been "freed" (the optimistic formula for unemployment) because of technical advances will ultimately find jobs, either by directing themselves toward industries with manpower shortages or through the fact that new inventions will produce new opportunities of employment and new vocations. The standard example cited in defense of this thesis is that of the vocational opportunities connected with the invention of the automobile. Admittedly, this technological device did suppress a certain number of vocations, but it brought innumerable others into being, with the final result that a vast number of persons are now employed by the servicing required by this industry. Hence the machine in question has actually created employment.

All of this is indeed true. It is, nevertheless, a terribly heartless view of the situation, because it neglects to mention the *interim* period. It is all very well to say that the worker rendered jobless will, *with the lapse of a certain time*, again find employment . . . and that, after he has been reclassified, unemployment will die out. But, humanly speaking, what is the situation of the unemployed worker in the interim? Here the problem of spreading out unemployment in time is posed.

In the Soviet Union, unemployment of a technological nature (which not only exists but springs from the same sources) is spread out in space. By this I mean that when in one place new machines are applied and workers "liberated" the affected workers will, without having to wait very long, receive a work card which tells them in effect: "Two thousand kilometers from here a job has been assigned to you; you are hereby directed to remove yourself to such and such a factory." In one way, such a procedure seems a little less inhuman; but, in another way, it seems just as inhuman as the time procedure of the capitalists, since no account is taken of one's attachments to family, friends, locality, and so on. The human being is only a pawn to be moved about. It is hard to tell, between the capitalist and the socialist ways of handling the problem, which solution presents the worse indecencies.

A further example of the inseparable mingling of good and bad effects is furnished by the noteworthy study of the American sociological historian, J. U. Nef, concerning "industry and war."[13] Nef shows how industrialism—that is, the development of industry taken as a whole—necessarily prods industrialized societies in the direction of war. His analysis has nothing to do with the inner essence of industrialism; the phenomena described by him lie purely at the level of the human being.

First, industrialism gives an increasing population the means to live. It is a law sociologically irrefutable that, the denser the population, the greater the number of wars. This phenomenon is, of course, well known as a *practical* matter to all sociologists, but only Nef has studied it carefully.

Second, industrialism creates the media of the press, transmission of information, and transport, and finally the means of making war, all of which make it more and more difficult and even almost impossible to distinguish between the aggressor and the aggressed. At the present, no one knows (and perhaps no one can know) which side has commenced hostilities, a fact not solely due to armaments, but also to facility of transport. The extraordinary rapidity of transport allows an aggression to be launched within twenty-four hours, or even less, without anyone being able to foresee it. Here the influence of the press is extremely important, since the press function is to confuse and addle the facts so that no one is able to gain any correct intelligence of them.

Finally, Nef indicates that the new means of destruction created by industrialism have greatly reduced the trouble, the difficulties, and the anguish implied in the act of killing people. A bombardier or artillerist has no feeling at all of killing anyone; he is, in fact, able to reach the conclusion that he has killed someone only with the aid of a syllogism. In hand-to-hand combat all the tiresome difficulties of conscience about the evil of murder keep obtruding themselves. In such ways, then, positive elements of industry result essentially (by very complex expedients) in favoring war and even in provoking it, even if no one has the *intention* of using Technique "badly."

Let us consider, as a final example of the relation between good effects and bad effects, the press and information.

It seems to be a simple matter, for example, to distinguish between information and propaganda. But closer study of the problem reveals that it is practically impossible to make such a distinction. Considering but a few elements of the situation, the problem of information is today no longer that of the necessity of transmitting *honest* information —everybody agrees on this point. On the moral level it is a commonplace that we ought to transmit true information. I merely inquire, How do we get it? To remain on the *moral* level is simply not to understand the situation. The *concrete* situation, to take but a single example, is something like the following: Over the wires and into the offices of the Associated Press pass daily up to three hundred thousand words of world news, approximately equal to an enormous volume of one thousand pages. From this mass of words, it is necessary for the Associated Press, in competition with all the other world agencies, to choose, cut, and re-expedite as quickly as possible, perhaps a twentieth part of the whole to its subscribers. How is it possible to select from such a flood just what should be retained, what is true, what is possibly false, and so on? The editors have no criteria, they are at the mercy of whatever comes in, and (even when they judge in good faith and knowledge) they must essentially judge subjectively. Then again, even if the editor had only true news, how should he assign it a coefficient of importance? To do so is his business, and here the stereotypes of the editor are true enough: The Catholic editor will deem the news of the latest Vatican Council of great significance, information which has not the slightest importance to the Communist editor. What we have to do with here is not a question of bad faith, but of a difference of perspective on the world. The result is that we never know, even under the most favorable circumstances, if a given piece of information is subjective. And we must always bear in mind that this information, whatever it is, has been worked over by at least four or five different pairs of hands.

My reasons for maintaining that good effects are inseparable from bad are now, I trust, clear. And, as communications improve, the freer will be the flow of the news and the more available to all agencies concerned. These factors will play an ever greater role, making the difficulties of editing proportionately more difficult, and the chance of selecting absurd rather than sound news ever greater.

4. ALL TECHNICAL PROGRESS CONTAINS UNFORESEEABLE EFFECTS

The final aspect of the ambiguity of technical progress resides in the following state of affairs: When scientists carry out their researches in one or another discipline and hit upon new technical means, they generally see clearly in what sphere the new technique will be applicable. Certain results are expected and gotten. *But* there are always secondary effects which had not been anticipated, which in the primary stage of the technical progress in question could not *in principle* have been anticipated. This unpredictability arises from the fact that predictability implies complete possibility of experimenting in *every* sphere, an inconceivable state of affairs.

The most elementary example is furnished by drugs. You have a cold in the head; you take an aspirin. The headache disappears, but aspirin has other actions besides doing away with headaches. In the beginning we were totally oblivious of these side effects; but, I should imagine, by now everyone has read articles warning against the use of aspirin because of its possible dangerous effects, say, on the blood picture. Grave hemorrhages have appeared in people who habitually took two or three aspirins daily. Yet aspirin was thought the perfect remedy a scant ten years ago—on the ground that no side effects were to be feared. Now such effects begin to appear even in what was, and is, probably the most harmless of all drugs.

Another spectacular example is that of DDT, a chemical which in 1945 was thought to be a prodigiously successful means for the destruction of all kinds of vermin and insects. One of the most admirable things about DDT was that it was said to be completely innocuous toward human beings. DDT was sprinkled over the whole surface of the globe. Then, by accident, it was discovered that in certain areas veal cattle were wasting away and dying. Research revealed that DDT in oily solution causes anemia. Cattle had been dusted with DDT in order to get rid of insects; they had subsequently licked themselves clean and ingested the DDT. The chemical in question passed into their milk and by this route found its way into oily solution, that is, in the milk fat. Calves suckled by such cows died of anemia, and it is needless to add that the same milk was ingested by human infants. Identical problems are potentially raised by *all* chemicals consumed by animals or men. Recall the recent example of thalidomide.

This is an example of the so-called secondary effects, effects which are essentially unpredictable and only revealed after the technique in question has been applied on a grand scale—that is, when it is no longer possible to retrace one's steps.

Another interesting example is furnished by the psychosociological studies of the particular psychology of big city dwellers, where, once more, we are confronted with the effect of the technical environment on the human being. One of the principal elements of big city life is the feeling of isolation, loneliness, absence of human contacts, and so on. One of the leading ideas of Le Corbusier in his *Maison des Hommes* was the admission that "big city dwellers do not know one another." "Let us create," said Le Corbusier, "great blocks of dwellings where people will meet one another as they did in the village, with everything (grocer, baker, butcher) included in the block so that people will get to know one another and a community will come into being." The result of Le Corbusier's creation was exactly the opposite of what had been planned; problems of loneliness and isolation in such blocks of dwellings proved to be much more tragic than in the normal and traditional city.

Then it was held (and this is the penultimate word in city planning) that it was necessary to rediscover groupings *on a human scale*, not on the scale of a block with, say, five thousand separate dwelling units. In the works and writings of sociologists and of city planners of perhaps seven or eight years ago we read: "At bottom, the only ones who understood what a city was were the people of the Middle Ages, who knew how to create a true city corresponding to the demands of a genuine city-planning technique, that is, a human community centered about a small square surrounded by small houses, toward which converged the (straight) city streets, etc." The new city planners, in keeping with these theories, applied them to the suburbs of Chicago and, in particular, to the well-known "village" of

Park Forest. There, it was thought, was to be found the distinctively human formula, one which really allows the human being his full scope. But, the most recent sociological and psychological analyses show this model community to represent nothing less than a new and unexpected difficulty. This time, people are traumatized because they are perpetually under the eyes and under the surveillance of their neighbors. The affected group is indeed much reduced in size; but no one dares to budge, because everybody knows just what everybody else is up to, a frightfully constricting situation, to say the least. It is clear that, even with the best intentions and with the application of hyper-modern and profound research in psychology and sociology, we only succeed in coming to results in every case which could not possibly have been anticipated.

I shall give one last example of these unforeseeable effects, this time from agriculture—namely, the massive cultivation of certain plants like corn and cotton. The cultivation of these plants in the "new countries" seems to represent undeniable progress. The deforestation of land too heavily forested is a felicitous operation, profitable from every point of view, and, consequently, represents technical progress. But it could not have been anticipated that corn and cotton are plants which not only impoverish the soil, but even annihilate it by the twofold action of removing certain natural elements and destroying the relation between the humus and the soil particles. Both these last are destroyed by the roots of cotton and corn to the degree that, after thirty or forty years of cultivation of these agricultural products, the soil is transformed into a veritable dust bowl. A strong wind need only to pass over it to reduce it to bare rock.

The phenomenon is world wide, and is to be encountered in the United States, Brazil, and Russia, among others. It is a bone of contention between Khrushchev and certain Soviet agricultural specialists. Khrushchev essentially emphasizes the cultivation of corn, as is well known; but many Soviet specialists insist that this emphasis is a very dangerous one. It allows a very rapid economic progress for, say, twenty years, only to be followed by a destruction of hitherto fertile lands which may last for centuries.

The inquiries of Castro and Vogt have shown that, at the present, in certain regions 20 percent of cultivated land is threatened with destruction in this way. If this factor is considered in connection with that of population growth, a very considerable difficulty seems to lurk in the offing. If arable land continues to diminish in extent beyond possibility of recovery, our chances of survival diminish correspondingly. Here we have an example of typical and unpredictable secondary effects, effects which in corn and cotton agriculture do not reveal themselves except after thirty years of experience. It is again impossible, therefore, to say whether technical progress is in essence good or bad.

We are launched into a world of an astonishing degree of complexity; at every step we let loose new problems and raise new difficulties. We succeed progressively in solving these difficulties, but only in such a way that when one has been resolved we are confronted by another. Such is the progress of technology in our society. All I have been able to do is to give a few fragmentary examples. What would be necessary in order to comprehend the problem in its entirety is a systematic and detailed study of all these points.

—Translated from the French
by John Wilkinson

Emmanuel G. Mesthene (1920-) is Professor of Philosophy and Director of Harvard University's Program of Studies in Technology and Society. This program has produced some outstanding publications, and has influenced American thinking on the relationship between science, technology, and society during the past decade.

How Technology Will Shape the Future

Emmanuel G. Mesthene

THERE ARE two ways, at least, to approach an understanding of how technology will affect the future. One, which I do not adopt here, is to try to predict the most likely technological developments of the future along with their most likely social effects.[1] The other way is to identify some respects in which technology entails change and to suggest the kinds or patterns of change that, by its nature, it brings about in society. It is along the latter lines that I speculate in what follows, restricting myself largely to the contemporary American scene.

New Technology Means Change

It is widely and ritually repeated these days that a technological world is a world of change. To the extent that this statement is meaningful at all, it would seem to be true only of a world characterized by a more or less continuous development of new technologies. There is no inherent impetus toward change in tools as such, no matter how many or how sophisticated they may be. When new tools emerge and displace older ones, however, there is a strong presumption that there will be changes in nature and in society.

I see no such necessity in the technology-culture or technology-society relationship as we associate with the Marxist tradition, according to which changes in the technology of production are inevitably and univocally determinative of culture and social structure. But I do see, in David Hume's words, a rather "constant conjunction" between technological change and social change as well as a number of good reasons why there should be one, after we discount for the differential time lags that characterize particular cases of social change consequent to the introduction of new technologies.

The traditional Marxist position has been thought of as asserting a strict or hard determinism. By contrast, I would defend a position that William James once called a "soft" determinism, although he used the phrase in a different context. (One may also call it a *probabilistic* determinism and thus avoid the trap of strict causation.) I would hold that the development and adoption of new technologies make for changes in social organization and values by virtue of creating new possibilities for human action and thus altering the mix of options available to men. They may not do so necessarily, but I suggest they do so frequently and with a very high probability.

Technology Creates New Possibilities

One of the most obvious characteristics of new technology is that it brings about or inhibits changes in *physical* nature, including changes in the patterns of physical objects or processes. By virtue of enhancing our ability to measure and predict, moreover, technology can, more specifically, lead to controlled or directed change. Thus the plow changes the texture of the soil in a specifiable way; the wheel speeds up the mobility (change in relative position) of people or objects; and the smokebox (or icebox) inhibits some processes of decay. It would be equally accurate to say that these technologies respectively *make possible* changes in soil texture, speed of transport, and so forth.

In these terms, we can define any new (nontrivial) technological change as one which (1) makes possible a new way of inducing a physical change, or (2) creates a wholly new physical possibility that simply did not exist before. A better mousetrap or faster airplane are examples of new ways and the Salk vaccine or the moon rocket are instances of new possibilities. Either kind of technological change will extend the range of what man *can* do, which is what technology is all about.

There is nothing in the nature or fact of a new tool, of course, that requires its use. As Lynn White has observed, "A new device merely opens a door; it does not compel one to enter."[2] I would add, however, that a newly opened door does *invite* one to enter.[3] A house in which a number of new doors have been installed is different from what it was before and the behavior of its inhabitants is very likely to change as a result. Possibility as such does not imply actuality (as a strict determinism would have to hold), but there is a high probability of realization of new possibilities that have been deliberately created by technological development, and therefore of change consequent to that realization.

Technology Alters the Mix of Choices

A correlative way in which new technology makes for change is by removing some options previously available. This consequence of technology is derivative, indirect, and more difficult to anticipate than the generation of new options. It is derivative in that old options are removed only after technology has created new ones. It is indirect, analogously, because the removal of options is not the result of the new technology, but of the act of choosing the new options that the technology has created.[4] It is more difficult to anticipate, finally, to the degree that the positive consequences *for* which a technology is developed and applied are seen as part of the process of decision to develop and apply, whereas other (often negative) consequences of the development are usually seen only later if at all.

Examples abound. Widespread introduction of modern plumbing can contribute to convenience and to public health, but it also destroys the kind of society that we associate with the village pump. Exploitation of industrial technology removes many of the options and values peculiar to an agricultural society. The automobile and airplane provide mobility, but often at the expense of stabilities and constancies that mobility can disturb.

Opportunity costs are involved in exploiting any opportunity, in other words, and therefore also the opportunities newly created by technology. Insofar as the new options are chosen and the new possibilities are exploited, older possibilities are displaced and older options are precluded or prior choices are reversed. The presumption, albeit not the necessity, that most of the new options will be chosen is therefore at the same time a presumption that the choice will be made to pay the new costs. Thus, whereas technology begins by simply adding to the options available to man, it ends by altering the spectrum of his options and the mix or hierarchy of his social choices.

Social Change

The first-order effect of technology is thus to multiply and diversify material possibilities and thereby offer new and altered opportunities to man. Different societies committed to different values can react differently (positively or negatively, or simply differently) to the same new possibilities, of course. This is part of the explanation, I believe, for the phenomenon currently being referred to as the "technological gap" between Western Europe and the United States. Moreover, as with all opportunities when badly handled, the ones created by technology can turn into new opportunities to make mistakes. None of this alters the fact that technology creates opportunity.

Since new possibilities and new opportunities generally require new organizations of human effort to realize and exploit them, technology generally has second-order effects that take the form of social change. There have been instances in which changes in technology and in the material culture of a society have not been accompanied by social change, but such cases are rare and exceptional.[5] More generally,

> over the millennia cultures in different environments have changed tremendously, and these changes are basically traceable to new adaptations required by changing technology and productive arrangements. Despite occasional cultural barriers, the useful arts have spread extremely widely, and the instances in which they have not been accepted because of preexisting cultural patterns are insignificant.[6]

While social change does not necessarily follow upon technological change, it almost always does in fact, thus encouraging the presumption that it generally will. The role of the heavy plow in the organization of rural society and that of the stirrup in the rise of feudalism provide fascinating medieval examples of a nearly direct technology-society relationship.[7] The classic case in our era, of course—which it was Karl Marx's contribution to see so clearly, however badly he clouded his perception and blinded many of his disciples by tying it at once to a rigid determinism and to a form of Hegelian absolutism—is the Industrial Revolution, whose social effects continue to proliferate.

When social change does result from the introduction of a new technology, it must, at least in some of its aspects, be of a sort conducive to exploitation of the new opportunities or possibilities created by that technology. Otherwise, it makes no sense even to speak of the social effects of technological change. Social consequences need not be and surely are not uniquely and univocally determined by the character of innovation, but they cannot be entirely independent of that character and still be accounted consequences. (Therein lies the distinction, ultimately, between a hard and soft determinism.) What the advent of nuclear weapons altered was the military organization of the country, not the structure of its communications industry, and the launching of satellites affects international relations much more directly than it does the institutions of organized sport. (A change in international relations may affect international competition in sports, of course, but while everything may be connected with everything else in the last analysis, it is not so in the first.)

There is a congruence between technology and its social effects that serves as intellectual ground for all inquiry into the technology-society relationship. This congruence has two aspects. First, the subset of social changes that can result from a given technological innovation is smaller than the set of all possible social changes and the changes that do in fact result are a still smaller subset of those that can result—that is, they are a sub-subset. In relation to any given innovation, the spectrum of all possible social changes can be divided into those that cannot follow as consequences, those that can (are made possible by the new technology), and those that do (the actual consequences).

It is the congruence of technology and its social consequences in this sense that provides the theoretical warrant for the currently fashionable art of "futurology." The more responsible practitioners of this art insist that they do not predict unique future events but rather identify and assess the likelihood of possible future events or situations. The effort is warranted by the twin facts that technology constantly alters the mix of possibilities and that any given technological change may have several consequences.

The second aspect of the congruence between technology and its social consequences is a certain "one-wayness" about the relationship—that is, the determinative element in it, however "soft." It is, after all, only technology that creates new *physical* possibilities (though it is not technology *alone* that does so, since science, knowledge, social organization, and other factors are also necessary to the process). To be sure, what technologies will be developed at any particular time is dependent on the social institutions and values that prevail at that particular time. I do not depreciate the interaction between technology and society, especially in our society which is learning to create scientific knowledge to order and develop new technologies for already established purposes. Nevertheless, once a new technology is created, it is the impetus for the social and institutional changes that follow it. This is especially so since a social decision to develop a particular technology is made in the principal expectation of its predicted first-order effects, whereas evaluation of the technology after it is developed and in operation usually takes account also of its less foreseeable second- and even third-order effects.

The "one-wayness" of the technology-society relationship that I am seeking to identify may be evoked by allusion to the game of dice. The initiative for throwing the dice lies with the player, but the "social" consequences that follow the throw are initiated *by the dice* and depend on how the dice fall. Similarly, the initiative for development of technology in any given instance lies with people, acting individually or as a public, deliberately or in response to such pressures as wars or revolutions. But the material initiative remains with the technology and the social adaptation to it remains its consequence. Where the analogy is weak, the point is strengthened. For the rules of the game remain the same no matter how the dice fall, but technology has the effect of adding new faces to the dice, thus inducing changes in society's rules so that it can take advantage of the new combinations that are created thereby. That is why new technology generally means change in society as well as in nature.

Technology and Values

New technology also means a high probability of change in individual and social values, because it alters the conditions of choice. It is often customary to distinguish rather sharply between individual and social values and, in another dimension, between tastes or preferences, which are usually taken to be relatively short-term, trivial, and localized, and values, which are seen as higher-level, relatively long-term, and extensive in scope. However useful for some purposes, these distinctions have no standing in logic, as Kenneth Arrow points out:

> One might want to reserve the term "values" for a specially elevated or noble set of choices. Perhaps choices in general might be referred to as "tastes." We do not ordinarily think of the preference for additional bread over additional beer as being a value worthy of philosophic inquiry. I believe, though, that the distinction cannot be made logically.[8]

The *logical* equivalence of preferences and values, whether individual or social, derives from the fact that all of them are rooted in choice behavior. If values be taken in the contemporary American sociologist's sense of broad dominant commitments that account for the cohesion of a society and the maintenance of its identity through time, their relation to choice can be seen both in their genesis (historically in the society, not psychologically in the individual) and in their exemplification (where they function as criteria for choice).

Since values in this sense are rather high-level abstractions, it is unlikely that technological change can be seen to influence them directly. We need, rather, to explore what difference technology makes for the choice behaviors that the values are abstractions from.

What we choose, whether individually or as a society (in whatever sense a society may be said to choose, by public action or by resultant of private actions), is limited, at any given time, by the options available. (Preferences and values are in this respect different from aspirations or ideals in that the latter can attach to imaginative constructs. To confuse the two is to confuse morality and fantasy.) When we say that technology makes possible what was not possible before, we say that we now have more options to choose from than we did before. Our old value clusters, whose hierarchical ordering was determined in the sense of being delimited by antecedent conditions of material possibility, are thus now subject to change because technology has altered the material conditions.

By making available new options, new technology can, and generally will, lead to a restructuring of the hierarchy of values, either by providing the means for bringing previously unattainable ideals within the realm of choice and therefore of realizable values, or by altering the relative ease with which different values can be implemented—that is, by changing the costs associated with realizing them. Thus the economic affluence that technological advance can bring may enhance the values we associate with leisure at the relative expense of the value of work and achievement, and the development of pain-killing and pleasure-producing drugs can make the value of material comfort relatively easier of achievement than the values we associate with maintaining a stiff upper lip during pain or adversity.

One may argue further that technological change leads to value change of a particular sort in exact analogy to the subset of possible social changes that a new technology may augur (as distinct from both the wider set that includes the impossible and the narrower, actual subset). There are two reasons for this. First, certain attitudes and values are more conducive than others to most effective exploitation of the potentialities of new tools or technologies. Choice behavior must be somehow attuned to the new options that technology creates, so that they will in fact be chosen. Thus to transfer or adapt industrial technologies to underdeveloped nations is only part of the problem of economic development; the more important part consists in altering value predispositions and attitudes so that the technologies can flourish. In more advanced societies, such as ours, people who hold values well adapted to exploitation of major new technologies will tend to grow rich and occupy elite positions in society, thus serving to reinforce those same values in the society at large.

Second, whereas technological choices will be made according to the values prevailing in society at any given time, those choices will, as previously noted, be based on the foreseeable consequences of the new technology. The essence of technology as creative of new possibility, however, means that there is an irreducible element of uncertainty—of unforeseeable consequence—in any innovation.[9] Techniques of the class of systems analysis are designed to anticipate as much of this uncertainty as possible, but it is in the nature of the case that they can never be more than partially successful, partly because a new technology will enter into interaction with a growing number and variety of ongoing processes as societies become more complex, and partly—at least in democratic societies—because the unforeseeable consequences of technological innovation may take the form of negative *political* reaction by certain groups in the society.

Since there is an irreducible element of uncertainty that attends every case of technological innovation, therefore, there is need for two evaluations: one before and one after the innovation. The first is an evaluation of prospects (of ends-in-view, as John Dewey called them). The second is an evaluation of results (of outcomes actually attained).[10] The uncertainty inherent in technological innovation means there will usually be a difference between the results of these two evaluations. To that extent new technology will lead to value change.

Contemporary Patterns of Change

Our own age is characterized by a deliberate fostering of technological change and, in general, by the growing social role of knowledge. "Every society now lives by innovation and growth; and it is theoretical knowledge that has become the matrix of innovation."[11]

In a modern industrialized society, particularly, there are a number of pressures that conspire toward this result. First, economic pressures argue for the greater efficiency implicit in a new technology. The principal example of this is the continuing process of capital modernization in industry. Second, there are political pressures that seek the greater absolute effectiveness of a new technology, as in our latest weapons, for example. Third, we turn more and more to the promise of new technology for help in dealing with our social problems. Fourth, there is the spur to action inherent in the mere availability of a technology: space vehicles spawn moon programs. Finally, political and industrial interests engaged in developing a new technology have the vested interest and powerful means needed to urge its adoption and widespread use irrespective of social utility.

If this social drive to develop ever more new technology is taken in conjunction with the very high probability that new technology will result in physical, social, and value changes, we have the conditions for a world whose defining characteristic is change, the kind of world I once described as Heraclitean, after the pre-Socratic philosopher Heraclitus, who saw change as the essence of being.[12]

When change becomes that pervasive in the world, it must color the ways in which we understand, organize, and evaluate the world. The sheer fact of change will have an impact on our sensibilities and ideas, our institutions and practices, our politics and values. Most of these have to date developed on the assumption that stability is more characteristic of the world than change—that is, that change is but a temporary perturbation of stability or a transition to a new (and presumed better or higher) stable state. What happens to them when that fundamental metaphysical assumption is undermined? The answer is implicit in a number of intellectual, social, and political trends in present-day American society.

Intellectual Trends

I have already noted the growing social role of knowledge.[13] Our society values the production and inculcation of knowledge more than ever before, as is evidenced by sharply rising research, development, and education expenditures over the last twenty years. There is an increasing devotion, too, to the systematic use of information in public and private decision making, as is exemplified by the President's Council of Economic Advisers, by various scientific advisory groups in and out of government, by the growing number of research and analysis organizations, by increasing appeal to such techniques as program planning and budgeting, and by the recent concern with assembling and analyzing a set of "social indicators" to help gauge the social health of the nation.[14]

A changing society must put a relatively strong accent on knowledge in order to offset the unfamiliarity and uncertainty that change implies. Traditional ways (beliefs, institutions, procedures, attitudes) may be adequate for dealing with the existent and known. But new technology can be generated and assimilated only if there is technical knowledge about its operation and capabilities, and economic, sociological, and political knowledge about the society into which it will be introduced.

This argues, in turn, for the importance of the social sciences. It is by now reasonably well established that policy making in many areas can be effective only if it takes account of the findings and potentialities of the natural sciences and of their associated technologies. Starting with economics, we are gradually coming to a similar recognition of the importance of the social sciences to public policy. Research and education in the social sciences are being increasingly supported by public funds, as the natural sciences have been by the military services and the National Science Foundation for the last quarter of a century. Also, both policy makers and social scientists are seeking new mechanisms of cooperation and are exploring the modifications these will require in their respective assumptions and procedures. This trend toward more applied social science is likely to be noticeable in any highly innovative society.

The scientific mores of such a society will also be influenced by the interest in applying technology that defines it. Inquiry is likely to be motivated by and focused on problems of the society rather than centering mainly around the unsolved puzzles of the scientific disciplines themselves. This does not mean, although it can, if vigilance against political interference is relaxed, (1) that the resulting science will be any less pure than that proceeding from disinterested curiosity, or (2) that there cannot therefore be any science motivated by curiosity, or (3) that the advancement of scientific knowledge may not be dependent on there always being some. The research into the atomic structure of matter that is undertaken in the interest of developing new materials for supersonic flight is no less basic or pure than the same research undertaken in pursuit of a new and intriguing particle, even though the research strategy may be different in the two cases. Even more to the point, social research into voting behavior is not *ipso facto* less basic or pure because it is paid for by an aspiring candidate rather than by a foundation grant.

There is a serious question, in any event, about just how pure is pure in scientific research. One need not subscribe to such an out-and-out Marxism as Hessen's postulation of exclusively social and economic origins for Newton's research interests, for example, in order to recognize "the demonstrable fact that the thematics of science in seventeenth century England were in large part determined by the social structure of the time."[15] Nor should we ignore the fashions in science, such as the strong emphasis on physics in recent years that was triggered by the military interest in physics-based technologies, or the very similar present-day passion for computers and computer science. An innovative society is one in which there is a strong interest in bringing the best available knowledge to bear on ameliorating society's problems and on taking advantage of its opportunities. It is not surprising that scientific objectives and choices in that society should be in large measure determined by what those problems and opportunities are, which does not mean, however, that scientific objectives are identical with or must remain tied to social objectives.

Another way in which a society of change influences its patterns of inquiry is by putting a premium on the formulation of new questions and, in general, on the synthetic aspects of knowing. Such a society is by description one that probes at scientific and intellectual frontiers, and a scientific frontier, according to the biologist C. H. Waddington, is where "we encounter problems about which we cannot yet ask sensible questions."[16] When change is prevalent, in other words, we are frequently in the position of not knowing just what we need to know. A goodly portion of the society's intellectual effort must then be devoted to formulating new research questions or reformulating old ones in the light of changed circumstances and needs so that inquiry can remain pertinent to the social problems that knowledge can alleviate.

Three consequences follow. First, there is a need to reexamine the knowledge already available for its meaning in the context of the new questions. This is the synthetic aspect of knowing. Second, the need to formulate new questions coupled with the problem orientation, as distinct from the discipline orientation, discussed above requires that answers be sought from the intersection of several disciplines. This is the impetus for current emphases on the importance of interdisciplinary or cross-disciplinary inquiry as a supplement to the academic research aimed at expanding knowledge and training scientists. Third, there is a need for further institutionalization of the function of transferring scientific knowledge to social use. This process, which began in the late nineteenth century with the creation of large central research laboratories in the chemical, electrical, and communications industries, now sees universities spawning problem- or area-oriented institutes, which surely augur eventual organizational change, and new policy-oriented research organizations arising at the borderlines of industry, government, and universities, and in a new no-man's land between the public and private sectors of our society.

A fundamental intellectual implication of a world of change is the greater theoretical utility of the concept of process over that of structure in sociological and cultural analysis. Equilibrium theories of various sorts imply ascription of greater reality to stable sociocultural patterns than to social change. But as the anthropologist Evon Vogt argues,

> Change is basic in social and cultural systems . . . Leach [E. R.] is fundamentally correct when he states that "every real society is a process in time." Our problem becomes one of describing, conceptualizing, and explaining a set of ongoing processes,

but none of the current approaches is satisfactory "in providing a set of conceptual tools for the description and analysis of the *changing* social and cultural *systems* that we observe."[17]

There is no denial of structure: "Once the processes are understood, the structures manifested at given time-points will emerge with even greater clarity," and Vogt goes on to distinguish between short-run "recurrent processes" and long-range and cumulative "directional processes." The former are the repetitive "structural dynamics" of a society. The latter "involve alterations in the structures of social and cultural systems."[18] It is clear that the latter, for Vogt, are more revelatory of the essence of culture and society as changing.[19]

Social Trends

Heraclitus' philosophy of universal change was a generalization from his observation of physical nature, as is evident from his appeal to the four elements of ancient physics (fire, earth, water, and air) in support of it.[20] Yet he offered it as a metaphysical generalization. Change, flux, is the essential characteristic of all of existence, not of matter only. We should expect to find it central also, therefore, to societies, institutions, values, population patterns, and personal careers.

We do. Among the effects of technological change that we are beginning to understand fairly well even now are those (1) on our principal institutions: industry, government, universities; (2) on our production processes and occupational patterns; and (3) on our social and individual environment: our values, educational requirements, group affiliations, physical locations, and personal identities. All of these are in movement. Most are also in process; that is, there is direction or pattern to the changes they are undergoing, and the direction is, moreover, recognizable as a consequence of the growing social role of knowledge induced by proliferation of new technology.

It used to be that industry, government, and universities operated almost independently of one another. They no longer do, because technical knowledge is increasingly necessary to the successful operation of industry and government, and because universities, as the principal sources and repositories of knowledge, find that they are adding a dimension of social service to their traditional roles of research and teaching. This conclusion is supported (1) by the growing importance of research, development, and systematic planning in industry; (2) by the proliferation and growth of knowledge-based industries; (3) by the changing role of the executive, who increasingly performs sifting, rearranging, and decision operations on ideas that are generated and come to him from below; (4) by the entry of technical experts into policy making at all levels of government; (5) by the increasing dependence of effective government on availability of information and analysis of data; (6) by the importance of education and training to successful entry into the society and to maintenance of economic growth; and (7) by the growth, not only of problem-oriented activities on university campuses, but also of the social role (as consultants, advisory boards, and so forth) of university faculties.

The economic affluence that is generated by modern industrial technology accelerates such institutional mixing-up by blurring the heretofore relatively clear distinction between the private and public sectors of society. Some societies, like the Scandinavian, can put a strong emphasis on the acquisition of public or social goods even in the absence of a highly productive economy. In our society affluence is a precondition of such an emphasis. Thus, as we dispose increasingly of resources not required for production of traditional consumer goods and services, they tend to be devoted to providing such public goods as education, urban improvement, clean air, and so forth.

What is more, goods and services once considered private more and more move into the public sector, as in scientific research and graduate education or the delivery of medical care. As we thus "socialize" an increasing number of goods once considered private, however, we tend also to farm out their procurement to private institutions, through such devices as government grants and contracts. As a result,

> Our national policy assumes that a great deal

of our new enterprise is likely to follow from technological developments financed by the government and directed in response to government policy; and many of our most dynamic industries are largely or entirely dependent on doing business with the government through a subordinate relationship that has little resemblance to the traditional market economy.[21]

Another observer says:

> Increasingly it will be recognized that the mature corporation, as it develops, becomes part of the larger administrative complex associated with the state. In time the line between the two will disappear.[22]

This fluidity of institutions and social sectors is not unreminiscent of the more literal fluidity that Heraclitus immortalized: "You cannot step twice into the same river, for other waters are continually flowing on."[23] But, also like the waters of a river, the institutional changes of a technologically active age are not aimless; they have direction, as noted, toward an enhancement of the use of knowledge in society:

> Perhaps it is not too much to say that if the business firm was the key institution of the past hundred years, because of its role in organizing production for the mass creation of products, the university will become the central institution of the next hundred years because of its role as the new (*sic*) source of innovation and knowledge.[24]

One should recall, in this connection, that the university is the portal through which more and more people enter into productive roles in society and that it increasingly provides the training necessary for leadership in business and government as well as in education.

The considerable debate of the last few years about the implications of technological change for employment and the character of work has not been in vain. Positions originally so extreme as to be untenable have been tempered in the process. Few serious students of the subject believe any longer that the progress of mechanization and automation in industry must lead to an irreversible increase in the level of *involuntary* unemployment in the society—whether in the form of unavailability of employment, or of a shortening work week, or of lengthening vacations, or of an extension of the period of formal schooling. These developments may occur, either voluntarily, because people choose to take some of their increased productivity in the form of leisure, or as a result of inadequate education, poor social management, or failure to ameliorate our race problem. But reduction of the overall level of employment is not a necessary consequence of new industrial technology.

Too much is beginning to be known about what the effects of technology on work and employment in fact are, on the other hand, for them to be adequately dealt with as merely transitional disruptions consequent on industrialization. A number of economists are, therefore, beginning to move away from explanation in terms of transition (which is typical of traditional equilibrium theory) to multilevel "steady state" models, or to dynamic theories of one sort or another, as more adequate to capturing the reality of constant change in the economy.[25]

The fact is that technological development has provided substitutes for human muscle power and mechanical skills for most of history. Developments in electronic computers are providing mechanical substitutes for at least some human mental operations. No technology as yet promises to duplicate human creativity, especially in the artistic sense, if only because we do not yet understand the conditions and functioning of creativity. (This is not to deny that computers can be useful aids to creative activity.) Nor are there in the offing mechanical equivalents for the initiatives inherent in human emotions, although emotions can of course be affected and modified by drugs or electrical means. For the foreseeable future, therefore, one may hazard the prediction that distinctively human work will be less and less of the "muscle and elementary mental" kind, and more and more of the "intellectual, artistic, and emotional" kind. (This need *not* mean that only highly inventive or artistic people will be employable in the future. There is much sympathy needed in the world, for example, and the provision of it is neither mechanizable nor requisite of genius. It is illustrative of what I think of as an "emotional" service.)

While advancing technology may not displace people by reducing employment in the aggregate, therefore, it unquestionably displaces some jobs by rendering them more efficiently performed by machines than by people. There devolves on the society, as a result, a major responsibility for inventing and adopting mechanisms and procedures of occupational innovation. These may range from financial and organizational innovations for diverting resources to neglected public needs to social policies which no longer treat human labor as a market commodity. Whatever the form of solution, however, the problem is more than a "transitional" one. It represents a qualitative and permanent alteration in the nature of human society consequent to perception of the ubiquity of change.

This perception and the anticipatory attitude that it implies have some additional consequences, which are not less important but are as yet less well understood even than those for institutional change and occupational patterns. For example, lifetime constancy of trade or profession has been a basis of personal identity and of the sense of individuality. Other bases for this same sense have been identification over time with a particular social group or set of groups as well as with physical or geographical location.

All of these are now subject to Heraclitean flux. The incidence of lifelong careers will inevitably lessen, as employing institutions and job contents both change. More than one career per lifetime is likely to be the norm henceforth. Group identities will shift as a result: every occupational change will involve the individual with new professional colleagues, and will often mean a sundering from old friends and cultivation of new ones. Increasing geographical mobility (already so characteristic of advanced industrial society) will not only reinforce these impermanencies, but also shake the sense of identity traditionally associated with ownership and residence upon a piece of land. Even the family will lose influence as a bastion of personality, as its loss of economic *raison d'être* is supplemented by a weakening of its educational and socializing functions and even of its prestige as the unit of reproduction.[26]

I have alluded elsewhere to the implications for education of a world seen as essentially changing.[27] Education has traditionally had the function of preparing youth to assume full membership in society (1) by imparting a sense for the history and accumulated knowledge of the race, (2) by imbuing the young with a sense of the culture, mores, practices, and values of the group, and (3) by teaching a skill or set of skills necessary to a productive social role. Philosophies of education have accordingly been elaborated on the assumption of stability of values and mores, and on the up-to-now demonstrable principle that one good set of skills well learned could serve a man through a productive lifetime.

This principle is undermined by contemporary and foreseeable occupational trends, and the burden of my general argument similarly disputes the assumption of unalterable cultural stabilities. There are significant implications for the enterprise of education. They include: (1) a decline in the importance of manual skills, (2) a consequent rising emphasis on general techniques of analysis and evaluation of alternatives, (3) training in occupational flexibility, (4) development of management skills, and (5) instruction in the potentialities and use of modern intellectual tools. The major problem of contemporary education at the primary and secondary levels is that the educational establishment is by and large unprepared, unequipped, and poorly organized to provide education consonant with these realities.

Above all, perhaps, higher education especially will need to attend more deliberately and systematically than it has in recent decades to developing the reflective, synthetic, speculative, and even the contemplative capacities of men, for understanding may be at a relatively greater premium henceforth than particular knowledge. When we can no longer lean on the world's stabilities, we must be able to rely on new abilities to cope with change and be comfortable with it.

There is an analogous implication for social values and for the human enterprise of valuing. There is concern expressed in many quarters these days about the threat of technology to values. Some writers go so far as to assert an incompatibility between technology and values and to warn that technological progress is tantamount to dehumanization and the destruction of all value.[28]

There is no question, as noted earlier, that technological change alters the mix of choices available to man and that choices made *ipso facto* preclude other choices that might have been made. Some values are destroyed in this process, which can thus involve punishing traumata of adjustment that it would be immoral to ignore. It is also unquestionably the case that some of the choices made are constrained by the very technology that makes them available. In such cases, the loss of value can be tragic, and justly regretted and inveighed against. It is in the hope of anticipating such developments that we are currently investigating means to assess and control technological development in the public interest.

On the other hand, I find no justification for the contention that technological progress must of necessity mean a progressive destruction of value. Such fears seem rather to be based partly on psychological resistance to change and partly on a currently fashionable literary mythology that interprets as a loss the fact that the average man today does not share the values that were characteristic of some tiny elites centuries ago. To the extent that it is more than that, this contention is based, I think, on a fundamental misunderstanding of the nature of value.

The values of a society change more slowly, to be sure, than the realities of human experience; their persistence is inherent in their emergence as values in the first place and in their function as criteria, which means that their adequacy will tend to be judged later rather than earlier. But values do change, as a glance at any history will show. They can change more quickly, moreover, the more quickly or extensively a society develops and introduces new technology. Since technological change is so prominent a characteristic of our own society, we tend to note inadequacies in our received values more quickly than might have been the case in other times. When that perception is coupled with the conviction of some that technology and value are inherently inimical to each other, the opinion is reinforced that the advance of technology must mean the decline of value and of the amenities of distinctively human civilization.

While particular values may vary with particular times and particular societies, however, the activity of valuing and the social function of values do not change. That is the source of the stability so necessary to human moral experience. It is not to be found, nor should it be sought, exclusively in the familiar values of the past. As the world and society are seen increasingly as processes in constant change under the impact of new technology, value analysis will have to concentrate on process, too: on the process of valuation in the individual and on the process of value formation and value change in the society. The emphasis will have to shift, in other words, from values to valu*ing*. For it is not particular familiar values as such that are valuable, but the human ability to extract values from experience and to use and cherish them. And that value is not threatened by technology; it is only challenged by it to remain adequate to human experience by guiding us in the reformulation of our ends to fit our new means and opportunities.

Political Trends

There are a number of respects in which technological change and the intellectual and social changes it brings with it are likely to alter the conditions and patterns of government. I construe *government* in this connection in the broadest possible sense of the term, as governance (with a small *g*) of a *polity*. Better yet, I take the word as equivalent to govern*ing*, since the participle helps to banish both visions of statism and connotations of public officialdom. What I seek to encompass by the term, in other words, is the social decision-making function in general, whether exemplified by small or large or public or private groups. I include in decision making, moreover, both the values and criteria that govern it and the institutions, mechanisms, procedures, and information by means of which it operates.

One notes that, as in other social sectors and institutions, the changes that technology purports for government are of a determinate sort—they have direction: they enhance the role of government in society and they enhance the role of knowledge in government.

The importance of decision making will tend to grow relative to other social functions (relative to production, for example, in an affluent society): (1) partly because the frequency with which new possibilities are created in a technologically active age will provide many opportunities for new choices; (2) partly because continuing alteration of the spectrum of available choice alternatives will shorten the useful life of decisions previously made; (3) partly because decisions in areas previously thought to be unrelated are increasingly found to impinge on and alter each other; and (4) partly because the economic affluence consequent on new technology will increase the scope of deliberate public decision making at the expense relatively, of the largely automatic and private charting of society's course by market forces. It is characteristic of our time that the market is increasingly distrusted as a goal-setting mechanism for society, although there is of course no question of its effectiveness as a signaling and controlling device for the formulation of economic policy.

Some of the ways in which knowledge increasingly enters the fabric of government have been amply noted, both above and in what is by now becoming a fairly voluminous literature on various aspects of the relation of science and public policy.[29] There are other ways, in addition, in which knowledge (information, technology, science) is bound to have fundamental impacts on the structures and processes of decision making that we as yet know little about.

The newest information-handling equipments and techniques find their way quickly into the agencies of federal and local government and into the operations of

industrial organizations, because there are many jobs that they can perform more efficiently than the traditional rows of clerks. But it is notorious that adopting new means in order to better accomplish old ends very often results in the substitution of new ends for old ones.[30] Computers and associated intellectual tools can thus, for example, make our public decisions more informed, efficient, and rational, and less subject to lethargy, partisanship, and ignorance. Yet that possibility seems to imply a degree of expertise and sophistication of policy-making and implementing procedures that may leave the public forever ill informed, blur the lines between executive and legislature (and private bureaucracies) as all increasingly rely on the same experts and sources of information, and chase the idea of federalism into the history books close on the heels of the public-private separation.

There is, in general, the problem of what happens to traditional relationships between citizens and government, to such prerogatives of the individual as personal privacy, electoral consent, and access to the independent social criticism of the press, and to the ethics of and public controls over a new elite of information keepers, when economic, military, and social policies become increasingly technical, long-range, machine-processed, information-based, and expert-dominated.

An exciting possibility that is however so dimly seen as perhaps to be illusory is that knowledge can widen the area of political consensus. There is no question here of a naive rationalism such as we associate with the eighteenth-century Enlightenment. No amount of reason will ever triumph wholly over irrationality, certainly, nor will vested interest fully yield to love of wisdom. Yet there are some political disputes and disagreements, surely, that derive from ignorance of information bearing on an issue or from lack of the means to analyze fully the probable consequences of alternative courses of action. Is it too much to expect that better knowledge may bring about greater political consensus in such cases as these? Is the democratic tenet that an informed public contributes to the commonweal pure political myth? The sociologist S. M. Lipset suggests not:

> Insofar as most organized participants in the political struggle accept the authority of experts in economics, military affairs, interpretations of the behavior of foreign nations and the like, it becomes increasingly difficult to challenge the views of opponents in moralistic "either/or" terms. Where there is some consensus among the scientific experts on specific issues, these tend to be removed as possible sources of intense controversy.[31]

Robert E. Lane of Yale has made the point more generally:

> If we employ the term "ideology" to mean a comprehensive, passionately believed, self-activating view of society, usually organized as a social movement . . . it makes sense to think of a domain of knowledge distinguishable from a domain of ideology, despite the extent to which they may overlap. Since knowledge and ideology serve somewhat as functional equivalents in orienting a person toward the problems he must face and the policies he must select, the growth of the domain of knowledge causes it to impinge on the domain of ideology.[32]

Harvey Brooks, finally, draws a similar conclusion from consideration of the extent to which scientific criteria and techniques have found their way into the management of political affairs. He finds an

> Increasing relegation of questions which used to be matters of political debate to professional cadres of technicians and experts which function almost independently of the democratic political process. . . . The progress which is achieved, while slower, seems more solid, more irreversible, more capable of enlisting a wide consensus.[33]

I raise this point as fundamental to the technology-polity relationship, not by way of hazarding a prediction. I ignore neither the possibility that *value* conflicts in political debate may become sharper still as factual differences are muted by better knowledge, nor the fact that decline of political ideology does not *ipso facto* mean a decline of political disagreement, nor the fear of some that the hippie movement, literary anti-intellectualism, and people's fears of genuine dangers implicit in continued technological advance may in fact augur an imminent retreat from rationality and an interlude—perhaps a long interlude—either of political know-nothingism reminiscent of Joseph McCarthy or of social concentration on contemplative or religious values.

Yet if the technology-values dualism is unwarranted, as I argued above, it is equally plausible to find no more warrant in principle in a sharp separation between knowledge and political action. Like all dualisms, this one too may have had its origins in the analytic abhorrence of uncertainty. (One is reminded in this connection of the radical dualism that Descartes arrived at as a result of his determination to base his philosophy on the only certain and self-evident principle he could discover.) There certainly is painfully much in political history and political experience to render uncertain a positive correlation between knowledge and political consensus. The correlation is not necessarily absent, therefore, and to find it and lead society to act on it may be the greatest challenge yet to political inquiry and political action.

To the extent that technological change expands and alters the spectrum of what man can do, it multiples the choices that society will have to make. These choices will increasingly have to be deliberate social choices, moreover, rather than market reflections of innumerable individual consumer choices, and will therefore have to be made by political means. Since it is unlikely—despite futurists and technological forecasters—that we will soon be able to predict future opportunities (and their attendant opportunity costs) with any significant degree of reliability in detail, it becomes important to investigate the conditions of a political system (I use the term in the wide sense I assigned to "government" above) with the flexibility and value presuppositions necessary to evaluate alternatives and make choices among them as they continue to emerge.

This prescription is analogous, for governance of a changing society, to those advanced above for educational policy and for our approach to the analysis of values. In all three cases, the emphasis shifts from allegiance to the known, stable, formulated, and familiar, to a *posture* of expectation of change and readiness to deal with it. It is this kind of shift, occurring across many elements of society, that is the hallmark of a truly Heraclitean age. It is what Vogt seeks to formalize in stressing processual as against structural analysis of culture and society. The mechanisms, values, attitudes, and procedures called for by a social posture of readiness will be different in kind from those characteristic of a society that sees itself as mature, "arrived," and in stable equilibrium. The most fundamental *political* task of a technological world, in other words, is that of systematizing and institutionalizing the

social expectation of the changes that technology will continue to bring about.

I see that task as a precondition of profiting from our accumulating knowledge of the effects of technological change. To understand those effects is an intellectual problem, but to do something about them and profit from the opportunities that technology offers is a political one. We need above all, in other words, to gauge the effects of technology on the *polity*, so that we can derive some social value from our knowledge. This, I suppose, is the twentieth-century form of the perennial ideal of wedding wisdom and government.[34]

6

Thomas Henry Huxley, SCIENCE AND CULTURE

Henry David Thoreau, WALKING

1. Why do you suppose that Huxley found it necessary to go to such great length to justify a place for physical science in a university curriculum?

2. Do you believe that Huxley would be satisfied with the state of familiarity with the physical sciences in the current United States population? Why?

3. Describe Huxley's attack on Arnold's contention that literature contains all that one needs to know about life. Do you detect any polemic in the argument?

4. Do you see any evidence that the debate on the relative merit of literature and science, begun by men like Huxley and Arnold, has continued into the late twentieth century?

5. Thoreau presents a view of nature and man's relationship to nature which is quite different from that of modern western society. How does a society's view of the man-nature relationship affect its technological growth?

6. Why do you think that Thoreau is accorded a place in a theme entitled "Science, Technology, and Society"?

7. What, for Thoreau, is the distinction between "useful knowledge" and "useful ignorance"? Relate this distinction to your own view of liberal education and, if you can, to Cardinal Newman's view of liberal education.

8. What special significance did the compass directions, East and West, hold for Thoreau?

Thomas Henry Huxley (1825-1895) was not only a distinguished English scientist, but also a significant interpreter of science and its impact to a non-scientific public. An ardent supporter of Darwin, Huxley had a major role in the debate between the proponents of evolution and those who opposed it for religious reasons.

SCIENCE AND CULTURE[1]

SIX years ago, as some of my present hearers may remember, I had the privilege of addressing a large assemblage of the inhabitants of this city, who had gathered together to do honor to the memory of their famous townsman, Joseph Priestley; and, if any satisfaction attaches to posthumous glory, we may hope that the manes of the burnt-out philosopher were then finally appeased.

No man, however, who is endowed with a fair share of common sense, and not more than a fair share of vanity, will identify either contemporary or posthumous fame with the highest good; and Priestley's life leaves no doubt that he, at any rate, set a much higher value upon the advancement of knowledge, and the promotion of that freedom of thought which is at once the cause and the consequence of intellectual progress.

Hence I am disposed to think that, if Priestley could be amongst us to-day, the occasion of our meeting would afford him even greater pleasure than the proceedings which celebrated the centenary of his chief discovery. The kindly heart would be moved, the high sense of social duty would be satisfied, by the spectacle of well-earned wealth, neither squandered in tawdry luxury and vainglorious show, nor scattered with the careless charity which blesses neither him that gives nor him that takes, but expended in the execution of a well-considered plan for the aid of present and future generations of those who are willing to help themselves.

We shall all be of one mind thus far. But it is needful to share Priestley's keen interest in physical science; and to have learned, as he had learned, the value of scientific training in fields of inquiry apparently far remote from physical science; in order to appreciate, as he would have appreciated, the value of the noble gift which Sir

[1] Originally delivered as an address, in 1880, at the opening of Mason College, Birmingham, England, now the University of Birmingham.

Josiah Mason has bestowed upon the inhabitants of the Midland district.

For us children of the nineteenth century, however, the establishment of a college under the conditions of Sir Josiah Mason's trust has a significance apart from any which it could have possessed a hundred years ago. It appears to be an indication that we are reaching the crisis of the battle, or rather of the long series of battles, which have been fought over education in a campaign which began long before Priestley's time, and will probably not be finished just yet.

In the last century, the combatants were the champions of ancient literature, on the one side, and those of modern literature on the other; but, some thirty years[2] ago, the contest became complicated by the appearance of a third army, ranged round the banner of physical science.

I am not aware that any one has authority to speak in the name of this new host. For it must be admitted to be somewhat of a guerilla force, composed largely of irregulars, each of whom fights pretty much for his own hand. But the impressions of a full private, who has seen a good deal of service in the ranks, respecting the present position of affairs and the conditions of a permanent peace, may not be devoid of interest; and I do not know that I could make a better use of the present opportunity than by laying them before you.

From the time that the first suggestion to introduce physical science into ordinary education was timidly whispered, until now, the advocates of scientific education have met with opposition of two kinds. On the one hand, they have been poohpoohed by the men of business who pride themselves on being the representatives of practicality; while, on the other hand, they have been excommunicated by the classical scholars, in their capacity of Levites in charge of the ark of culture and monopolists of liberal education.

The practical men believed that the idol whom they worship—rule of thumb—has been the source of the past prosperity, and will suffice for the future welfare of the arts and manufactures. They were of opinion that science is speculative rubbish; that theory and practice have nothing to do with one another; and that the scientific

[2] The advocacy of the introduction of physical science into general education by George Combe and others commenced a good deal earlier; but the movement had acquired hardly any practical force before the time to which I refer.

habit of mind is an impediment, rather than an aid, in the conduct of ordinary affairs.

I have used the past tense in speaking of the practical men—for although they were very formidable thirty years ago, I am not sure that the pure species has not been extirpated. In fact, so far as mere argument goes, they have been subjected to such a *feu d'enfer* that it is a miracle if any have escaped. But I have remarked that your typical practical man has an unexpected resemblance to one of Milton's angels. His spiritual wounds, such as are inflicted by logical weapons, may be as deep as a well and as wide as a church door, but beyond shedding a few drops of ichor, celestial or otherwise, he is no whit the worse. So, if any of these opponents be left, I will not waste time in vain repetition of the demonstrative evidence of the practical value of science; but knowing that a parable will sometimes penetrate where syllogisms fail to effect an entrance, I will offer a story for their consideration.

Once upon a time, a boy, with nothing to depend upon but his own vigorous nature, was thrown into the thick of the struggle for existence in the midst of a great manufacturing population. He seems to have had a hard fight, inasmuch as, by the time he was thirty years of age, his total disposable funds amounted to twenty pounds. Nevertheless, middle life found him giving proof of his comprehension of the practical problems he had been roughly called upon to solve, by a career of remarkable prosperity.

Finally, having reached old age with its well-earned surroundings of "honor, troops of friends," the hero of my story bethought himself of those who were making a like start in life, and how he could stretch out a helping hand to them.

After long and anxious reflection this successful practical man of business could devise nothing better than to provide them with the means of obtaining "sound, extensive, and practical scientific knowledge." And he devoted a large part of his wealth and five years of incessant work to this end.

I need not point the moral of a tale which, as the solid and spacious fabric of the Scientific College assures us, is no fable, nor can anything which I could say intensify the force of this practical answer to practical objections.

We may take it for granted then, that, in the opinion of those best qualified to judge, the diffusion of thorough scientific education is an absolutely essential condition of industrial progress; and that the college which has been opened to-day will confer an inestimable boon upon those whose livelihood is to be gained by the practice of the arts and manufactures of the district.

The only question worth discussion is, whether the conditions, under which the work of the college is to be carried out, are such as to give it the best possible chance of achieving permanent success.

Sir Josiah Mason, without doubt most wisely, has left very large freedom of action to the trustees, to whom he proposes ultimately to commit the administration of the college, so that they may be able to adjust its arrangements in accordance with the changing conditions of the future. But, with respect to three points, he has laid most explicit injunctions upon both administrators and teachers.

Party politics are forbidden to enter into the minds of either, so far as the work of the college is concerned; theology is as sternly banished from its precincts; and finally, it is especially declared that the college shall make no provision for "mere literary instruction and education."

It does not concern me at present to dwell upon the first two injunctions any longer than may be needful to express my full conviction of their wisdom. But the third prohibition brings us face to face with those other opponents of scientific education, who are by no means in the moribund condition of the practical man, but alive, alert, and formidable.

It is not impossible that we shall hear this express exclusion of "literary instruction and education" from a college which, nevertheless, professes to give a high and efficient education, sharply criticised. Certainly the time was that the Levites of culture would have sounded their trumpets against its walls as against an educational Jericho.

How often have we not been told that the study of physical science is incompetent to confer culture; that it touches none of the higher problems of life; and, what is worse, that the continual devotion to scientific studies tends to generate a narrow and bigoted belief in the applicability of scientific methods to the search after truth of all kinds. How frequently one has reason to observe that no reply to a troublesome argument tells so well as calling its author a "mere scientific specialist." And, as I am afraid it is not permissible to speak of this form of opposition to scientific education in the past tense; may we not expect to be told that this, not only omission, but prohibition, of "mere literary instruction and education" is a patent example of scientific narrow-mindedness?

I am not acquainted with Sir Josiah Mason's reasons for the action which he has taken; but if, as I apprehend is the case, he refers to the ordinary classical course of our schools and universities by the name of "mere literary instruction and education," I venture to offer sundry reasons of my own in support of that action.

For I hold very strongly by two convictions. The first is, that neither the discipline nor the subject-matter of classical education is of such direct value to the student of physical science as to justify the expenditure of valuable time upon either; and the second is, that for the purpose of attaining real culture, an exclusively scientific education is at least as effectual as an exclusively literary education.

I need hardly point out to you that these opinions, especially the latter, are diametrically opposed to those of the great majority of educated Englishmen, influenced as they are by school and university traditions. In their belief, culture is obtainable only by a liberal education; and a liberal education is synonymous, not merely with education and instruction in literature, but in one particular form of literature, namely, that of Greek and Roman antiquity. They hold that the man who has learned Latin and Greek, however little, is educated; while he who is versed in other branches of knowledge, however deeply, is a more or less respectable specialist, not admissible into cultured caste. The stamp of the educated man, the university degree, is not for him.

I am too well acquainted with the generous catholicity of spirit, the true sympathy with scientific thought, which pervades the writings of our chief apostle of culture to identify him with these opinions; and yet one may cull from one and another of those epistles to the Philistines, which so much delight all who do not answer to that name, sentences which lend them some support.

Mr. Arnold tells us that the meaning of culture is "to know the best that has been thought and said in the world." It is the criticism of life contained in literature. That criticism regards "Europe as being, for intellectual and spiritual purposes, one great confederation, bound to a joint action and working to a common result; and whose members have, for their common outfit, a knowledge of Greek, Roman, and Eastern antiquity, and of one another. Special, local, and temporary advantages being put out of account, that modern nation will in the intellectual and spiritual sphere make most progress, which most thoroughly carries out this programme. And what is that but saying that we too, all of us, as individuals, the more thoroughly we carry it out, shall make the more progress?"

We have here to deal with two distinct propositions. The first, that a criticism of life is the essence of culture; the second, that literature contains the materials which suffice for the construction of such a criticism.

I think that we must all assent to the first proposition. For culture certainly means something quite different from learning or technical skill. It implies the possession of an ideal, and the habit of critically estimating the value of things by comparison with a theoretic standard. Perfect culture should apply a complete theory of life, based upon a clear knowledge alike of its possibilities and of its limitations.

But we may agree to all this, and yet strongly dissent from the assumption that literature alone is competent to supply this knowledge. After having learnt all that Greek, Roman, and Eastern antiquity have thought and said, and all that modern literatures have to tell us, it is not self-evident that we have laid a sufficiently broad and deep foundation for the criticism of life which constitutes culture.

Indeed, to any one acquainted with the scope of physical science, it is not at all evident. Considering progress only in the "intellectual and spiritual sphere," I find myself wholly unable to admit that either nations or individuals will really advance, if their common outfit draws nothing from the stores of physical science. I should say that an army, without weapons of precision, and with no particular base of operations, might more hopefully enter upon a campaign on the Rhine, than a man, devoid of a knowledge of what physical science has done in the last century, upon a criticism of life.

When a biologist meets with an anomaly, he instinctively turns to the study of development to clear it up. The rationale of contradictory opinions may with equal confidence be sought in history.

It is, happily, no new thing that Englishmen should employ their wealth in building and endowing institutions for educational purposes. But, five or six hundred years ago, deeds of foundation expressed or implied conditions as nearly as possible contrary to those which have been thought expedient by Sir Josiah Mason. That is to say, physical science was practically ignored, while a certain literary training was enjoined as a means to the acquirement of knowledge which was essentially theological.

The reason of this singular contradiction between the actions of men alike animated by a strong and disinterested desire to promote the welfare of their fellows, is easily discovered.

At that time, in fact, if any one desired knowledge beyond such as could be obtained by his own observation, or by common conversation, his first necessity was to learn the Latin language, inasmuch as all the higher knowledge of the western world was contained in works written in that language. Hence, Latin grammar, with logic and rhetoric, studied through Latin, were the fundamentals of education. With respect to the substance of the knowledge imparted through this channel, the Jewish and Christian Scriptures, as interpreted and supplemented by the Romish Church, were held to contain a complete and infallibly true body of information.

Theological dicta were, to the thinkers of those days, that which the axioms and definitions of Euclid are to the geometers of these. The business of the philosophers of the Middle Ages was to deduce from the data furnished by the theologians, conclusions in accordance with ecclesiastical decrees. They were allowed the high privilege of showing, by logical process, how and why that which the Church said was true, must be true. And if their demonstrations fell short of or exceeded this limit, the Church was maternally ready to check their aberrations, if need be, by the help of the secular arm.

Between the two, our ancestors were furnished with a compact and complete criticism of life. They were told how the world began, and how it would end; they learned that all material existence was but a base and insignificant blot upon the fair face of the spiritual world, and that nature was, to all intents and purposes, the playground of the devil; they learned that the earth is the centre of the visible universe, and that man is the cynosure of things terrestrial; and more especially is it inculcated that the course of nature had no fixed order, but that it could be, and constantly was, altered by the agency of innumerable spiritual beings, good and bad, according as they were moved by the deeds and prayers of men. The sum and substance of the whole doctrine was to produce the conviction that the only thing really worth knowing in this world was how to secure that place in a better, which, under certain conditions, the Church promised.

Our ancestors had a living belief in this theory of life, and acted upon it in their dealings with education, as in all other matters. Culture meant saintliness—after the fashion of the saints of those days; the education that led to it was, of necessity, theological; and the way to theology lay through Latin.

That the study of nature—further than was requisite for the satisfaction of everyday wants—should have any bearing on human life was far from the thoughts of men thus trained. Indeed, as nature had been cursed for man's sake, it was an obvious conclusion that those who meddled with nature were likely to come into pretty close contact with Satan. And, if any born scientific investigator followed his instincts, he might safely reckon upon earning the reputation, and probably upon suffering the fate, of a sorcerer.

Had the western world been left to itself in Chinese isolation, there is no saying how long this state of things might have endured. But, happily, it was not left to itself. Even earlier than the thirteenth century, the development of Moorish civilization in Spain and the great movement of the Crusades had introduced the leaven which, from that day to this, has never ceased to work. At first, through the intermediation of Arabic translations, afterwards by the study of the originals, the western nations of Europe became acquainted with the writings of the ancient philosophers and poets, and, in time, with the whole of the vast literature of antiquity.

Whatever there was of high intellectual aspiration or dominant capacity in Italy, France, Germany, and England, spent itself for centuries in taking possession of the rich inheritance left by the dead civilization of Greece and Rome. Marvelously aided by the invention of printing, classical learning spread and flourished. Those who possessed it prided themselves on having attained the highest culture then within the reach of mankind.

And justly. For, saving Dante on his solitary pinnacle, there was no figure in modern literature at the time of the Renaissance to compare with the men of antiquity; there was no art to compete with their sculpture; there was no physical science but that which Greece had created. Above all, there was no other example of perfect intellectual freedom—of the unhesitating acceptance of reason as the sole guide to truth and the supreme arbiter of conduct.

The new learning necessarily soon exerted a profound influence upon education. The language of the monks and schoolmen seemed little better than gibberish to scholars fresh from Vergil and Cicero, and the study of Latin was placed upon a new foundation. Moreover, Latin itself ceased to afford the sole key to knowledge. The student who sought the highest thought of antiquity found only a second-hand reflection of it in Roman literature, and turned his face to the full light of the Greeks. And after a battle, not altogether dissimilar to that which is at present being fought over the teaching of physical science, the study of Greek was recognized as an essential element of all higher education.

Thus the humanists, as they were called, won the day; and the great reform which they effected was of incalculable service to mankind. But the Nemesis of all reformers is finality; and the reformers of education, like those of religion, fell into the profound, however common, error of mistaking the beginning for the end of the work of reformation.

The representatives of the humanists in the nineteenth century take their stand upon classical education as the sole avenue to culture, as firmly as if we were still in the age of Renaissance. Yet, surely, the present intellectual relations of the modern and the ancient worlds are profoundly different from those which obtained three centuries ago. Leaving aside the existence of a great and characteristically modern literature, of modern painting, and, especially, of modern music, there is one feature of the present state of the civilized world which separates it more widely from the Renaissance than the Renaissance was separated from the Middle Ages.

This distinctive character of our own times lies in the vast and constantly increasing part which is played by natural knowledge. Not only is our daily life shaped by it, not only does the prosperity of millions of men depend upon it, but our whole theory of life has long been influenced, consciously or unconsciously, by the general conceptions of the universe, which have been forced upon us by physical science.

In fact, the most elementary acquaintance with the results of scientific investigation shows us that they offer a broad and striking contradiction to the opinions so implicitly credited and taught in the Middle Ages.

The notions of the beginning and the end of the world entertained by our forefathers are no longer credible. It is very certain that the earth is not the chief body in the material universe, and that the world is not subordinated to man's use. It is even more certain that nature is the expression of a definite order with which nothing interferes, and that the chief business of mankind is to learn that order and govern themselves accordingly. Moreover this scientific "criticism of life" presents itself to us with different credentials from any other. It appeals not to authority, nor to what anybody may have thought or said, but to nature. It admits that all our interpretations of natural fact are more or less imperfect and symbolic, and bids the learner seek for truth not among words but among things. It warns us that the assertion which outstrips evidence is not only a blunder but a crime.

The purely classical education advocated by the representatives of the humanists in our day gives no inkling of all this. A man may be a better scholar than Erasmus, and know no more of the chief causes of the present intellectual fermentation than Erasmus did. Scholarly and pious persons, worthy of all respect, favor us with allocutions upon the sadness of the antagonism of science to their mediæval way of thinking, which betray an ignorance of the first principles of scientific investigation, an incapacity for understanding what a man of science means by veracity, and an unconsciousness of the weight of established scientific truths, which is almost comical.

There is no great force in the *tu quoque* argument, or else the advocates of scientific education might fairly enough retort upon the modern humanists that they may be learned specialists, but that they possess no such sound foundation for a criticism of life as deserves the name of culture. And, indeed, if we were disposed to be cruel, we might urge that the humanists have brought this reproach upon themselves, not because they are too full of the spirit of the ancient Greek, but because they lack it.

The period of the Renaissance is commonly called that of the "Revival of Letters," as if the influences then brought to bear upon the mind of Western Europe had been wholly exhausted in the field of literature. I think it is very commonly forgotten that the revival of science, effected by the same agency, although less conspicuous, was not less momentous.

In fact, the few and scattered students of nature of that day picked up the clew to her secrets exactly as it fell from the hands of the Greeks a thousand years before. The foundations of mathematics were so well laid by them that our children learn their geometry from a book written for the schools of Alexandria two thousand years ago. Modern astronomy is the natural continuation and development of the work of Hipparchus and of Ptolemy; modern physics of that of Democritus and of Archimedes; it was long before modern biological science outgrew the knowledge bequeathed to us by Aristotle, by Theophrastus, and by Galen.

We cannot know all the best thoughts and sayings of the Greeks unless we know what they thought about natural phenomena. We cannot fully apprehend their criticism of life unless we understand the extent to which that criticism was affected by scientific conceptions. We falsely pretend to be the inheritors of their culture, unless we are penetrated, as the best minds among them were, with an unhesitating faith that the free employment of reason, in accordance with scientific method, is the sole method of reaching truth.

Thus I venture to think that the pretensions of our modern humanists to the possession of the monopoly of culture and to the exclusive inheritance of the spirit of antiquity must be abated, if not abandoned. But I should be very sorry that anything I have said should be taken to imply a desire on my part to depreciate the value of classical education, as it might be and as it sometimes is. The native capacities of mankind vary no less than their opportunities; and while culture is one, the road by which one man may best reach it is widely different from that which is most advantageous to another. Again, while scientific education is yet inchoate and tentative, classical education is thoroughly well organized upon the practical experience of generations of teachers. So that, given ample time for learning and destination for ordinary life, or for a literary career, I do not think that a young Englishman in search of culture can do better than follow the course usually marked out for him, supplementing its deficiencies by his own efforts.

But for those who mean to make science their serious occupation; or who intend to follow the profession of medicine; or who have to enter early upon the business of life; for all these, in my opinion, classical education is a mistake; and it is for this reason that I am glad to see "mere literary education and instruction" shut out from the curriculum of Sir Josiah Mason's college, seeing that its inclusion would probably lead to the introduction of the ordinary smattering of Latin and Greek.

Nevertheless, I am the last person to question the importance of genuine literary education, or to suppose that intellectual culture can be complete without it. An exclusively scientific training will bring about a mental twist as surely as an exclusive literary training. The value of the cargo does not compensate for a ship's being out of trim; and I should be very sorry to think that the Scientific College would turn out none but lop-sided men.

There is no need, however, that such a catastrophe should happen. Instruction in English, French, and German is provided, and thus the three greatest literatures of the modern world are made accessible to the student.

French and German, and especially the latter language, are absolutely indispensable to those who desire full knowledge in any department of science. But even supposing that the knowledge of these languages acquired is not more than sufficient for purely scientific purposes, every Englishman has, in his native tongue, an almost perfect instrument of literary expression; and, in his own literature, models of every kind of literary excellence. If an Englishman cannot get literary culture out of his Bible, his Shakespeare, his Milton, neither, in my belief, will the profoundest study of Homer and Sophocles, Vergil and Horace, give it to him.

Thus, since the constitution of the college makes sufficient provision for literary as well as for scientific education, and since artistic instruction is also contemplated, it seems to me that a fairly complete culture is offered to all who are willing to take advantage of it.

But I am not sure that at this point the "practical" man, scotched but not slain, may ask what all this talk about culture has to do with an institution, the object of which is defined to be "to promote the prosperity of the manufactures and the industry of the country." He may suggest that what is wanted for this end is not culture, nor even a purely scientific discipline, but simply a knowledge of applied science.

I often wish that this phrase, "applied science," had never been invented. For it suggests that there is a sort of scientific knowledge of direct practical use, which can be studied apart from another sort of scientific knowledge, which is of no practical utility, and which is termed "pure science." But there is no more complete fallacy than this. What people call applied science is nothing but the application of pure science to particular classes of problems. It consists of deductions from those general principles, established by reasoning and observation, which constitute pure science. No one can safely make these deductions until he has a firm grasp of the principles; and he can obtain that grasp only by personal experience of the operations of observation and of reasoning on which they are founded.

Almost all the processes employed in the arts and manufactures fall within the range either of physics or of chemistry. In order to improve them one must thoroughly understand them; and no one has a chance of really understanding them, unless he has obtained that mastery of principles and that habit of dealing with facts which is given by long-continued and well-directed purely scientific training in the physical and chemical laboratory. So that there really is no question as to the necessity of purely scientific discipline, even if the work of the college were limited by the narrowest interpretation of its stated aims.

And, as to the desirableness of a wider culture than that yielded by science alone, it is to be recollected that the improvement of manufacturing processes is only one of the conditions which contribute to the prosperity of industry. Industry is a means and not an end; and mankind work only to get something which they want. What that something is depends partly on their innate, and partly on their acquired, desires.

If the wealth resulting from prosperous industry is to be spent upon the gratification of unworthy desires, if the increasing perfection of manufacturing processes is to be accompanied by an increasing debasement of those who carry them on, I do not see the good of industry and prosperity.

Now it is perfectly true that men's views of what is desirable depend upon their characters; and that the innate proclivities to which we give that name are not touched by any amount of instruction. But it does not follow that even mere intellectual education may not, to an indefinite extent, modify the practical manifestation of the characters of men in their actions, by supplying them with motives unknown to the ignorant. A pleasure-loving character will have pleasure of some sort; but, if you give him the choice, he may prefer pleasures which do not degrade him to those which do. And this choice is offered to every man who possesses in literary or artistic culture a never-failing source of pleasures, which are neither withered by age, nor staled by custom, nor embittered in the recollection by the pangs of self-reproach.

If the institution opened to-day fulfils the intention of its founder, the picked intelligences among all classes of the population of this district will pass through it. No child born in Birmingham, henceforward, if he have the capacity to profit by the opportunities offered to him, first in the primary and other schools, and afterward in the Scientific College, need fail to obtain, not merely the instruction, but the culture most appropriate to the conditions of his life.

Within these walls the future employer and the future artisan may sojourn together for awhile, and carry, through all their lives, the stamp of the influences then brought to bear upon them. Hence, it is not beside the mark to remind you that the prosperity of industry depends not merely upon the improvement of manufacturing processes, not merely upon the ennobling of the individual character, but upon a third condition, namely, a clear understanding of the conditions of social life on the part of both the capitalist and the operative, and their agreement upon common principles of social action. They must learn that social phenomena are as much the expression of natural laws as any others; that no social arrangements can be permanent unless they harmonize with the requirements of social statics and dynamics; and that, in the nature of things, there is an arbiter whose decisions execute themselves.

But this knowledge is only to be obtained by the application of the methods of investigation adopted in physical researches to the investigation of the phenomena of society. Hence, I confess I should like to see one addition made to the excellent scheme of education propounded for the college, in the shape of provision for the teaching of sociology. For though we are all agreed that party politics are to have no place in the instruction of the college; yet in this country, practically governed as it is now by universal suffrage, every man who does his duty must exercise political functions. And, if the evils which are inseparable from the good of political liberty are to be checked, if the perpetual oscillation of nations between anarchy and despotism is to be replaced by the steady march of self-restraining freedom; it will be because men will gradually bring themselves to deal with political, as they now deal with scientifical questions; to be as ashamed of undue haste and partisan prejudice in the one case as in the other; and to believe that the machinery of society is at least as delicate as that of a spinning-jenny, and as little likely to be improved by the meddling of those who have not taken the trouble to master the principles of its action.

In conclusion, I am sure that I make myself the mouthpiece of all present in offering to the venerable founder of the institution, which now commences its beneficent career, our congratulations on the completion of his work; and in expressing the conviction that the remotest posterity will point to it as a crucial instance of the wisdom which natural piety leads all men to ascribe to their ancestors.

Henry David Thoreau (1817-1862) was a highly individualistic nineteenth-century American philosopher and naturalist. Rejecting the trappings of contemporary New England society, Thoreau sought a solitary existence which afforded him ample time to observe his natural surroundings, to ponder his place in nature, and to record his thoughts for generations of delighted readers.

WALKING

[1862]

I WISH to speak a word for Nature, for absolute freedom and wildness, as contrasted with a freedom and culture merely civil, —to regard man as an inhabitant, or a part and parcel of Nature, rather than a member of society. I wish to make an extreme statement, if so I may make an emphatic one, for there are enough champions of civilization: the minister and the school committee and every one of you will take care of that.

I have met with but one or two persons in the course of my life who understood the art of Walking, that is, of taking walks,—who had a genius, so to speak, for *sauntering:* which word is beautifully derived from "idle people who roved about the country, in the Middle Ages, and asked charity, under pretence of going *à la Sainte Terre,*" to the Holy Land, till the children exclaimed, "There goes a *Sainte-Terrer,*" a Saunterer, a Holy-Lander. They who never go to the Holy Land in their walks, as they pretend, are indeed mere idlers and vagabonds; but they who do go there are saunterers in the good sense, such as I mean. Some, however, would derive the word from *sans terre,* without land or a home, which, therefore, in the good sense, will mean, having no particular home, but equally at home everywhere. For this is the secret of successful sauntering. He who sits still in a house all the time may be the greatest vagrant of all; but the saunterer, in the good sense, is no more vagrant than the meandering river, which is all the while sedulously seeking the shortest course to the sea. But I prefer the first, which, indeed, is the most probable derivation. For every walk is a sort of crusade, preached by some Peter the Hermit in us, to go forth and reconquer this Holy Land from the hands of the Infidels.

It is true, we are but faint-hearted crusaders, even the walkers, nowadays, who undertake no persevering, never-ending enterprizes.

Our expeditions are but tours, and come round again at evening to the old hearth-side from which we set out. Half the walk is but retracing our steps. We should go forth on the shortest walk, perchance, in the spirit of undying adventure, never to return—prepared to send back our embalmed hearts only as relics to our desolate kingdoms. If you are ready to leave father and mother, and brother and sister, and wife and child and friends, and never see them again —if you have paid your debts, and made your will, and settled all your affairs, and are a free man, then you are ready for a walk.

To come down to my own experience, my companion and I, for I sometimes have a companion, take pleasure in fancying ourselves knights of a new, or rather an old, order—not Equestrians or Chevaliers, not Ritters or riders, but Walkers, a still more ancient and honourable class, I trust. The chivalric and heroic spirit which once belonged to the Rider seems now to reside in, or perchance to have subsided into, the Walker,—not the Knight, but Walker, Errant. He is a sort of fourth estate, outside of Church and State and People.

We have felt that we almost alone hereabouts practised this noble art; though, to tell the truth, at least, if their own assertions are to be received, most of my townsmen would fain walk sometimes, as I do, but they cannot. No wealth can buy the requisite leisure, freedom, and independence which are the capital in this profession. It comes only by the grace of God. It requires a direct dispensation from Heaven to become a walker. You must be born into the family of the Walkers. *Ambulator nascitur, non fit.* Some of my townsmen, it is true, can remember and have described to me some walks which they took ten years ago, in which they were so blessed as to lose themselves for half an hour in the woods; but I know very well that they have confined themselves to the highway ever since, whatever pretensions they may make to belong to this select class. No doubt they were elevated for a moment as by the reminiscence of a previous state of existence, when even they were foresters and outlaws.

"When he came to grene wode,
In a mery mornynge,
There he herde the notes small
Of byrdes mery syngynge.

WALKING

"It is ferre gone, sayd Robyn,
That I was last here;
Me lyste a lytell for to shote
At the donne dere."

I think that I cannot preserve my health and spirits, unless I spend four hours a day at least,—and it is commonly more than that,—sauntering through the woods and over the hills and fields, absolutely free from all worldly engagements. You may safely say, A penny for your thoughts, or a thousand pounds. When sometimes I am reminded that the mechanics and shopkeepers stay in their shops not only all the forenoon, but all the afternoon too, sitting with crossed legs, so many of them—as if the legs were made to sit upon, and not to stand or walk upon—I think that they deserve some credit for not having all committed suicide long ago.

I, who cannot stay in my chamber for a single day without acquiring some rust, and when sometimes I have stolen forth for a walk at the eleventh hour or four o'clock in the afternoon, too late to redeem the day, when the shades of night were already beginning to be mingled with the daylight, have felt as if I had committed some sin to be atoned for,—I confess that I am astonished at the power of endurance, to say nothing of the moral insensibility, of my neighbours who confine themselves to shops and offices the whole day for weeks and months, ay, and years almost together. I know not what manner of stuff they are of—sitting there now at three o'clock in the afternoon, as if it were three o'clock in the morning. Bonaparte may talk of the three-o'clock-in-the-morning courage, but it is nothing to the courage which can sit down cheerfully at this hour in the afternoon over against one's self whom you have known all the morning, to starve out a garrison to whom you are bound by such strong ties of sympathy. I wonder that about this time, or say between four and five o'clock in the afternoon, too late for the morning papers and too early for the evening ones, there is not a general explosion heard up and down the street, scattering a legion of antiquated and house-bred notions and whims to the four winds for an airing—and so the evil cure itself.

How womankind, who are confined to the house still more than men, stand it I do not know; but I have ground to suspect that most of them do not *stand* it at all. When, early in a summer afternoon, we have been shaking the dust of the village from the skirts of our garments, making haste past those houses with purely Doric or Gothic fronts, which have such an air of repose about them, my companion whispers that probably about these times their occupants are all gone to bed. Then it is that I appreciate the beauty and the glory of architecture, which itself never turns in, but forever stands out and erect, keeping watch over the slumberers.

No doubt temperament, and, above all, age, have a good deal to do with it. As a man grows older, his ability to sit still and follow indoor occupations increases. He grows vespertinal in his habits as the evening of life approaches, till at last he comes forth only just before sundown, and gets all the walk that he requires in half an hour.

But the walking of which I speak has nothing in it akin to taking exercise, as it is called, as the sick take medicine at stated hours—as the swinging of dumb-bells or chairs; but is itself the enterprise and adventure of the day. If you would get exercise, go in search of the springs of life. Think of a man's swinging dumb-bells for his health, when those springs are bubbling up in far-off pastures unsought by him!

Moreover, you must walk like a camel, which is said to be the only beast which ruminates when walking. When a traveller asked Wordsworth's servant to show him her master's study, she answered, "Here is his library, but his study is out of doors."

Living much out of doors, in the sun and wind, will no doubt produce a certain roughness of character—will cause a thicker cuticle to grow over some of the finer qualities of our nature, as on the face and hands, or as severe manual labor robs the hands of some of their delicacy of touch. So staying in the house, on the other hand, may produce a softness and smoothness, not to say thinness of skin, accompanied by an increased sensibility to certain impressions. Perhaps we should be more susceptible to some influences important to our intellectual and moral growth, if the sun had shone and the wind blown on us a little less; and no doubt it is a nice matter to proportion rightly the thick and thin skin. But methinks that is a scurf that will fall off fast enough—that the natural remedy is to be found in the proportion which the night bears to the day, the winter to the summer, thought to experience. There will be so much the more air and sunshine in our thoughts. The callous palms of the laborer are conversant with finer tissues of self-respect and heroism, whose touch thrills the heart, than the languid fingers of idleness. That is mere sentimentality that lies abed by day and thinks itself white, far from the tan and callus of experience.

When we walk, we naturally go to the fields and woods: what would become of us if we walked only in a garden or a mall? Even some sects of philosophers have felt the necessity of importing the woods to themselves, since they did not go to the woods. "They planted groves and walks of Platanes," where they took *subdiales ambulationes* in porticos open to the air. Of course it is of no use to direct our steps to the woods, if they do not carry us thither. I am alarmed when it happens that I have walked a mile into the woods bodily, without getting there in spirit. In my afternoon walk I would fain forget all my morning occupations and my obligations to society. But it sometimes happens that I cannot easily shake off the village. The thought of some work will run in my head and I am not where my body is—I am out of my senses. In my walks I would fain return to my senses. What business have I in the woods, if I am thinking of something out of the woods? I suspect myself, and cannot help a shudder, when I find myself so implicated even in what are called good works—for this may sometimes happen.

My vicinity affords many good walks; and though for so many years I have walked almost every day, and sometimes for several days together, I have not yet exhausted them. An absolutely new prospect is a great happiness, and I can still get this any afternoon. Two or three hours' walking will carry me to as strange a country as I expect ever to see. A single farmhouse which I had not seen before is sometimes as good as the dominions of the King of Dahomey. There is in fact a sort of harmony discoverable between the capabilities of the landscape within a circle of ten miles' radius, or the limits of an afternoon walk, and the threescore years and ten of human life. It will never become quite familiar to you.

Nowadays almost all man's improvements, so called, as the building of houses, and the cutting down of the forest and of all large trees, simply deform the landscape, and make it more and more tame and cheap. A people who would begin by burning the fences and let the forest stand! I saw the fences half consumed, their ends lost in the middle of the prairie, and some worldly miser with a surveyor looking after his bounds, while heaven had taken place around him, and he did not see the angels going to and fro, but was looking for an old post-hole in the midst of paradise. I looked again, and saw him standing in the middle of a boggy stygian fen, surrounded by devils, and he had found his bounds without a doubt, three little stones, where a stake had been driven, and looking nearer, I saw that the Prince of Darkness was his surveyor.

I can easily walk ten, fifteen, twenty, any number of miles, commencing at my own door, without going by any house, without crossing a road except where the fox and the mink do: first along by the river, and then the brook, and then the meadow and the wood-side. There are square miles in my vicinity which have no inhabitant. From many a hill I can see civilisation and the abodes of man afar. The farmers and their works are scarcely more obvious than woodchucks and their burrows. Man and his affairs, church and state and school, trade and commerce, and manufactures and agriculture, even politics, the most alarming of them all,—I am pleased to see how little space they occupy in the landscape. Politics is but a narrow field, and that still narrower highway yonder leads to it. I sometimes direct the traveller thither. If you would go to the political world, follow the great road—follow that market-man, keep his dust in your eyes, and it will lead you straight to it; for it, too, has its place merely, and does not occupy all space. I pass from it as from a bean-field into the forest, and it is forgotten. In one half-hour I can walk off to some portion of the earth's surface where a man does not stand from one year's end to another, and there, consequently, politics are not, for they are but as the cigar-smoke of a man.

The village is the place to which the roads tend, a sort of expansion of the highway, as a lake of a river. It is the body of which roads are the arms and legs—a trivial or quadrivial place, the thoroughfare and ordinary of travellers. The word is from the Latin *villa,* which together with *via,* a way, or more anciently *ved* and *vella,* Varro derives from *veho,* to carry, because the villa is the place to and from which things are carried. They who got their living by team-ing were said *vellaturam facere.* Hence, too, apparently, the Latin word *vilis* and our vile; also *villain.* This suggests what kind of degeneracy villagers are liable to. They are wayworn by the travel that goes by and over them, without travelling themselves.

Some do not walk at all; others walk in the highways; a few walk across lots. Roads are made for horses and men of business. I do not travel in them much, comparatively, because I am not in a hurry to get to any tavern or grocery or livery-stable or depot to which they lead. I am a good horse to travel, but not from choice a roadster. The landscape-painter uses the figures of men to mark a road. He would not make that use of my figure. I walk out into a Nature such as the old prophets and poets, Menu, Moses, Homer, Chaucer, walked in. You may name it America, but it is not America: neither Americus Vespucius, nor Columbus, nor the rest were the discoverers of it. There is a truer account of it in mythology than in any history of America, so called, that I have seen.

However, there are a few old roads that may be trodden with profit, as if they led somewhere now that they are nearly discontinued. There is the Old Marlborough Road, which does not go to Marlborough now, methinks, unless that is Marlborough where it carries me. I am the bolder to speak of it here, because I presume that there are one or two such roads in every town.

THE OLD MARLBOROUGH ROAD

Where they once dug for money,
Where sometimes Martial Miles.
But never found any;
Singly files,
And Elijah Wood,
I fear for no good:
No other man,
Save Elisha Dugan,—
O man of wild habits,
Partridges and rabbits,
Who hast no cares
Only to set snares,
Who liv'st all alone,
Close to the bone,
And where life is sweetest
Constantly eatest.

THOREAU

When the spring stirs my blood
  With the instinct to travel
  I can get enough gravel
On the Old Marlborough Road.
    Nobody repairs it,
    For nobody wears it;
    It is a living way,
    As the Christians say.
Not many there be
  Who enter therein,
Only the guests of the
  Irishman Quin.
What is it, what is it,
  But a direction out there,
And the bare possibility
    Of going somewhere?
    Great guide-boards of stone.
    But travellers none;
    Cenotaphs of the towns
    Named on their crowns.
    It is worth going to see
    Where you *might* be.
    What king
    Did the thing,
    I am still wondering;
    Set up how or when,
    By what selectmen,
    Gourgas or Lee,
    Clark or Darby?
    They're a great endeavor
    To be something forever;
    Blank tablets of stone,
    Where a traveller might groan,
    And in one sentence
    Grave all that is known;
    Which another might read,
    In his extreme need.
    I know one or two
    Lines that would do,
    Literature that might stand
    All over the land,
    Which a man could remember
    Till next December,
    And read again in the Spring,

After the thawing.
If with fancy unfurled
You leave your abode,
You may go round the world
By the Old Marlborough Road.

At present, in this vicinity, the best part of the land is not private property; the landscape is not owned, and the walker enjoys comparative freedom. But possibly the day will come when it will be partitioned off into so-called pleasure-grounds, in which a few will take a narrow and exclusive pleasure only,—when fences shall be multiplied, and man-traps and other engines invented to confine men to the *public* road, and walking over the surface of God's earth shall be construed to mean trespassing on some gentleman's grounds. To enjoy a thing exclusively is commonly to exclude yourself from the true enjoyment of it. Let us improve our opportunities, then, before the evil days come.

What is it that makes it so hard sometimes to determine whither we will walk? I believe that there is a subtle magnetism in Nature, which, if we unconsciously yield to it, will direct us aright. It is not indifferent to us which way we walk. There is a right way; but we are very liable from heedlessness and stupidity to take the wrong one. We would fain take that walk, never yet taken by us through this actual world, which is perfectly symbolical of the path which we love to travel in the interior and ideal world; and sometimes, no doubt, we find it difficult to choose our direction, because it does not yet exist distinctly in our idea.

When I go out of the house for a walk, uncertain as yet whither I will bend my steps, and submit myself to my instinct to decide for me, I find, strange and whimsical as it may seem, that I finally and inevitably settle southwest, toward some particular wood or meadow or deserted pasture or hill in that direction. My needle is slow to settle,—varies a few degrees, and does not always point due southwest, it is true, and it has good authority for this variation, but it always settles between west and south-southwest. The future lies that way to me, and the earth seems more unexhausted and richer on that side. The outline which would bound my walks would be, not a circle, but a parabola, or rather like one of those cometary orbits which have been thought to be non-returning curves, in this case opening westward, in which my house occupies the place of the sun. I turn round and round irresolute sometimes for a quarter of an hour, until I decide, for a thousandth time, that I will walk into the southwest or west. Eastward I go only by force; but westward I go free. Thither no business leads me. It is hard for me to believe that I shall find fair landscapes or sufficient wildness and freedom behind the eastern horizon. I am not excited by the prospect of a walk thither; but I believe that the forest which I see in the western horizon stretches uninterruptedly toward the setting sun, and there are no towns nor cities in it of enough consequence to disturb me. Let me live where I will, on this side is the city, on that the wilderness, and ever I am leaving the city more and more, and withdrawing into the wilderness. I should not lay so much stress on this fact, if I did not believe that something like this is the prevailing tendency of my countrymen. I must walk toward Oregon, and not toward Europe. And that way the nation is moving, and I may say that mankind progress from east to west. Within a few years we have witnessed the phenomenon of a south-eastward migration, in the settlement of Australia; but this affects us as a retrograde movement, and, judging from the moral and physical character of the first generation of Australians, has not yet proved a successful experiment. The eastern Tartars think that there is nothing west beyond Thibet. "The world ends there," say they; "beyond there is nothing but a shoreless sea." It is unmitigated East where they live.

We go eastward to realise history and study the works of art and literature, retracing the steps of the race; we go westward as into the future, with a spirit of enterprise and adventure. The Atlantic is a Lethean stream, in our passage over which we have had an opportunity to forget the Old World and its institutions. If we do not succeed this time, there is perhaps one more chance for the race left before it arrives on the banks of the Styx; and that is in the Lethe of the Pacific, which is three times as wide.

I know not how significant it is, or how far it is an evidence of singularity, that an individual should thus consent in his pettiest walk with the general movement of the race; but I know that something akin to the migratory instinct in birds and quadrupeds,—which, in some instances, is known to have affected the squirrel tribe, impelling them to a general and mysterious movement, in which they were seen, say some, crossing the broadest rivers, each on its particular chip, with its tail raised for a sail, and bridging narrower streams with their dead,—that something like the *furor* which affects the domestic cattle in the spring, and which is referred to a worm in their tails,—affects both nations and individuals, either perennially or from time to time. Not a flock of wild geese cackles over our town, but it to some extent unsettles the value of real estate here, and if I were a broker I should probably take that disturbance into account.

> "Than longen folk to gon on pilgrimages,
> And palmeres for to seken strange strondes."

Every sunset which I witness inspires me with the desire to go to a West as distant and as fair as that into which the sun goes down. He appears to migrate westward daily, and tempts us to follow him. He is the Great Western Pioneer whom the nations follow. We dream all night of those mountain-ridges in the horizon, though they may be of vapor only, which were last gilded by his rays. The island of Atlantis, and the islands and gardens of the Hesperides, a sort of terrestrial paradise, appear to have been the Great West of the ancients, enveloped in mystery and poetry. Who has not seen in imagination, when looking into the sunset sky, the gardens of the Hesperides, and the foundation of all those fables?

Columbus felt the westward tendency more strongly than any before. He obeyed it, and found a New World for Castile and Leon. The herd of men in those days scented fresh pastures from afar.

> "And now the sun had stretched out all the hills,
> And now was dropped into the western bay;
> At last *he* rose, and twitched his mantle blue;
> To-morrow to fresh woods and pastures new."

Where on the globe can there be found an area of equal extent with that occupied by the bulk of our States, so fertile and so rich and varied in its productions, and at the same time so habitable by the European, as this is? Michaux, who knew but part of them, says that "the species of large trees are much more numerous in North America than in Europe; in the United States there are more than one hundred and forty species that exceed thirty feet in height; in France there are but thirty that attain this size." Later botanists more than confirm his observations. Humboldt came to America to realize his youthful dreams of a tropical vegetation, and he beheld it in its greatest perfection in the primitive forests of the Amazon, the most gigantic wilderness on the earth, which he has so eloquently described. The geographer Guyot, himself a European, goes farther —farther than I am ready to follow him; yet not when he says: "As the plant is made for the animal, as the vegetable world is made for the animal world, America is made for the man of the Old World. . . . The man of the Old World sets out upon his way. Leaving the highlands of Asia, he descends from station to station towards Europe. Each of his steps is marked by a new civilization superior to the preceding, by a greater power of development. Arrived at the Atlantic, he pauses on the shore of this unknown ocean, the bounds of which he knows not, and turns upon his footprints for an instant." When he has exhausted the rich soil of Europe, and reinvigorated himself, "then recommences his adventurous career westward as in the earliest ages." So far Guyot.

From this western impulse coming in contact with the barrier of the Atlantic sprang the commerce and enterprise of modern times. The younger Michaux, in his "Travels West of the Alleghanies in 1802," says that the common inquiry in the newly settled West was, " 'From what part of the world have you come?' As if these vast and fertile regions would naturally be the place of meeting and common country of all the inhabitants of the globe."

To use an obsolete Latin word, I might say, *Ex Oriente lux; ex Occidente* FRUX. From the East light; from the West fruit.

Sir Francis Head, an English traveller and a Governor-General of Canada, tells us that "in both the northern and southern hemispheres of the New World, Nature has not only outlined her works on a larger scale, but has painted the whole picture with brighter and more costly colors than she used in delineating and in beautifying the Old World. . . . The heavens of America appear infinitely higher, the sky is bluer, the air is fresher, the cold is intenser, the moon looks larger, the stars are brighter, the thunder is louder, the lightning is vivider, the wind is stronger, the rain is heavier, the mountains are higher, the rivers longer, the forests bigger, the plains broader." This statement will do at least to set against Buffon's account of this part of the world and its productions.

Linnæus said long ago, "Nescio quæ facies *læta, glabra* plantis Americanis: I know not what there is of joyous and smooth in the aspect of American plants;" and I think that in this country there are no, or at most very few, *Africanæ bestiæ,* African beasts, as the Romans called them, and that in this respect also it is peculiarly fitted for the habitation of man. We are told that within three miles of the centre of the East Indian city of Singapore, some of the inhabitants are annually carried off by tigers; but the traveller can lie down in the woods at night almost anywhere in North America without fear of wild beasts.

These are encouraging testimonies. If the moon looks larger here than in Europe, probably the sun looks larger also. If the heavens of America appear infinitely higher, and the stars brighter, I trust that these facts are symbolical of the height to which the philosophy and poetry and religion of her inhabitants may one day soar. At length, perchance, the immaterial heaven will appear as much higher to the American mind, and the intimations that star it as much brighter. For I believe that climate does thus react on man—as there is something in the mountain air that feeds the spirit and inspires. Will not man grow to greater perfection intellectually as well as physically under these influences? Or is it unimportant how many foggy days there are in his life? I trust that we shall be more imaginative, that our thoughts will be clearer, fresher, and more ethereal, as our sky—our understanding more comprehensive and broader, like our plains—our intellect generally on a grander scale, like our thunder and lightning, our rivers and mountains and forests—and our hearts shall even correspond in breadth and depth and grandeur to our inland seas. Perchance there will appear to the traveller something, he knows not what, of *læta* and *glabra,* of joyous and serene, in our very faces. Else to what end does the world go on, and why was America discovered?

To Americans I hardly need to say—

"Westward the star of empire takes its way."

As a true patriot, I should be ashamed to think that Adam in paradise was more favourably situated on the whole than the backwoodsman in this country.

Our sympathies in Massachusetts are not confined to New England; though we may be estranged from the South, we sympathize with the West. There is the home of the younger sons, as among the Scandinavians they took to the sea for their inheritance. It is too late to be studying Hebrew; it is more important to understand even the slang of to-day.

Some months ago I went to see a panorama of the Rhine. It was like a dream of the Middle Ages. I floated down its historic stream in something more than imagination, under bridges built by the Romans, and repaired by later heroes, past cities and castles whose very names were music to my ears, and each of which was the subject of a legend. There were Ehrenbreitstein and Rolandseck and Coblentz, which I knew only in history. They were ruins that interested me chiefly. There seemed to come up from its waters and its vine-clad hills and valleys a hushed music as of Crusaders departing for the Holy Land. I floated along under the spell of enchantment, as if I had been transported to an heroic age, and breathed an atmosphere of chivalry.

Soon after, I went to see a panorama of the Mississippi, and as I worked my way up the river in the light of to-day and saw the steamboats wooding up, counted the rising cities, gazed on the fresh ruins of Nauvoo, beheld the Indians moving west across the stream, and, as before I had looked up the Moselle, now looked up the Ohio and the Missouri and heard the legends of Dubuque and of Wenona's Cliff,—still thinking more of the future than of the past or present, —I saw that this was a Rhine stream of a different kind; that the foundations of castles were yet to be laid, and the famous bridges were yet to be thrown over the river; and I felt that *this was the heroic age itself,* though we know it not, for the hero is commonly the simplest and obscurest of men.

The West of which I speak is but another name for the Wild; and what I have been preparing to say is, that in Wildness is the preservation of the World. Every tree sends its fibres forth in search of the Wild. The cities import it at any price. Men plough and sail for it. From the forest and wilderness come the tonics and barks which brace mankind. Our ancestors were savages. The story of Romulus and Remus being suckled by a wolf is not a meaningless fable. The founders of every state which has risen to eminence have drawn their nourishment and vigor from a similar wild source. It was because the children of the Empire were not suckled by the wolf that they were conquered and displaced by the children of the northern forests who were.

I believe in the forest, and in the meadow, and in the night in which the corn grows. We require an infusion of hemlock-spruce or arbor-vitæ in our tea. There is a difference between eating and drinking for strength and from mere gluttony. The Hottentots eagerly devour the marrow of the koodoo and other antelopes raw, as a matter of course. Some of our Northern Indians eat raw the marrow of the Arctic reindeer, as well as various other parts, including the summits of the antlers, as long as they are soft. And herein, perchance, they have stolen a march on the cooks of Paris. They get what usually goes to feed the fire. This is probably better than stall-fed beef and slaughter-house pork to make a man of. Give me a wildness whose glance no civilization can endure—as if we lived on the marrow of koodoos devoured raw.

There are some intervals which border the strain of the wood-thrush, to which I would migrate,—wild lands where no settler has squatted; to which, methinks, I am already acclimated.

The African hunter Cumming tells us that the skin of the eland, as well as that of most other antelopes just killed, emits the most delicious perfume of trees and grass. I would have every man so much like a wild antelope, so much a part and parcel of Nature, that his very person should thus sweetly advertise our senses of his presence, and remind us of those parts of Nature which he most haunts. I feel no disposition to be satirical, when the trapper's coat emits the odour of musquash even; it is a sweeter scent to me than that which commonly exhales from the merchant's or the scholar's garments. When I go into their wardrobes and handle their vestments, I am reminded of no grassy plains and flowery meads which they have frequented, but of dusty merchants' exchanges and libraries rather.

A tanned skin is something more than respectable, and perhaps olive is a fitter color than white for a man—a denizen of the woods. "The pale white man!" I do not wonder that the African pitied him. Darwin the naturalist says, "A white man bathing by the side of a Tahitian was like a plant bleached by the gardener's art, compared with a fine, dark green one, growing vigorously in the open fields."

Ben Jonson exclaims,—

"How near to good is what is fair!"

So I would say—

How near to good is what is *wild!*

Life consists with wildness. The most alive is the wildest. Not yet subdued to man, its presence refreshes him. One who pressed forward incessantly and never rested from his labors, who grew fast and made infinite demands on life, would always find himself in a new country or wilderness, and surrounded by the raw material of life. He would be climbing over the prostrate stems of primitive forest trees.

Hope and the future for me are not in lawns and cultivated fields, not in towns and cities, but in the impervious and quaking swamps. When, formerly, I have analyzed my partiality for some farm which I had contemplated purchasing, I have frequently found that I was attracted solely by a few square rods of impermeable and unfathomable bog—a natural sink in one corner of it. That was the jewel which dazzled me. I derive more of my subsistence from the swamps which surround my native town than from the cultivated gardens in the village. There are no richer parterres to my eyes than the dense beds of dwarf andromeda (*Cassandra calyculata*) which cover these tender places on the earth's surface. Botany cannot go farther than tell me the names of the shrubs which grow there—the high-blueberry, panicled andromeda, lamb-kill, azalea, and rhodora—all standing in the quaking sphagnum. I often think that I should like to have my house front on this mass of dull red bushes, omitting other flower plots and borders, transplanted spruce and trim box; even graveled walks—to have this fertile spot under my windows, not a few imported barrow-fulls of soil only to cover the sand which was thrown out in digging the cellar. Why not put my house, my parlour, behind this plot, instead of behind that meagre assemblage of curiosities, that poor apology for a Nature and Art, which I call my front-yard? It is an effort to clear up and make a decent appearance when the carpenter and mason have departed, though done as much for the passer-by as the dweller within. The most tasteful front-yard fence was never an agreeable object of study to me; the most elaborate ornaments, acorn-tops, or what not, soon wearied and disgusted me. Bring your sills up to the very edge of the swamp, then (though it may not be the best place for a dry cellar), so that there be no access on that side to citizens. Front-yards are not made to walk in, but, at most, through, and you could go in the back way.

Yes, though you may think me perverse, if it were proposed to me to dwell in the neighborhood of the most beautiful garden that ever human art contrived, or else of a Dismal Swamp, I should certainly decide for the swamp. How vain, then, have been all your labors, citizens, for me!

My spirits infallibly rise in proportion to the outward dreariness. Give me the ocean, the desert, or the wilderness! In the desert, pure air and solitude compensate for want of moisture and fertility. The traveller Burton says of it—"Your *morale* improves; you become frank and cordial, hospitable and single-minded. . . . In the desert, spirituous liquors excite only disgust. There is a keen enjoyment in a mere animal existence." They who have been travelling long on the steppes of Tartary say: "On reëntering cultivated lands, the agitation, perplexity, and turmoil of civilization oppressed and suffocated us; the air seemed to fail us, and we felt every moment as if about to die of asphyxia." When I would recreate myself, I seek the darkest wood, the thickest and most interminable and, to the citizen, most dismal swamp. I enter a swamp as a sacred place,— a *sanctum sanctorum*. There is the strength, the marrow of Nature. The wild-wood covers the virgin mould,—and the same soil is good for men and for trees. A man's health requires as many acres of meadow to his prospect as his farm does loads of muck. There are the strong meats on which he feeds. A town is saved, not more by the righteous men in it than by the woods and swamps that surround it. A township where one primitive forest waves above while another primitive forest rots below,—such a town is fitted to raise not only corn and potatoes, but poets and philosophers for the coming ages. In such a soil grew Homer and Confucius and the rest, and out of such a wilderness comes the Reformer eating locusts and wild honey.

To preserve wild animals implies generally the creation of a forest for them to dwell in or resort to. So it is with man. A hundred years ago they sold bark in our streets peeled from our own woods. In the very aspect of those primitive and rugged trees there was, methinks, a tanning principle which hardened and consolidated the fibres of men's thoughts. Ah! already I shudder for these comparatively degenerate days of my native village, when you cannot collect a load of bark of good thickness; and we no longer produce tar and turpentine.

The civilised nations—Greece, Rome, England—have been sustained by the primitive forests which anciently rotted where they stand. They survive as long as the soil is not exhausted. Alas for human culture! little is to be expected of a nation, when the vegetable mould is exhausted, and it is compelled to make manure of the bones of its fathers. There the poet sustains himself merely by his own superfluous fat, and the philosopher comes down on his marrowbones.

It is said to be the task of the American "to work the virgin soil," and that "agriculture here already assumes proportions unknown everywhere else." I think that the farmer displaces the Indian even because he redeems the meadow, and so makes himself stronger and in some respects more natural. I was surveying for a man the other day a single straight line one hundred and thirty-two rods long, through a swamp, at whose entrance might have been written the words which Dante read over the entrance to the infernal regions,—"Leave all hope, ye that enter,"—that is, of ever getting out again; where at one time I saw my employer actually up to his neck and swimming for his life in his property, though it was still winter. He had another similar swamp which I could not survey at all, because it was completely under water, and nevertheless, with regard to a third swamp, which I did *survey* from a distance, he remarked to me, true to his instincts, that he would not part with it for any consideration, on account of the mud which it contained. And that man intends to put a girdling ditch round the whole in the course of forty months, and so redeem it by the magic of his spade. I refer to him only as the type of a class.

The weapons with which we have gained our most important victories, which should be handed down as heirlooms from father to son, are not the sword and the lance, but the bush-whack, the turf-cutter, the spade, and the bog-hoe, rusted with the blood of many a meadow, and begrimed with the dust of many a hard-fought field. The very winds blew the Indian's corn-field into the meadow, and pointed out the way which he had not the skill to follow. He had no better implement with which to intrench himself in the land than a clam-shell. But the farmer is armed with plough and spade.

In Literature it is only the wild that attracts us. Dulness is but another name for tameness. It is the uncivilized free and wild thinking in "Hamlet" and the "Iliad," in all the Scriptures and Mythologies, not learned in the schools, that delights us. As the wild duck is more swift and beautiful than the tame, so is the wild—the mallard—thought, which 'mid falling dews wings its way above the fens. A truly good book is something as natural, and as unexpectedly and unaccountably fair and perfect, as a wild flower discovered on the prairies of the West or in the jungles of the East. Genius is a light which makes the darkness visible, like the lightning's flash, which perchance shatters the temple of knowledge itself,—and not a taper lighted at the hearthstone of the race, which pales before the light of common day.

English literature, from the days of the minstrels to the Lake Poets,—Chaucer and Spenser and Milton, and even Shakespeare, included—breathes no quite fresh and, in this sense, wild strain. It is an essentially tame and civilized literature, reflecting Greece and Rome. Her wilderness is a green wood, her wild man, a Robin Hood. There is plenty of genial love of Nature, but not so much of Nature herself. Her chronicles inform us when her wild animals, but not when the wild man in her, became extinct.

The science of Humboldt is one thing, poetry is another thing. The poet to-day, notwithstanding all the discoveries of science, and the accumulated learning of mankind, enjoys no advantage over Homer.

Where is the literature which gives expression to Nature? He would be a poet who could impress the winds and streams into his service, to speak for him; who nailed words to their primitive senses, as farmers drive down stakes in the spring, which the frost has heaved; who derived his words as often as he used them—transplanted them to his page with earth adhering to their roots; whose words were so true and fresh and natural that they would appear to expand like the buds at the approach of spring, though they lay half-smothered between two musty leaves in a library,—aye, to bloom and bear fruit there, after their kind, annually, for the faithful reader, in sympathy with surrounding Nature.

I do not know of any poetry to quote which adequately expresses this yearning for the Wild. Approached from this side, the best poetry is tame. I do not know where to find in any literature, ancient or modern, any account which contents me of that Nature with which even I am acquainted. You will perceive that I demand something which no Augustan nor Elizabethan age, which no *culture,* in short, can give. Mythology comes nearer to it than anything. How much more fertile a Nature, at least, has Grecian mythology its root in than English literature! Mythology is the crop which the Old World bore before its soil was exhausted, before the fancy and imagination were affected with blight; and which it still bears, wherever its pristine vigor is unabated. All other literatures endure only as the elms which overshadow our houses; but this is like the great dragon-tree of the Western Isles, as old as mankind, and, whether that does or not, will endure as long; for the decay of other literatures makes the soil in which it thrives.

The West is preparing to add its fables to those of the East. The valleys of the Ganges, the Nile, and the Rhine having yielded their crop, it remains to be seen what the valleys of the Amazon, the Platte, the Orinoco, the St. Lawrence, and the Mississippi will produce. Perchance, when, in the course of ages, American liberty has become a fiction of the past—as it is to some extent a fiction of the present—the poets of the world will be inspired by American mythology.

The wildest dreams of wild men, even, are not the less true, though they may not recommend themselves to the sense which is most common among Englishmen and Americans to-day. It is not every truth that recommends itself to the common sense. Nature has a place for the wild clematis as well as for the cabbage. Some expressions of truth are reminiscent—others merely *sensible,* as the phrase is,—others prophetic. Some forms of disease, even, may prophesy forms of health. The geologist has discovered that the figures of serpents, griffins, flying dragons, and other fanciful embellishments of heraldry, have their prototypes in the forms of fossil species which were extinct before man was created, and hence "indicate a faint and shadowy knowledge of a previous state of organic existence." The Hindoos dreamed that the earth rested on an elephant, and the elephant on a tortoise, and the tortoise on a serpent; and though it may be an unimportant coincidence, it will not be out of place here to state, that a fossil tortoise has lately been discovered in Asia large enough to support an elephant. I confess that I am partial to these wild fancies, which transcend the order of time and development. They are the sublimest recreation of the intellect. The partridge loves peas, but not those that go with her into the pot.

In short, all good things are wild and free. There is something in a strain of music, whether produced by an instrument or by the human voice,—take the sound of a bugle in a summer night, for instance,—which by its wildness, to speak without satire, reminds me of the cries emitted by wild beasts in their native forests. It is so much of their wildness as I can understand. Give me for my friends and neighbors wild men, not tame ones. The wildness of the savage is but a faint symbol of the awful ferity with which good men and lovers meet.

I love even to see the domestic animals reassert their native rights, —any evidence that they have not wholly lost their original wild habits and vigor; as when my neighbor's cow breaks out of her pasture early in the spring and boldly swims the river, a cold, grey tide, twenty-five or thirty rods wide, swollen by the melted snow. It is the buffalo crossing the Mississippi. This exploit confers some dignity on the herd in my eyes—already dignified. The seeds of instinct are preserved under the thick hides of cattle and horses, like seeds in the bowels of the earth, an indefinite period.

Any sportiveness in cattle is unexpected. I saw one day a herd of a dozen bullocks and cows running about and frisking in unwieldy sport, like huge rats, even like kittens. They shook their heads, raised their tails, and rushed up and down a hill, and I perceived by their horns, as well as by their activity, their relation to the deer tribe. But, alas! a sudden loud *Whoa!* would have damped their ardor at once, reduced them from venison to beef, and stiffened their sides and sinews like the locomotive. Who but the Evil One has cried, "Whoa!" to mankind? Indeed, the life of cattle, like that of many men, is but a sort of locomotiveness; they move a side at a time, and man, by his machinery, is meeting the horse and the ox half-way. Whatever part the whip has touched is thenceforth palsied. Who would ever think of a *side* of any of the supple cat tribe, as we speak of a *side* of beef?

I rejoice that horses and steers have to be broken before they can be made the slaves of men, and that men themselves have some wild oats still left to sow before they become submissive members of society. Undoubtedly, all men are not equally fit subjects for civilization; and because the majority, like dogs and sheep, are tame by inherited disposition, this is no reason why the others should have their natures broken that they may be reduced to the same level. Men are in the main alike, but they were made several in order that they might be various. If a low use is to be served, one man will do nearly or quite as well as another; if a high one, individual excellence is to be regarded. Any man can stop a hole to keep the wind away, but no other man could serve so rare a use as the author of this illustration did. Confucius says—"The skins of the tiger and the leopard, when they are tanned, are as the skins of the dog and the sheep tanned." But it is not the part of a true culture to tame tigers, any more than it is to make sheep ferocious; and tanning their skins for shoes is not the best use to which they can be put.

When looking over a list of men's names in a foreign language, as of military officers, or of authors who have written on a particular subject, I am reminded once more that there is nothing in a name. The name Menschikoff, for instance, has nothing in it to my ears more human than a whisker, and it may belong to a rat. As the names of the Poles and Russians are to us, so are ours to them. It is as if they had been named by the child's rigmarole—*Iery wiery ichery van, tittle-tol-tan.* I see in my mind a herd of wild creatures swarming over the earth, and to each the herdsman has affixed some barbarous sound in his own dialect. The names of men are of course as cheap and meaningless as *Bose* and *Tray,* the names of dogs.

Methinks it would be some advantage to philosophy, if men were named merely in the gross, as they are known. It would be necessary only to know the genus and perhaps the race or variety, to know the individual. We are not prepared to believe that every private soldier in a Roman army had a name of his own,—because we have not supposed that he had a character of his own.

At present our only true names are nicknames. I knew a boy who, from his peculiar energy, was called "Buster" by his playmates, and this rightly supplanted his Christian name. Some travellers tell us that an Indian had no name given him at first, but earned it, and his name was his fame: and among some tribes he acquired a new name with every new exploit. It is pitiful when a man bears a name for convenience merely, who has earned neither name nor fame.

I will not allow mere names to make distinctions for me, but still see men in herds for all them. A familiar name cannot make a man less strange to me. It may be given to a savage who retains in secret his own wild title earned in the woods. We have a wild savage in us, and a savage name is perchance somewhere recorded as ours. I see that my neighbor, who bears the familiar epithet William, or Edwin, takes it off with his jacket. It does not adhere to him when asleep or in anger, or aroused by any passion or inspiration. I seem to hear pronounced by some of his kin at such a time his original wild name in some jaw-breaking or else melodious tongue.

Here is this vast, savage, howling mother of ours, Nature, lying all around, with such beauty, and such affection for her children, as the leopard; and yet we are so early weaned from her breast to society, to that culture which is exclusively an interaction of man on man—a sort of breeding in and in, which produces at most a merely English nobility, a civilization destined to have a speedy limit.

In society, in the best institutions of men, it is easy to detect a certain precocity. When we should still be growing children, we are already little men. Give me a culture which imports much muck from the meadows, and deepens the soil—not that which trusts to heating manures, and improved implements and modes of culture only!

Many a poor sore-eyed student that I have heard of would grow faster, both intellectually and physically, if, instead of sitting up so very late, he honestly slumbered a fool's allowance.

There may be an excess even of informing light. Niépce, a Frenchman, discovered "actinism," that power in the sun's rays which produces a chemical effect; that granite rocks, and stone structures, and statues of metal, "are all alike destructively acted upon during the hours of sunshine, and, but for provisions of Nature no less wonderful, would soon perish under the delicate touch of the most subtile of the agencies of the universe." But he observed that "those bodies which underwent this change during the daylight possessed the power of restoring themselves to their original conditions during the hours of night, when this excitement was no longer influencing them." Hence it has been inferred that "the hours of darkness are as necessary to the inorganic creation as we know night and sleep are to the organic kingdom." Not even does the moon shine every night, but gives place to darkness.

I would not have every man nor every part of a man cultivated, any more than I would have every acre of earth cultivated: part will be tillage, but the greater part will be meadow and forest, not only serving an immediate use, but preparing a mould against a distant future, by the annual decay of the vegetation which it supports.

There are other letters for the child to learn than those which Cadmus invented. The Spaniards have a good term to express this wild and dusky knowledge—*Gramática parda,* tawny grammar—a kind of mother-wit derived from that same leopard to which I have referred.

We have heard of a Society for the Diffusion of Useful Knowledge. It is said that knowledge is power; and the like. Methinks there is equal need of a Society for the Diffusion of Useful Ignorance, what we will call Beautiful Knowledge, a knowledge useful in a higher sense: for what is most of our boasted so-called knowledge but a conceit that we know something, which robs us of the advantage of our actual ignorance? What we call knowledge is often our positive ignorance; ignorance our negative knowledge. By long years of patient industry and reading of the newspapers,—for what are the libraries of science but files of newspapers?—a man accumulates a myriad facts, lays them up in his memory, and then when in some spring of his life he saunters abroad into the Great Fields of thought, he, as it were, goes to grass like a horse and leaves all his harness behind in the stable. I would say to the Society for the Diffusion of Useful Knowledge, sometimes,—Go to grass. You have eaten hay long enough. The spring has come with its green crop. The very cows are driven to their country pastures before the end of May; though I have heard of one unnatural farmer who kept his cow in the barn and fed her on hay all the year round. So, frequently, the Society for the Diffusion of Useful Knowledge treats its cattle.

A man's ignorance some times is not only useful, but beautiful,—while his knowledge, so called, is oftentimes worse than useless, besides being ugly. Which is the best man to deal with,—he who knows nothing about a subject, and, what is extremely rare, knows that he knows nothing, or he who really knows something about it, but thinks that he knows all?

My desire for knowledge is intermittent; but my desire to bathe my head in atmospheres unknown to my feet is perennial and constant. The highest that we can attain to is not Knowledge, but Sympathy with Intelligence. I do not know that this higher knowledge amounts to anything more definite than a novel and grand surprise on a sudden revelation of the insufficiency of all that we called Knowledge before,—a discovery that there are more things in heaven and earth than are dreamed of in our philosophy. It is the lighting up of the mist by the sun. Man cannot *know* in any higher sense than this, any more than he can look serenely and with impunity in the face of the sun: Ὡς τὶ νοῶν, οὐ κεῖνον νοήσεις,—"You will not perceive that, as perceiving a particular thing," say the Chaldean Oracles.

There is something servile in the habit of seeking after a law which we may obey. We may study the laws of matter at and for our convenience, but a successful life knows no law. It is an unfortunate discovery certainly, that of a law which binds us where we did not know before that we were bound. Live free, child of the mist,—and with respect to knowledge we are all children of the mist. The man who takes the liberty to live is superior to all the laws, by virtue of his relation to the law-maker. "That is active duty," says the Vishnu Purana, "which is not for our bondage; that is knowledge which is for our liberation: all other duty is good only unto weariness; all other knowledge is only the cleverness of an artist."

It is remarkable how few events or crises there are in our histories; how little exercised we have been in our minds; how few experiences we have had. I would fain be assured that I am growing apace and rankly, though my very growth disturb this dull equanimity,—though it be with struggle through long, dark, muggy nights or seasons of gloom. It would be well, if all our lives were a divine tragedy even, instead of this trivial comedy or farce. Dante, Bunyan, and others appear to have been exercised in their minds more than we: they were subjected to a kind of culture such as our district schools and colleges do not contemplate. Even Mahomet, though many may scream at his name, had a good deal more to live for, aye, and to die for, than they have commonly.

When, at rare intervals, some thought visits one, as perchance he is walking on a railroad, then indeed the cars go by without his hearing them. But soon, by some inexorable law, our life goes by and the cars return.

"Gentle breeze, that wanderest unseen,
And bendest the thistles round Loira of storms,
Traveller of the windy glens,
Why hast thou left my ear so soon?"

While almost all men feel an attraction drawing them to society, few are attracted strongly to Nature. In their relation to Nature men appear to me for the most part, notwithstanding their arts, lower than the animals. It is not often a beautiful relation, as in the case of the animals. How little appreciation of the beauty of the landscape there is among us! We have to be told that the Greeks called the world Κόσμος, Beauty, or Order, but we do not see clearly why they did so, and we esteem it at best only a curious philological fact.

For my part, I feel that with regard to Nature I live a sort of border life, on the confines of a world into which I make occasional and transient forays only, and my patriotism and allegiance to the State into whose territories I seem to retreat are those of a moss-trooper. Unto a life which I call natural I would gladly follow even a will-o'-the-wisp through bogs and sloughs unimaginable, but no moon nor firefly has shown me the causeway to it. Nature is a personality so vast and universal that we have never seen one of her features. The walker in the familiar fields which stretch around my native town sometimes finds himself in another land than is described in their owners' deeds, as it were in some far-away field on the confines of the actual Concord, where her jurisdiction ceases, and the idea which the word Concord suggests ceases to be suggested. These farms which I have myself surveyed, these bounds which I have set up, appear dimly still as through a mist; but they have no chemistry to fix them; they fade from the surface of the glass; and the picture which the painter painted stands out dimly from beneath. The world with which we are commonly acquainted leaves no trace, and it will have no anniversary.

I took a walk on Spaulding's Farm the other afternoon. I saw the setting sun lighting up the opposite side of a stately pine wood. Its golden rays straggled into the aisles of the wood as into some noble hall. I was impressed as if some ancient and altogether admirable and shining family had settled there in that part of the land called Concord, unknown to me,—to whom the sun was servant,—who had not gone into society in the village,—who had not been called on. I saw their park, their pleasure-ground, beyond through the wood, in Spaulding's cranberry-meadow. The pines furnished them with gables as they grew. Their house was not obvious to vision; the trees grew through it. I do not know whether I heard the sounds of a suppressed hilarity or not. They seemed to recline on the sunbeams. They have sons and daughters. They are quite well. The farmer's cart-path, which leads directly through their hall, does not in the least put them out, as the muddy bottom of a pool is sometimes seen through the reflected skies. They never heard of Spaulding, and do not know that he is their neighbor,—notwithstanding I heard him whistle as he drove his team through the house. Nothing can equal the serenity of their lives. Their coat of arms is simply a lichen. I saw it painted on the pines and oaks. Their attics were in the tops of the trees. They are of no politics. There was no noise of labor. I did not perceive that they were weaving or spinning. Yet I did detect, when the wind lulled and hearing was done away, the finest imaginable sweet musical hum,—as of a distant hive in May, which perchance was the sound of their thinking. They had no idle thoughts, and no one without could see their work, for their industry was not as in knots and excrescences embayed.

But I find it difficult to remember them. They fade irrevocably out of my mind even now while I speak and endeavor to recall them and recollect myself. It is only after a long and serious effort to recollect my best thoughts that I become again aware of their cohabitancy. If it were not for such families as this, I think I should move out of Concord.

We are accustomed to say in New England that few and fewer pigeons visit us every year. Our forests furnish no mast for them. So, it would seem, few and fewer thoughts visit each growing man from year to year, for the grove in our minds is laid waste,—sold to feed unnecessary fires of ambition, or sent to mill, and there is scarcely a twig left for them to perch on. They no longer build nor breed with us. In some more genial season, perchance, a faint shadow flits across the landscape of the mind, cast by the *wings* of some thought in its vernal or autumnal migration, but, looking up, we are unable to detect the substance of the thought itself. Our winged thoughts are turned to poultry. They no longer soar, and they attain only to a Shanghai and Cochin-China grandeur. Those *gra-a-ate thoughts,* those *gra-a-ate men* you hear of!

We hug the earth,—how rarely we mount! Methinks we might elevate ourselves a little more. We might climb a tree, at least. I found my account in climbing a tree once. It was a tall white pine, on the top of a hill; and though I got well pitched, I was well paid for it, for I discovered new mountains in the horizon which I had never seen before,—so much more of the earth and the heavens. I might have walked about the foot of the tree for three-score years and ten, and yet I certainly should never have seen them. But, above all, I discovered around me,—it was near the end of June,—on the ends of the topmost branches only, a few minute and delicate red cone-like blossoms, the fertile flower of the white pine looking heavenward. I carried straightway to the village the topmost spire, and showed it to stranger jurymen who walked the streets,—for it was court-week,—and to farmers and lumber-dealers and wood-choppers and hunters, and not one had ever seen the like before, but wondered as at a star dropped down. Tell of ancient architects finishing their works on the tops of columns as perfectly as on the lower and more visible parts! Nature has from the first expanded the minute blossoms of the forest only toward the heavens, above men's heads and unobserved by them. We see only the flowers that are under our feet in the meadows. The pines have developed their delicate blossoms on the highest twigs of the wood every summer for ages, as well over the heads of Nature's red children as of her white ones; yet scarcely a farmer or hunter in the land has ever seen them.

Above all, we cannot afford not to live in the present. He is blessed over all mortals who loses no moment of the passing life in remembering the past. Unless our philosophy hears the cock crow in every barn-yard within our horizon, it is belated. That sound commonly reminds us that we are growing rusty and antique in our employments and habits of thought. His philosophy comes down to a more recent time than ours. There is something suggested by it that is a newer testament,—the gospel according to this moment. He has not fallen astern; he has got up early and kept up early, and to be where he is is to be in season, in the foremost rank of time. It is an expression of the health and soundness of Nature, a brag for all the world,—healthiness as of a spring burst forth, a new fountain of the Muses, to celebrate this last instant of time.

Where he lives no fugitive slave laws are passed. Who has not betrayed his master many times since last he heard that note?

The merit of this bird's strain is in its freedom from all plaintiveness. The singer can easily move us to tears or to laughter, but where is he who can excite in us a pure morning joy? When, in doleful dumps, breaking the awful stillness of our wooden sidewalk on a Sunday, or, perchance, a watcher in the house of mourning, I hear a cockerel crow far or near, I think to myself, "There is one of us well, at any rate,"—and with a sudden gush return to my senses.

We had a remarkable sunset one day last November. I was walking in a meadow, the source of a small brook, when the sun at last, just before setting, after a cold gray day, reached a clear stratum in the horizon, and the softest, brightest morning sunlight fell on the dry grass and on the stems of the trees in the opposite horizon and on the leaves of the shrub-oaks on the hillside, while our shadows stretched long over the meadow eastward, as if we were the only motes in its beams. It was such a light as we could not have imagined a moment before, and the air also was so warm and serene that nothing was wanting to make a paradise of that meadow. When we reflected that this was not a solitary phenomenon, never to happen again, but that it would happen forever and ever an infinite number of evenings, and cheer and reassure the latest child that walked there, it was more glorious still.

The sun sets on some retired meadow, where no house is visible, with all the glory and splendor that it lavishes on cities, and perchance as it has never set before,—where there is but a solitary marsh-hawk to have his wings gilded by it, or only a musquash looks out from his cabin, and there is some little black-veined brook in the midst of the marsh, just beginning to meander, winding slowly round a decaying stump. We walked in so pure and bright a light, gilding the withered grass and leaves, so softly and serenely bright, I thought I had never bathed in such golden flood, without a ripple or a murmur to it. The west side of every wood and rising ground gleamed like the boundary of Elysium, and the sun on our backs seemed like a gentle herdsman driving us home at evening.

So we saunter toward the Holy Land, till one day the sun shall shine more brightly than ever he has done, shall perchance shine into our minds and hearts, and light up our whole lives with a great awakening light, as warm and serene and golden as on a bankside in autumn.

7

Henry Adams, MONT-SAINT-MICHEL AND CHARTRES
THE EDUCATION OF HENRY ADAMS
"PRAYER TO THE VIRGIN OF CHARTRES'

1. Why did Adams believe the Gothic cathedrals were important to twentieth-century America?
2. Why did the people in twelth-century France spend such huge amounts of money on cathedrals like the one at Chartres?
3. What did the virgin demand from the people?
4. In our society of today, what might fill the role or take the place of the Virgin of Chartres?
5. Why did Adams believe the dynamo represented his age?
6. Why did the people of early twentieth-century United States spend so much money on dynamos?
7. What did the dynamo demand from the people?
8. What do the Virgin and the Dynamo represent to their respective societies?
9. What does Adams mean by the "Abyss of Ignorance"?
10. Adams refers to his age as the Age of Multiplicity and to the twelth century as the Age of Unity. Why does he do this? How does the Virgin create unity and the Dynamo, multiplicity?
11. Do you think Adams successfully applies the laws of physics to human history? Why or why not?
12. In which image did Adams see salvation, the Virgin or the Dynamo?
13. What does Adams fear about the Dynamo?

14. Does it make any sense to pray to the Dynamo?

15. Why was the Virgin Mary the Dynamo of the Middle Ages?

16. What was the difference between the power of the Virgin and the power of the Dynamo?

17. Was the Dynamo moral?

18. How does Adams see the future of the world?

19. Does the poem make you sad, happy, or what emotion?

Henry Brooks Adams (1838-1918) was a grandson of President John Quincy Adams and a great-grandson of President John Adams. A philosopher-historian who taught for a time at Harvard, Adams was particularly impressed with the rise of science and technology. Living under the burden of being an Adams, Henry's life was filled with personal tragedy and a constant feeling that he was not living up to the family name.

Adams was tremendously impressed with the magnificent beauty and overpowering force of the great Gothic cathedrals of Europe, particularly the Cathedral of Chartres. In this excerpt from the book, *Mont-Saint-Michel and Chartres (1905),* Adams attempts to explain the power of the "Virgin of Chartres".

The Education of Henry Adams was written in 1906 as a sequel to *Mont-Saint-Michel and Chartres*. It was originally distributed only to Henry Adams' friends for comment. In the excerpt reproduced below, Adams concludes that the power of his age was represented by the great dynamos which powered American industry. Adams goes on to apply the laws of physics to the movement of human history.

Adams puts the image of the Virgin together with the image of the Dynamo in these poems which are two prayers, one to the Virgin of Chartres and one to the modern Dynamo, with a postscript prayer to the Virgin once again.

CHAPTER VI

THE VIRGIN OF CHARTRES

WE must take ten minutes to accustom our eyes to the light, and we had better use them to seek the reason why we come to Chartres rather than to Rheims or Amiens or Bourges, for the cathedral that fills our ideal. The truth is, there are several reasons; there generally are, for doing the things we like; and after you have studied Chartres to the ground, and got your reasons settled, you will never find an antiquarian to agree with you; the architects will probably listen to you with contempt; and even these excellent priests, whose kindness is great, whose patience is heavenly, and whose good opinion you would so gladly gain, will turn from you with pain, if not with horror. The Gothic is singular in this; one seems easily at home in the Renaissance; one is not too strange in the Byzantine; as for the Roman, it is ourselves; and we could walk blindfolded through every chink and cranny of the Greek mind; all these styles seem modern, when we come close to them; but the Gothic gets away. No two men think alike about it, and no woman agrees with either man. The Church itself never agreed about it, and the architects agree even less than the priests. To most minds it casts too many shadows; it wraps itself in mystery; and when people talk of mystery, they commonly mean fear. To others, the Gothic seems hoary with age and decrepitude, and its shadows mean death. What is curious to watch is the fanatical conviction of the Gothic enthusiast, to whom the twelfth century means exuberant youth, the eternal child of Wordsworth, over whom its immortality broods like the day; it is so simple and yet so complicated; it sees so much and so little; it loves so many toys and cares for so few necessities; its youth is so young, its age so old, and its youthful yearning for old thought is so disconcerting, like the mysterious senility of the baby that —

> Deaf and silent, reads the eternal deep,
> Haunted forever by the eternal mind.

One need not take it more seriously than one takes the baby itself. Our amusement is to play with it, and to catch its meaning in its smile; and whatever Chartres may be now, when young it was a smile. To the Church, no doubt, its cathedral here has a fixed and administrative meaning, which is the same as that of every other bishop's seat and with which we have nothing whatever to do. To us, it is a child's fancy; a toy-house to please the Queen of Heaven, — to please her so much that she would be happy in it, — to charm her till she smiled.

The Queen Mother was as majestic as you like; she was absolute; she could be stern; she was not above being angry; but she was still a woman, who loved grace, beauty, ornament, — her toilette, robes, jewels; — who considered the arrangements of her palace with attention, and liked both light and colour; who kept a keen eye on her Court, and exacted prompt and willing obedience from king and archbishops as well as from beggars and drunken priests. She protected her friends and punished her enemies. She required space, beyond what was known in the Courts of kings, because she was liable at all times to have ten thousand people begging her for favours — mostly inconsistent with law — and deaf to refusal. She was extremely sensitive to neglect, to disagreeable impressions, to want of intelligence in her surroundings. She was the greatest artist, as she was the greatest philosopher and musician and theologist, that ever lived on earth, except her Son, Who, at Chartres, is still an Infant under her guardianship. Her taste was infallible; her sentence eternally final. This church was built for her in this spirit of simple-minded, practical, utilitarian faith, — in this singleness of thought, exactly as a little girl sets up a doll-house for her favourite blonde doll. Unless you can go back to your dolls, you are out of place here. If you can go back to them, and get rid for one small hour of the weight of custom, you shall see Chartres in glory.

The palaces of earthly queens were hovels compared with these palaces of the Queen of Heaven at Chartres, Paris, Laon, Noyon, Rheims, Amiens, Rouen, Bayeux, Coutances, — a list that might be stretched into a volume. The nearest approach we have made to a palace was the Merveille at Mont-Saint-Michel, but no Queen had a palace equal to that. The Merveille was built, or designed, about the year 1200; toward the year 1500, Louis XI built a great castle at Loches in Touraine, and there Queen Anne de Bretagne had apartments which still exist, and which we will visit. At Blois you shall see the residence which served for Catherine de Medicis till her death in 1589. Anne de Bretagne was trebly queen, and Catherine de Medicis took her standard of comfort from the luxury of Florence. At Versailles you can see the apartments which the queens of the Bourbon line occupied through their century of magnificence. All put together, and then trebled in importance, could not rival the splendour of any single cathedral dedicated to Queen Mary in the thirteenth century; and of them all, Chartres was built to be peculiarly and exceptionally her delight.

One has grown so used to this sort of loose comparison, this reckless waste of words, that one no longer adopts an idea unless it is driven in with hammers of statistics and columns of figures. With the irritating demand for literal exactness and perfectly straight lines which lights up every truly American eye, you will certainly ask when this exaltation of Mary began, and unless you get the dates, you will doubt the facts. It is your own fault if they are tiresome; you might easily read them all in the "Iconographie de la Sainte Vierge," by M. Rohault de Fleury, published in 1878. You can start at Byzantium with the Empress Helena in 326, or with the Council of Ephesus in 431. You will find the Virgin acting as the patron saint of Constantinople and of the Imperial residence, under as many names as Artemis or

Aphrodite had borne. As Godmother (Θεομητηρ), Deipara (Θεοτοκος), Pathfinder (Ὁδηγητρια), she was the chief favourite of the Eastern Empire, and her picture was carried at the head of every procession and hung on the wall of every hut and hovel, as it is still wherever the Greek Church goes. In the year 610, when Heraclius sailed from Carthage to dethrone Phocas at Constantinople, his ships carried the image of the Virgin at their mastheads. In 1143, just before the flèche on the Chartres clocher was begun, the Basileus John Comnenus died, and so devoted was he to the Virgin that, on a triumphal entry into Constantinople, he put the image of the Mother of God in his chariot, while he himself walked. In the Western Church the Virgin had always been highly honoured, but it was not until the crusades that she began to overshadow the Trinity itself. Then her miracles became more frequent and her shrines more frequented, so that Chartres, soon after 1100, was rich enough to build its western portal with Byzantine splendour. A proof of the new outburst can be read in the story of Citeaux. For us, Citeaux means Saint Bernard, who joined the Order in 1112, and in 1115 founded his Abbey of Clairvaux in the territory of Troyes. In him, the religious emotion of the half-century between the first and second crusades (1095–1145) centred as in no one else. He was a French precursor of Saint Francis of Assisi who lived a century later. If we were to plunge into the story of Citeaux and Saint Bernard we should never escape, for Saint Bernard incarnates what we are trying to understand, and his mind is further from us than the architecture. You would lose hold of everything actual, if you could comprehend in its contradictions the strange mixture of passion and caution, the austerity, the self-abandonment, the vehemence, the restraint, the love, the hate, the miracles, and the scepticism of Saint Bernard. The Cistercian Order, which was founded in 1098, from the first put all its churches under the special protection of the Virgin, and Saint Bernard in his time was regarded as the apple of the Virgin's eye. Tradition as old as the twelfth century, which long afterwards gave to Murillo the subject of a famous painting, told that once, when he was reciting before her statue the "Ave Maris Stella," and came to the words, "Monstra te esse Matrem," the image, pressing its breast, dropped on the lips of her servant three drops of the milk which had nourished the Saviour. The same miracle, in various forms, was told of many other persons, both saints and sinners; but it made so much impression on the mind of the age that, in the fourteenth century, Dante, seeking in Paradise for some official introduction to the foot of the Throne, found no intercessor with the Queen of Heaven more potent than Saint Bernard. You can still read Bernard's hymns to the Virgin, and even his sermons, if you like. To him she was the great mediator. In the eyes of a culpable humanity, Christ was too sublime, too terrible, too just, but not even the weakest human frailty could fear to approach his Mother. Her attribute was humility; her love and pity were infinite. "Let him deny your mercy who can say that he has ever asked it in vain."

Saint Bernard was emotional and to a certain degree mystical, like Adam de Saint-Victor, whose hymns were equally famous, but the emotional saints and mystical poets were not by any means allowed to establish exclusive rights to the Virgin's favour. Abélard was as devoted as they were, and wrote hymns as well. Philosophy claimed her, and Albert the Great, the head of scholasticism, the teacher of Thomas Aquinas, decided in her favour the question: "Whether the Blessed Virgin possessed perfectly the seven liberal arts." The Church at Chartres had decided it a hundred years before by putting the seven liberal arts next her throne, with Aristotle himself to witness; but Albertus gave the reason: "I hold that she did, for it is written, 'Wisdom has built herself a house, and has sculptured seven columns.' That house is the blessed Virgin; the seven columns are the seven liberal arts. Mary, therefore, had perfect mastery of science." Naturally she had also perfect mastery of economics, and most of her great churches were built in economic centres. The guilds were, if possible, more devoted to her than the monks; the bourgeoisie of Paris, Rouen. Amiens, Laon, spend money by millions to gain her favour. Most surprising of all, the great military class was perhaps the most vociferous. Of all inappropriate haunts for the gentle, courteous, pitying Mary, a field of battle seems to be the worst, if not distinctly blasphemous; yet the greatest French warriors insisted on her leading them into battle, and in the actual mêlée when men were killing each other, on every battle-field in Europe, for at least five hundred years, Mary was present, leading both sides. The battle-cry of the famous Constable du Guesclin was "Notre-Dame-Guesclin"; "Notre-Dame-Coucy" was the cry of the great Sires de Coucy; "Notre-Dame-Auxerre"; "Notre-Dame-Sancerre"; Notre-Dame-Hainault"; Notre-Dame-Gueldres"; "Notre-Dame-Bourbon"; "Notre-Dame-Bearn"; — all well-known battle-cries. The King's own battle at one time cried, "Notre-Dame-Saint-Denis-Montjoie"; the Dukes of Burgundy cried, "Notre-Dame-Bourgogne"; and even the soldiers of the Pope were said to cry, "Notre-Dame-Saint-Pierre."

The measure of this devotion, which proves to any religious American mind, beyond possible cavil, its serious and practical reality, is the money it cost. According to statistics, in the single century between 1170 and 1270, the French built eighty cathedrals and nearly five hundred churches of the cathedral class, which would have cost, according to an estimate made in 1840, more than five thousand millions to replace. Five thousand million francs is a thousand million dollars, and this covered only the great churches of a single century. The same scale of expenditure had been going on since the year 1000, and almost every parish in France had rebuilt its church in stone; to this day France is strewn with the ruins of this architecture, and yet the still preserved churches of the eleventh and twelfth centuries, among the churches that belong to the Romanesque and Transition period, are numbered by hundreds until they reach well into the thousands. The share of this capital which was — if one may use a commercial figure — invested in the Virgin cannot be fixed, any more than the total sum given to religious objects between 1000 and 1300; but in a spiritual and artistic sense, it was almost the whole, and expressed an intensity of conviction never again reached by any passion, whether of religion, of loyalty, of patriotism, or of wealth; perhaps never even parallelled by any single economic effort, except in war. Nearly every great church of the twelfth and thirteenth centuries belonged to Mary, until in France one asks for the church of Notre Dame as though it meant cathedral; but, not satisfied with this, she contracted the habit of requiring in all churches a chapel of her own, called in English the "Lady Chapel," which was apt to be as large as the church but was always meant to be handsomer; and there, behind the high altar, in her own private apartment, Mary sat, receiving her innumerable suppliants, and ready at any moment to step up upon the high altar itself to support the tottering authority of the local saint.

Expenditure like this rests invariably on an economic idea. Just as the French of the nineteenth century invested their surplus capital in a railway system in the belief that they would make money by it in this life, in the thirteenth they trusted their money to the Queen of Heaven because of their belief in her power to repay it with interest in the life to come. The investment was based on the power of Mary as Queen rather than on any orthodox Church conception of the Virgin's legitimate station. Papal Rome never greatly loved Byzantine empresses or French queens. The Virgin of Chartres was never wholly sympathetic to the Roman Curia. To this day the Church writers — like the Abbé Bulteau or M. Rohault de Fleury — are singularly shy of the true Virgin of majesty, whether at Chartres or at Byzantium or wherever she is seen. The fathers Martin and Cahier at Bourges alone felt her true value. Had the Church controlled her, the Virgin would perhaps have remained prostrate at the foot of the Cross. Dragged by a Byzantine Court, backed by popular insistence and impelled by overpowering self-interest, the Church accepted the Virgin throned and crowned, seated by Christ, the Judge throned and crowned; but even this did not wholly satisfy the French of the thirteenth century who seemed bent on absorbing Christ in His Mother, and making the Mother the Church, and Christ the Symbol.

The Church had crowned and enthroned her almost from the beginning, and could not have dethroned her if it would. In all Christian art — sculpture or mosaic, painting or poetry — the Virgin's rank was expressly asserted. Saint Bernard, like John Comnenus, and probably at the same time (1120–40), chanted hymns to the Virgin as Queen: —

O salutaris Virgo Stella Maris
Generans prolem, Æquitatis solem,
Lucis auctorem, Retinens pudorem,
Suscipe laudem!

Celi Regina Per quam medicina
Datur ægrotis, Gratia devotis,
Gaudium mœstis, Mundo lux cœlestis,
Spesque salutis;

Aula regalis, Virgo specialis,
Posce medelam Nobis et tutelam,
Suscipe vota, Precibusque cuncta
Pelle molesta!

O saviour Virgin, Star of Sea,
Who bore for child the Son of Justice,
The source of Light, Virgin always
Hear our praise!

Queen of Heaven who have given
Medicine to the sick, Grace to the devout,
Joy to the sad, Heaven's light to the world
And hope of salvation;

Court royal, Virgin typical,
Grant us cure and guard,
Accept our vows, and by prayers
Drive all griefs away!

As the lyrical poet of the twelfth century, Adam de Saint-Victor seems to have held rank higher if possible than that of Saint Bernard, and his hymns on the Virgin are certainly quite as emphatic an assertion of her majesty: —

Imperatrix supernorum!
Superatrix infernorum!
Eligenda via cœli,
Retinenda spe fideli,
Separatos a te longe
Revocatos ad te junge
Tuorum collegio!

Empress of the highest,
Mistress over the lowest,
Chosen path of Heaven,
Held fast by faithful hope,
Those separated from you far,
Recalled to you, unite
In your fold!

To delight in the childish jingle of the mediæval Latin is a sign of a futile mind, no doubt, and I beg pardon of you and of the Church for wasting your precious summer day on poetry which was regarded as mystical in its age and which now sounds like a nursery rhyme; but a verse or two of Adam's hymn on the Assumption of the Virgin completes the record of her rank, and goes to complete also the documentary proof of her majesty at Chartres: —

Salve, Mater Salvatoris!
Vas electum! Vas honoris!
  Vas cœlestis Gratiæ!
Ab æterno Vas provisum!
Vas insigne! Vas excisum
  Manu sapientiæ!

Mother of our Saviour, hail!
Chosen vessel! Sacred Grail!
  Font of celestial grace!
From eternity forethought!
By the hand of Wisdom wrought!
  Precious, faultless Vase!

Salve, Mater pietatis,
Et totius Trinitatis
  Nobile Triclinium!
Verbi tamen incarnati
Speciale majestati
  Præparans hospitium!

Hail, Mother of Divinity!
Hail, Temple of the Trinity!
  Home of the Triune God!
In whom the Incarnate Word had birth,
The King! to whom you gave on earth
  Imperial abode.

O Maria! Stella maris!
Dignitate singularis,
Super omnes ordinaris
  Ordines cœlestium!
In supremo sita poli
Nos commenda tuæ proli,
Ne terrores sive doli
  Nos supplantent hostium!

Oh, Maria! Constellation!
Inspiration! Elevation!
Rule and Law and Ordination
  Of the angels' host!
Highest height of God's Creation,
Pray your Son's commiseration,
Lest, by fear or fraud, salvation
  For our souls be lost!

Constantly — one might better say at once, officially, she was addressed in these terms of supreme majesty: "Imperatrix supernorum!" "Cœli Regina!" "Aula regalis!" but the twelfth century seemed determined to carry the idea out to its logical conclusion in defiance of dogma. Not only was the Son absorbed in the Mother, or represented as under her guardianship, but the Father fared no better, and the Holy Ghost followed. The poets regarded the Virgin as the "Templum Trinitatis"; "totius Trinitatis nobile Triclinium." She was the refectory of the Trinity — the "Triclinium" — because the refectory was the largest room and contained the whole of the members, and was divided in three parts by two rows of columns. She was the "Templum Trinitatis," the Church itself, with its triple aisle. The Trinity was absorbed in her.

This is a delicate subject in the Church, and you must feel it with delicacy, without brutally insisting on its necessary contradictions. All theology and all philosophy are full of contradictions quite as flagrant and far less sympathetic. This particular variety of religious faith is simply human, and has made its appearance in one form or another in nearly all religions; but though the twelfth century carried it to an extreme, and at Chartres you see it in its most charming expression, we have got always to make allowances for what was going on beneath the surface in men's minds, consciously or unconsciously, and for the latent scepticism which lurks behind all faith. The Church itself never quite accepted the full claims of what was called Mariolatry. One may be sure, too, that the bourgeois capitalist and the student of the schools, each from his own point of view, watched the Virgin with anxious interest. The bourgeois had put an enormous share of his capital into what was in fact an economical speculation, not unlike the South Sea Scheme, or the railway system of our own time; except that in one case the energy was devoted to shortening the road to Heaven; in the other, to shortening the road to Paris; but no serious schoolman could have felt entirely convinced that God would enter into a business partnership with man, to establish a sort of joint-stock society for altering the operation of divine and universal laws. The bourgeois cared little for the philosophical doubt if the economical result proved to be good, but he watched this result with his usual practical sagacity, and required an experience of only about three generations (1200–1300) to satisfy himself that relics were not certain in their effects; that the Saints were not always able or willing to help; that Mary herself could not certainly be bought or bribed; that prayer without money seemed to be quite as efficacious as prayer with money; and that neither the road to Heaven nor Heaven itself had been made surer or brought nearer by an investment of capital which amounted to the best part of the wealth of France. Economically speaking, he became satisfied that his enormous money-investment had proved to be an almost total loss, and the reaction on his mind was as violent as the emotion. For three hundred years it prostrated France. The efforts of the bourgeoisie and the peasantry to recover their property, so far as it was recoverable, have lasted to the present day and we had best take care not to get mixed in those passions.

If you are to get the full enjoyment of Chartres, you must, for the time, believe in Mary as Bernard and Adam did, and feel her presence as the architects did, in every stone they placed, and every touch they chiselled. You must try first to rid your mind of the traditional idea that the Gothic is an intentional expression of religious gloom. The necessity for light was the motive of the Gothic architects. They needed light and always more light, until they sacrificed safety and common sense in trying to get it. They converted their walls into windows, raised their vaults, diminished their piers, until their churches could no longer stand. You will see the limits at Beauvais; at Chartres we have not got so far, but even here, in places where the Virgin wanted it, — as above the high altar, — the architect has taken all the light there was to take. For the same reason, fenestration became the most important part of the Gothic architect's work, and at Chartres was uncommonly interesting because the architect was obliged to design a new system, which should at the same time satisfy the laws of construction and the taste and imagination of Mary. No doubt the first command of the Queen of Heaven was for light, but the second, at least equally imperative, was for colour. Any earthly queen, even though she were not Byzantine in taste, loved colour; and the truest of queens — the only true Queen of Queens — had richer and finer taste in colour than the queens of fifty earthly kingdoms, as you will see when we come to the immense effort to gratify her in the glass of her windows. Illusion for illusion, — granting for the moment that Mary was an illusion, — the Virgin Mother in this instance repaid to her worshippers a larger return for their money than the capitalist has ever been able to get, at least in this world, from any other illusion of wealth which he has tried to make a source of pleasure and profit.

The next point on which Mary evidently insisted was the arrangement for her private apartments, the apse, as distinguished from her throne-room, the choir; both being quite distinct from the hall, or reception-room of the public, which was the nave with its enlargements in the transepts. This arrangement marks the distinction between churches built as shrines for the deity and churches built as halls of worship for the public. The difference is chiefly in the apse, and the apse of Chartres is the most interesting of all apses from this point of view.

The Virgin required chiefly these three things, or, if you like, these four: space, light, convenience; and colour decoration to unite and harmonize the whole. This concerns the interior; on the exterior she required statuary, and the only complete system of decorative sculpture that existed seems to belong to her churches: — Paris, Rheims, Amiens, and Chartres. Mary required all this magnificence at Chartres for herself alone, not for the public. As far as one can see into the spirit of the builders, Chartres was exclusively intended for the Virgin, as the Temple of Abydos was intended for Osiris. The wants of man, beyond a mere roof-cover, and perhaps space to some degree, enter to no very great extent into the problem of Chartres. Man came to render homage or to ask favours. The Queen received him in her palace, where she alone was at home, and alone gave commands.

The artist's second thought was to exclude from his work everything that could displease Mary; and since Mary differed from living queens only in infinitely greater majesty and refinement, the artist could admit only what pleased the actual taste of the great ladies who dictated taste at the Courts of France and England, which surrounded the little Court of the Counts of Chartres. What they were — these women of the twelfth and thirteenth centuries — we shall have to see or seek in other directions; but Chartres is perhaps the most magnificent and permanent monument they left of their taste, and we can begin here with learning certain things which they were not.

In the first place, they were not in the least vague, dreamy, or mystical in a modern sense; — far from it! They seemed anxious only to throw the mysteries into a blaze of light; not so much physical, perhaps, — since they, like all women, liked moderate shadow for their toilettes, — but luminous in the sense of faith. There is nothing about Chartres that you would think mystical, who know your Lohengrin, Siegfried, and Parsifal. If you care to make a study of the whole literature of the subject, read M. Mâle's "Art Religieux du XIIIe Siècle en France," and use it for a guide-book. Here you need only note how symbolic and how simple the sculpture is, on the portals and porches. Even what seems a grotesque or an abstract idea is no more than the simplest child's personification. On the walls you may have noticed the *Ane qui vielle*, — the ass playing the lyre; and on all the old churches you can see "bestiaries," as they were called, of fabulous animals, symbolic or not; but the symbolism is as simple as the realism of the oxen at Laon. It gave play to the artist in his effort for variety of decoration, and it amused the people, — probably the Virgin also was not above being amused; — now and then it seems about to suggest what you would call an esoteric meaning, that is to say, a meaning which each one of us can consider private property reserved for our own amusement, and from which the public is excluded; yet, in truth, in the Virgin's churches the public is never excluded, but invited. The Virgin even had the additional charm to the public that she was popularly supposed to have no very marked fancy for priests as such; she was a queen, a woman, and a mother, functions, all, which priests could not perform. Accordingly, she seems to have had little taste for mysteries of any sort, and even the symbols that seem most mysterious were clear to every old peasant-woman in her church. The most pleasing and promising of them all is the woman's figure you saw on the front of the cathedral in Paris; her eyes bandaged; her head bent down; her crown falling; without cloak or royal robe; holding in her hand a guidon or banner with its staff broken in more than one place. On the opposite pier stands another woman, with royal mantle, erect and commanding. The symbol is so graceful that one is quite eager to know its meaning; but every child in the Middle Ages would have instantly told you that the woman with the falling crown meant only the Jewish Synagogue, as the one with the royal robe meant the Church of Christ.

Another matter for which the female taste seemed not much to care was theology in the metaphysical sense. Mary troubled herself little about theology except when she retired into the south transept with Pierre de Dreux. Even there one finds little said about the Trinity, always the most metaphysical subtlety of the Church. Indeed, you might find much amusement here in searching the cathedral for any distinct expression at all of the Trinity as a dogma recognized by Mary. One cannot take seriously the idea that the three doors, the three portals, and the three aisles express the Trinity, because, in the first place, there was no rule about it; churches might have what portals and aisles they pleased; both Paris and Bourges have five; the doors themselves are not allotted to the three members of the Trinity, nor are the portals; while another more serious objection is that the side doors and aisles are not of equal importance with the central, but mere adjuncts and dependencies, so that the architect who had misled the ignorant public into accepting so black a heresy would have deserved the stake, and would probably have gone to it. Even this suggestion of trinity is wanting in the transepts, which have only one aisle, and in the choir, which has five, as well as five or seven chapels, and, as far as an ignorant mind can penetrate, no triplets whatever. Occasionally, no doubt, you will discover in some sculpture or window, a symbol of the Trinity, but this discovery itself amounts to an admission of its absence as a controlling idea, for the ordinary worshipper must have been at least as blind as we are, and to him, as to us, it would have seemed a wholly subordinate detail. Even if the Trinity, too, is anywhere expressed, you will hardly find here an attempt to explain its metaphysical meaning — not even a mystic triangle.

The church is wholly given up to the Mother and the Son. The Father seldom appears; the Holy Ghost still more rarely. At least, this is the impression made on an ordinary visitor who has no motive to be orthodox; and it must have been the same with the thirteenth-century worshipper who came here with his mind absorbed in the perfections of Mary. Chartres represents, not the Trinity, but the identity of the Mother and Son. The Son represents the Trinity, which is thus absorbed in the Mother. The idea is not orthodox, but this is no affair of ours. The Church watches over its own.

The Virgin's wants and tastes, positive and negative, ought now to be clear enough to enable you to feel the artist's sincerity in trying to satisfy them; but first you have still to convince yourselves of the people's sincerity in employing the artists. This point is the easiest of all, for the evidence is express. In the year 1145 when the old flèche was begun, — the year before Saint Bernard preached the second crusade at Vézelay, — Abbot Haimon, of Saint-Pierre-sur-Dives in Normandy, wrote to the monks of Tutbury Abbey in England a famous letter to tell of the great work which the Virgin was doing in France and which began at the Church of Chartres. "Hujus sacræ institutionis ritus apud Carnotensem ecclesiam est inchoatus." From Chartres it had spread through Normandy, where it produced among other things the beautiful spire which we saw at Saint-Pierre-sur-Dives. "Postremo per totam fere Normanniam longe lateque convaluit ac loca per singula Matri misericordiæ dicata præcipue occupavit." The movement affected especially the places devoted to Mary, but ran through all Normandy, far and wide. Of all Mary's miracles, the best attested, next to the preservation of her church, is the building of it; not so much because it surprises us as because it surprised even more the people of the time and the men who were its instruments. Such deep popular movements are always surprising, and at Chartres the miracle seems to have occurred three times, coinciding more or less with the dates of the crusades, and taking the organization of a crusade, as Archibishop Hugo of Rouen described it in a letter to Bishop Thierry of Amiens. The most interesting part of this letter is the evident astonishment of the writer, who might be talking to us to-day, so modern is he: —

> The inhabitants of Chartres have combined to aid in the construction of their church by transporting the materials; our Lord has rewarded their humble zeal by miracles which have roused the Normans to imitate the piety of their neighbours. . . . Since then the faithful of our diocese and of other neighbouring regions have formed associations for the same object; they admit no one into their company unless he has been to confession, has renounced enmities and revenges, and has reconciled himself with his enemies. That done, they elect a chief, under whose direction they conduct their waggons in silence and with humility.

The quarries at Berchères-l'Evêque are about five miles from Chartres. The stone is excessively hard, and was cut in blocks of considerable size, as you can see for yourselves; blocks which required great effort to transport and lay in place. The work was done with feverish rapidity, as it still shows, but it is the solidist building of the age, and without a sign of weakness yet. The Abbot told, with more surprise than pride, of the spirit which was built into the cathedral with the stone: —

> Who has ever seen! — Who has ever heard tell, in times past, that powerful princes of the world, that men brought up in honour and in wealth, that nobles, men and women, have bent their proud and haughty necks to the harness of carts, and that, like beasts of burden, they have dragged to the abode of Christ these waggons, loaded with wines, grains, oil, stone, wood, and all that is necessary for the wants of life, or for the construction of the church? But while they draw these burdens, there is one thing admirable to observe; it is that often when a thousand persons and more are attached to the chariots, — so great is the difficulty, — yet they march in such silence that not a murmur is heard, and truly if one did not see the thing with one's eyes, one might believe that among such a multitude there was hardly a person present. When they halt on the road, nothing is heard but the confession of sins, and pure and suppliant prayer to God to obtain pardon. At the voice of the priests who exhort their hearts to peace, they forget all hatred, discord is thrown far aside, debts are remitted, the unity of hearts is established.

But if any one is so far advanced in evil as to be unwilling to pardon an offender, or if he rejects the counsel of the priest who has piously advised him, his offering is instantly thrown from the wagon as impure, and he himself ignominiously and shamefully excluded from the society of the holy. There one sees the priests who preside over each chariot exhort every one to penitence, to confession of faults, to the resolution of better life! There one sees old people, young people, little children, calling on the Lord with a suppliant voice, and uttering to Him, from the depth of the heart, sobs and sighs with words of glory and praise! After the people, warned by the sound of trumpets and the sight of banners, have resumed their road, the march is made with such ease that no obstacle can retard it. . . . When they have reached the church they arrange the wagons about it like a spiritual camp, and during the whole night they celebrate the watch by hymns and canticles. On each waggon they light tapers and lamps; they place there the infirm and sick, and bring them the precious relics of the Saints for their relief. Afterwards the priests and clerics close the ceremony by processions which the people follow with devout heart, imploring the clemency of the Lord and of his Blessed Mother for the recovery of the sick.

Of course, the Virgin was actually and constantly present during all this labour, and gave her assistance to it, but you would get no light on the architecture from listening to an account of her miracles, nor do they heighten the effect of popular faith. Without the conviction of her personal presence, men would not have been inspired; but, to us, it is rather the inspiration of the art which proves the Virgin's presence, and we can better see the conviction of it in the work than in the words. Every day, as the work went on, the Virgin was present, directing the architects, and it is this direction that we are going to study, if you have now got a realizing sense of what it meant. Without this sense, the church is dead. Most persons of a deeply religious nature would tell you emphatically that nine churches out of ten actually were dead-born, after the thirteenth century, and that church architecture became a pure matter of mechanism and mathematics; but that is a question for you to decide when you come to it; and the pleasure consists not in seeing the death, but in feeling the life.

Now let us look about!

CHAPTER XXV

The Dynamo and the Virgin (1900)

Until the Great Exposition of 1900 closed its doors in November, Adams haunted it, aching to absorb knowledge, and helpless to find it. He would have liked to know how much of it could have been grasped by the best-informed man in the world. While he was thus meditating chaos, Langley came by, and showed it to him. At Langley's behest, the Exhibition dropped its superfluous rags and stripped itself to the skin, for Langley knew what to study, and why, and how; while Adams might as well have stood outside in the night, staring at the Milky Way. Yet Langley said nothing new, and taught nothing that one might not have learned from Lord Bacon, three hundred years before; but though one should have known the "Advancement of Science" as well as one knew the "Comedy of Errors," the literary knowledge counted for nothing until some teacher should show how to apply it. Bacon took a vast deal of trouble in teaching King James I and his subjects, American or other, towards the year 1620, that true science was the development or economy of forces; yet an elderly American in 1900 knew neither the formula nor the forces: or even so much as to say to himself that his historical business in the Exposition concerned only the economies or developments of force since 1893, when he began the study at Chicago.

Nothing in education is so astonishing as the amount of ignorance it accumulates in the form of inert facts. Adams had looked at most of the accumulations of art in the storehouses called Art Museums; yet he did not know how to look at the art exhibits of 1900. He had studied Karl Marx and his doctrines of history with profound attention, yet he could not apply them at Paris. Langley, with the ease of a great master of experiment, threw out of the field every exhibit that did not reveal a new application of force, and naturally threw out, to begin with, almost the whole art exhibit. Equally, he ignored almost the whole industrial exhibit. He led his pupil directly to the forces. His chief interest was in new motors to make his airship feasible, and he taught Adams the astonishing complexities of the new Daimler motor, and of the automobile, which, since 1893, had become a nightmare at a hundred kilometres an hour, almost as destructive as the electric tram which was only ten years older; and threatening to become as terrible as the locomotive steam-engine itself, which was almost exactly Adams's own age.

Then he showed his scholar the great hall of dynamos, and explained how little he knew about electricity or force of any kind, even of his own special sun, which spouted heat in inconceivable volume, but which, as far as he knew, might spout less or more, at any time, for all the certainty he felt in it. To him, the dynamo itself was but an ingenious channel for conveying somewhere the heat latent in a few tons of poor coal hidden in a dirty engine-house carefully kept out of sight; but to Adams the dynamo became a symbol of infinity. As he grew accustomed to the great gallery of machines, he began to feel the forty-foot dynamos as a moral force, much as the early Christians felt the Cross. The planet itself seemed less impressive, in its old-fashioned, deliberate, annual or daily revolution, than this huge wheel, revolving within arm's-length at some vertiginous speed, and barely murmuring—scarcely humming an audible warning to stand a hair's-breadth further for respect of power—while it would not wake the baby lying close against its frame. Before the end, one began to pray to it; inherited instinct taught the natural expression of man before silent and infinite force. Among the thousand symbols of ultimate energy, the dynamo was not so human as some, but it was the most expressive.

Yet the dynamo, next to the steam-engine, was the most familiar of exhibits. For Adams's objects its value lay chiefly in its occult mechanism. Between the dynamo in the gallery of machines and the engine-house outside, the break of continuity amounted to abysmal fracture for a historian's objects. No more relation could he discover between the steam and the electric current than between the Cross and the cathedral. The forces were interchangeable if not reversible, but he could see only an absolute *fiat* in electricity as in faith. Langley could not help him. Indeed, Langley seemed to be worried by the same trouble, for he constantly repeated that the new forces were anarchical, and specially that he was not responsible for the new rays, that were little short of parricidal in their wicked spirit towards science. His own rays, with which he had doubled the solar spectrum, were altogether harmless and beneficent; but Radium denied its God—or, what was to Langley the same thing, denied the truths of his Science. The force was wholly new.

A historian who asked only to learn enough to be as futile as Langley or Kelvin, made rapid progress under this teaching, and mixed himself up in the tangle of ideas until he achieved a sort of Paradise of ignorance vastly consoling to his fatigued senses. He wrapped himself in vibrations and rays which were new, and he would have hugged Marconi and Branly had he met them, as he hugged the dynamo; while he lost his arithmetic in trying to figure out the equation between the discoveries and the economies of force. The economies, like the discoveries, were absolute, supersensual, occult; incapable of expression in horse-power. What mathematical equivalent could he suggest as the value of a Branly coherer? Frozen air, or the electric furnace, had some scale of measurement, no doubt, if somebody could invent a thermometer adequate to the purpose; but X-rays had played no part whatever in man's consciousness, and the atom itself had figured only as a fiction of thought. In these seven years man had translated himself into a new universe which had no common scale of measurement with the old. He had entered a supersensual world, in which he could measure nothing except by chance collisions of movements imperceptible to his senses, perhaps even imperceptible to his instruments, but perceptible to each other, and so to some known ray at the end of the scale. Langley seemed prepared for anything, even for an indeterminable number of universes interfused — physics stark mad in metaphysics.

Historians undertake to arrange sequences,— called stories, or histories — assuming in silence a relation of cause and effect. These assumptions, hidden in the depths of dusty libraries, have been astounding, but commonly unconscious and childlike; so much so, that if any captious critic were to drag them to light, historians would probably reply, with one voice, that they had never supposed themselves required to know what they were talking about. Adams, for one, had toiled in vain to find out what he meant. He had even published a dozen volumes of American history for no other purpose than to satisfy himself whether, by the severest process of stating, with the least possible comment, such facts as seemed sure, in such order as seemed rigorously consequent, he could fix for a familiar moment a necessary sequence of human movement. The result had satisfied him as little as at Harvard College. Where he saw sequence, other men saw something quite different, and no one saw the same unit of measure. He cared little about his experiments and less about his statesmen, who seemed to him quite as ignorant as himself and, as a rule, no more honest; but he insisted on a relation of sequence, and if he could not reach it by one method, he would try as many methods as science knew. Satisfied that the sequence of men led to nothing and that the sequence of their society could lead no further, while the mere sequence of time was artificial, and the sequence of thought was chaos, he turned at last to the sequence of force: and thus it happened that, after ten years' pursuit, he found himself lying in the Gallery of Machines at the Great Exposition of 1900, his historical neck broken by the sudden irruption of forces totally new.

Since no one else showed much concern, an elderly person without other cares had no need to betray alarm. The year 1900 was not the first to upset schoolmasters. Copernicus and Galileo had broken many professorial necks about 1600; Columbus had stood the world on its head towards 1500; but the nearest approach to the revolution of 1900 was that of 310, when Constantine set up the Cross. The rays that Langley disowned, as well as those which he fathered, were occult, supersensual, irrational; they were a revelation of mysterious energy like that of the Cross; they were what, in terms of mediæval science, were called immediate modes of the divine substance.

The historian was thus reduced to his last resources. Clearly if he was bound to reduce all these forces to a common value, this common value could have no measure but that of their attraction on his own mind. He must treat them as they had been felt; as convertible, reversible, interchangeable attractions on thought. He made up his mind to venture it; he would risk translating rays into faith. Such a reversible process would vastly amuse a chemist, but the chemist could not deny that he, or some of his fellow physicists, could feel the force of both. When Adams was a boy in Boston, the best chemist in the place had probably never heard of Venus except by way of scandal, or of the Virgin except as idolatry; neither had he heard of dynamos or automobiles or radium; yet his mind was ready to feel the force of all, though the rays were unborn and the women were dead.

Here opened another totally new education, which promised to be by far the most hazardous of all. The knife-edge along which he must crawl, like Sir Lancelot in the twelfth century, divided two kingdoms of force which had nothing in common but attraction. They were as different as a magnet is from gravitation, supposing one knew what a magnet was, or gravitation, or love. The force of the Virgin was still felt at Lourdes, and seemed to be as potent as X-rays; but in America neither Venus nor Virgin ever had value as force — at most as sentiment. No American had ever been truly afraid of either.

This problem in dynamics gravely perplexed an American historian. The Woman had once been supreme; in France she still seemed potent, not merely as a sentiment, but as a force. Why was she unknown in America? For evidently America was ashamed of her, and she was ashamed of herself, otherwise they would not have strewn fig-leaves so profusely all over her. When she was a true force, she was ignorant of fig-leaves, but the monthly-magazine-made American female had not a feature that would have been recognized by Adam. The trait was notorious, and often humorous, but any one brought up among Puritans knew that sex was sin. In any previous age, sex was strength. Neither art nor beauty was needed. Every one, even among Puritans, knew that neither Diana of the Ephesians nor any of the Oriental goddesses was worshipped for her beauty. She was goddess because of her force; she was the animated dynamo; she was reproduction — the greatest and most mysterious of all energies; all she needed was to be fecund. Singularly enough, not one of Adams's many schools of education had ever drawn his attention to the opening lines of Lucretius, though they were perhaps the finest in all Latin literature, where the poet invoked Venus exactly as Dante invoked the Virgin:—

"Quae quoniam rerum naturam *sola* gubernas."

The Venus of Epicurean philosophy survived in the Virgin of the Schools:—

"Donna, sei tanto grande, e tanto vali,
Che qual vuol grazia, e a te non ricorre,
Sua disianza vuol volar senz' ali."

All this was to American thought as though it had never existed. The true American knew something of the facts, but nothing of the feelings; he read the letter, but he never felt the law. Before this historical chasm, a mind like that of Adams felt itself helpless; he turned from the Virgin to the Dynamo as though he were a Branly coherer. On one side, at the Louvre and at Chartres, as he knew by the record of work actually done and still before his eyes, was the highest energy ever known to man, the creator of four-fifths of his noblest art, exercising vastly more attraction over the human mind than all the steam-engines and dynamos ever dreamed of; and yet this energy was unknown to the American mind. An American Virgin would never dare command; an American Venus would never dare exist.

The question, which to any plain American of the nineteenth century seemed as remote as it did to Adams, drew him almost violently to study, once it was posed; and on this point Langleys were as useless as though they were Herbert Spencers or dynamos. The idea survived only as art. There one turned as naturally as though the artist were himself a woman. Adams began to ponder, asking himself whether he knew of any American artist who had ever insisted on the power of sex, as every classic had always done; but he could think only of Walt Whitman; Bret Harte, as far as the magazines would let him venture; and one or two painters, for the flesh-tones. All the rest had used sex for sentiment, never for force; to them, Eve was a tender flower, and Herodias an unfeminine horror. American art, like the American language and American education, was as far as possible sexless. Society regarded this victory over sex as its greatest triumph, and the historian readily admitted it, since the moral issue, for the moment, did not concern one who was studying the relations of unmoral force. He cared nothing for the sex of the dynamo until he could measure its energy.

Vaguely seeking a clue, he wandered through the art exhibit, and, in his stroll, stopped almost every day before St. Gaudens's General Sherman, which had been given the central post of honor. St. Gaudens himself was in Paris, putting on the work his usual interminable last touches, and listening to the usual contradictory suggestions of brother sculptors. Of all the American artists who gave to American art whatever life it breathed in the seventies, St. Gaudens was perhaps the most sympathetic, but certainly the most inarticulate. General Grant or Don Cameron had scarcely less instinct of rhetoric than he. All the others — the Hunts, Richardson, John La Farge, Stanford White — were exuberant; only St. Gaudens could never discuss or dilate on an emotion, or suggest artistic arguments for giving to his work the forms that he felt. He never laid down the law, or affected the despot, or became brutalized like Whistler by the brutalities of his world. He required no incense; he was no egoist; his simplicity of thought was excessive; he could not imitate, or give any form but his own to the creations of his hand. No one felt more strongly than he the strength of other men, but the idea that they could affect him never stirred an image in his mind.

This summer his health was poor and his spirits were low. For such a temper, Adams was not the best companion, since his own gaiety was not *folle;* but he risked going now and then to the studio on Mont Parnasse to draw him out for a stroll in the Bois de Boulogne, or dinner as pleased his moods, and in return St. Gaudens sometimes let Adams go about in his company.

Once St. Gaudens took him down to Amiens, with a party of Frenchmen, to see the cathedral. Not until they found themselves actually studying the sculpture of the western portal, did it dawn on Adams's mind that, for his purposes, St. Gaudens on that spot had more interest to him than the cathedral itself. Great men before great monuments express great truths, provided they are not taken too solemnly. Adams never tired of quoting the supreme phrase of his idol Gibbon, before the Gothic cathedrals: "I darted a contemptuous look on the stately monuments of superstition." Even in the footnotes of his history, Gibbon had never inserted a bit of humor more human than this, and one would have paid largely for a photograph of the fat little historian, on the background of Notre Dame of Amiens, trying to persuade his readers — perhaps himself — that he was darting a contemptuous look on the stately monument, for which he felt in fact the respect which every man of his vast study and active mind always feels before objects worthy of it; but besides the humor, one felt also the relation. Gibbon ignored the Virgin, because in 1789 religious monuments were out of fashion. In 1900 his remark sounded fresh and simple as the green fields to ears that had heard a hundred years of other remarks, mostly no more fresh and certainly less simple. Without malice, one might find it more instructive than a whole lecture of Ruskin. One sees what one brings, and at that moment Gibbon brought the French Revolution. Ruskin brought reaction against the Revolution. St. Gaudens had passed beyond all. He liked the stately monuments much more than he liked Gibbon or Ruskin; he loved their dignity; their unity; their scale; their lines; their lights and shadows; their decorative sculpture; but he was even less conscious than they of the force that created it all — the Virgin, the Woman — by whose genius "the stately monuments of superstition" were built, through which she was expressed. He would have seen more meaning in Isis with the cow's horns, at Edfoo, who expressed the same thought. The art remained, but the energy was lost even upon the artist.

Yet in mind and person St. Gaudens was a survival of the 1500; he bore the stamp of the Renaissance, and should have carried an image of the Virgin round his neck, or stuck in his hat, like Louis XI. In mere time he was a lost soul that had strayed by chance into the twentieth century, and forgotten where it came from. He writhed and cursed at his ignorance, much as Adams did at his own, but in the opposite sense. St. Gaudens was a child of Benvenuto Cellini, smothered in an American cradle. Adams was a quintessence of Boston, devoured by curiosity to think like Benvenuto. St. Gaudens's art was starved from birth, and Adams's instinct was blighted from babyhood. Each had but half of a nature, and when they came together before the Virgin of Amiens they ought both to have felt in her the force that made them one; but it was not so. To Adams she became more than ever a channel of force; to St. Gaudens she remained as before a channel of taste.

For a symbol of power, St. Gaudens instinctively preferred the horse, as was plain in his horse and Victory of the Sherman monument. Doubtless Sherman also felt it so. The attitude was so American that, for at least forty years, Adams had never realized that any other could be in sound taste. How many years had he taken to admit a notion of what Michael Angelo and Rubens were driving at? He could not say; but he knew that only since 1895 had he begun to feel the Virgin or Venus as force, and not everywhere even so. At Chartres — perhaps at Lourdes — possibly at Cnidos if one could still find there the divinely naked Aphrodite of Praxiteles — but otherwise one must look for force to the goddesses of Indian mythology. The idea died out long ago in the German and English stock. St. Gaudens at Amiens was hardly less sensitive to the force of the female energy than Matthew Arnold at the Grande Chartreuse. Neither of them felt goddesses as power — only as reflected emotion, human expression, beauty, purity, taste, scarcely even as sympathy. They felt a railway train as power; yet they, and all other artists, constantly complained that the power embodied in a railway train could never be embodied in art. All the steam in the world could not, like the Virgin, build Chartres.

Yet in mechanics, whatever the mechanicians might think, both energies acted as interchangeable forces on man, and by action on man all known force may be measured. Indeed, few men of science measured force in any other way. After once admitting that a straight line was the shortest distance between two points, no serious mathematician cared to deny anything that suited his convenience, and rejected no symbol, unproved or unproveable, that helped him to accomplish work. The symbol was force, as a compass-needle or a triangle was force, as the mechanist might prove by losing it, and nothing could be gained by ignoring their value. Symbol or energy, the Virgin had acted as the greatest force the Western world ever felt, and had drawn man's activities to herself more strongly than any other power, natural or supernatural, had ever done; the historian's business was to follow the track of the energy; to find where it came from and where it went to; its complex source and shifting channels; its values, equivalents, conversions. It could scarcely be more complex than radium; it could hardly be deflected, diverted, polarized, absorbed more perplexingly than other radiant matter. Adams knew nothing about any of them, but as a mathematical problem of influence on human progress, though all were occult, all reacted on his mind, and he rather inclined to think the Virgin easiest to handle.

The pursuit turned out to be long and tortuous, leading at last into the vast forests of scholastic science. From Zeno to Descartes, hand in hand with Thomas Aquinas, Montaigne, and Pascal, one stumbled as stupidly as though one were still a German student of 1860. Only with the instinct of despair could one force one's self into this old thicket of ignorance after having been repulsed at a score of entrances more promising and more popular. Thus far, no path had led anywhere, unless perhaps to an exceedingly modest living. Forty-five years of study had proved to be quite futile for the pursuit of power; one controlled no more force in 1900 than in 1850, although the amount of force controlled by society had enormously increased. The secret of education still hid itself somewhere behind ignorance, and one fumbled over it as feebly as ever. In such labyrinths, the staff is a force almost more necessary than the legs; the pen becomes a sort of blind-man's dog, to keep him from falling into the gutters. The pen works for itself, and acts like a hand, modelling the plastic material over and over again to the form that suits it best. The form is never arbitrary, but is a sort of growth like crystallization, as any artist knows too well; for often the pencil or pen runs into side-paths and shapelessness, loses its relations, stops or is bogged. Then it has to return on its trail, and recover, if it can, its line of force. The result of a year's work depends more on what is struck out than on what is left in; on the sequence of the main lines of thought, than on their play or variety. Compelled once more to lean heavily on this support, Adams covered more thousands of pages with figures as formal as though they were algebra, laboriously striking out, altering, burning, experimenting, until the year had expired, the Exposition had long been closed, and winter drawing to its end, before he sailed from Cherbourg, on January 19, 1901, for home.

CHAPTER XXIX

The Abyss of Ignorance (1902)

The years hurried past, and gave hardly time to note their work. Three or four months, though big with change, come to an end before the mind can catch up with it. Winter vanished; spring burst into flower; and again Paris opened its arms, though not for long. Mr. Cameron came over, and took the castle of Inverlochy for three months, which he summoned his friends to garrison. Lochaber seldom laughs, except for its children, such as Camerons, McDonalds, Campbells and other products of the mist; but in the summer of 1902 Scotland put on fewer airs of coquetry than usual. Since the terrible harvest of 1879 which one had watched sprouting on its stalks on the Shropshire hillsides, nothing had equalled the gloom. Even when the victims fled to Switzerland, they found the Lake of Geneva and the Rhine not much gayer, and Carlsruhe no more restful than Paris; until at last, in desperation, one drifted back to the Avenue of the Bois de Boulogne, and, like the Cuckoo, dropped into the nest of a better citizen. Diplomacy has its uses. Reynolds Hitt, transferred to Berlin, abandoned his attic to Adams, and there, for long summers to come, he hid in ignorance and silence.

Life at last managed of its own accord to settle itself into a working arrangement. After so many years of effort to find one's drift, the drift found the seeker, and slowly swept him forward and back, with a steady progress oceanwards. Such lessons as summer taught, winter tested, and one had only to watch the apparent movement of the stars in order to guess one's declination. The process is possible only for men who have exhausted auto-motion. Adams never knew why, knowing nothing of Faraday, he began to mimic Faraday's trick of seeing lines of force all about him, where he had always seen lines of will. Perhaps the effect of knowing no mathematics is to leave the mind to imagine figures — images — phantoms; one's mind is a watery mirror at best; but, once conceived, the image became rapidly simple, and the lines of force presented themselves as lines of attraction. Repulsions counted only as battle of attractions. By this path, the mind stepped into the mechanical theory of the universe before knowing it, and entered a distinct new phase of education.

This was the work of the dynamo and the Virgin of Chartres. Like his masters, since thought began, he was handicapped by the eternal mystery of Force — the sink of all science. For thousands of years in history, he found that Force had been felt as occult attraction — love of God and lust for power in a future life. After 1500, when this attraction began to decline, philosophers fell back on some *vis a tergo* — instinct of danger from behind, like Darwin's survival of the fittest; and one of the greatest minds, between Descartes and Newton — Pascal — saw the master-motor of man in *ennui*, which was also scientific: "I have often said that all the troubles of man come from his not knowing how to sit still." Mere restlessness forces action. "So passes the whole of life. We combat obstacles in order to get repose, and, when got, the repose is insupportable; for we think either of the troubles we have, or of those that threaten us; and even if we felt safe on every side, *ennui* would of its own accord spring up from the depths of the heart where it is rooted by nature, and would fill the mind with its venom."

> "If goodness lead him not, yet weariness
> May toss him to My breast."

Ennui, like Natural Selection, accounted for change, but failed to account for direction of change. For that, an attractive force was essential; a force from outside; a shaping influence. Pascal and all the old philosophies called this outside force God or Gods. Caring but little for the name, and fixed only on tracing the Force, Adams had gone straight to the Virgin at Chartres, and asked her to show him God, face to face, as she did for St. Bernard. She replied, kindly as ever, as though she were still the young mother of to-day, with a sort of patient pity for masculine dulness: "My dear outcast, what is it you seek? This is the Church of Christ! If you seek him through me, you are welcome, sinner or saint; but he and I are one. We are Love! We have little or nothing to do with God's other energies which are infinite, and concern us the less because our interest is only in man, and the infinite is not knowable to man. Yet if you are troubled by your ignorance, you see how I am surrounded by the masters of the schools! Ask them!"

The answer sounded singularly like the usual answer of British science which had repeated since Bacon that one must not try to know the unknowable, though one was quite powerless to ignore it; but the Virgin carried more conviction, for her feminine lack of interest in all perfections except her own was honester than the formal phrase of science; since nothing was easier than to follow her advice, and turn to Thomas Aquinas, who, unlike modern physicists, answered at once and plainly: "To me," said St. Thomas, "Christ and the Mother are one Force — Love — simple, single, and sufficient for all human wants; but Love is a human interest which acts even on man so partially that you and I, as philosophers, need expect no share in it. Therefore we turn to Christ and the Schools who represent all other Force. We deal with Multiplicity and call it God. After the Virgin has redeemed by her personal Force as Love all that is redeemable in man, the Schools embrace the rest, and give it Form, Unity, and Motive."

This chart of Force was more easily studied than any other possible scheme, for one had but to do what the Church was always promising to do — abolish in one flash of lightning not only man, but also the Church itself, the earth, the other planets, and the sun, in order to clear the air; without affecting mediæval science. The student felt warranted in doing what the Church threatened — abolishing his solar system altogether — in order to look at God as actual; continuous movement, universal cause, and interchangeable force. This was pantheism, but the Schools were pantheist; at least as pantheistic as the *Energetik* of the Germans; and their duty was the ultimate energy, whose thought and act were one.

Rid of man and his mind, the universe of Thomas Aquinas seemed rather more scientific than that of Haeckel or Ernst Mach. Contradiction for contradiction, Attraction for attraction, Energy for energy, St. Thomas's idea of God had merits. Modern science offered not a vestige of proof, or a theory of connection between its forces, or any scheme of reconciliation between thought and mechanics; while St. Thomas at least linked together the joints of his machine. As far as a superficial student could follow, the thirteenth century supposed mind to be a mode of force directly derived from the intelligent prime motor, and the cause of all form and sequence in the universe—therefore the only proof of unity. Without thought in the unit, there could be no unity; without unity no orderly sequence or ordered society. Thought alone was Form. Mind and Unity flourished or perished together.

This education startled even a man who had dabbled in fifty educations all over the world; for, if he were obliged to insist on a Universe, he seemed driven to the Church. Modern science guaranteed no unity. The student seemed to feel himself, like all his predecessors, caught, trapped, meshed in this eternal drag-net of religion.

In practice the student escapes this dilemma in two ways: the first is that of ignoring it, as one escapes most dilemmas; the second is that the Church rejects pantheism as worse than atheism, and will have nothing to do with the pantheist at any price. In wandering through the forests of ignorance, one necessarily fell upon the famous old bear that scared children at play; but, even had the animal shown more logic than its victim, one had learned from Socrates to distrust, above all other traps, the trap of logic — the mirror of the mind. Yet the search for a unit of force led into catacombs of thought where hundreds of thousands of educations had found their end. Generation after generation of painful and honest-minded scholars had been content to stay in these labyrinths forever, pursuing ignorance in silence, in company with the most famous teachers of all time. Not one of them had ever found a logical highroad of escape.

Adams cared little whether he escaped or not, but he felt clear that he could not stop there, even to enjoy the society of Spinoza and Thomas Aquinas. True, the Church alone had asserted unity with any conviction, and the historian alone knew what oceans of blood and treasure the assertion had cost; but the only honest alternative to affirming unity was to deny it; and the denial would require a new education. At sixty-five years old a new education promised hardly more than the old.

Possibly the modern legislator or magistrate might no longer know enough to treat as the Church did the man who denied unity, unless the denial took the form of a bomb; but no teacher would know how to explain what he thought he meant by denying unity. Society would certainly punish the denial if ever any one learned enough to understand it. Philosophers, as a rule, cared little what principles society affirmed or denied, since the philosopher commonly held that though he might sometimes be right by good luck on some one point, no complex of individual opinions could possibly be anything but wrong; yet, supposing society to be ignored, the philosopher was no further forward. Nihilism had no bottom. For thousands of years every philosopher had stood on the shore of this sunless sea, diving for pearls and never finding them. All had seen that, since they could not find bottom, they must assume it. The Church claimed to have found it, but, since 1450, motives for agreeing on some new assumption of Unity, broader and deeper than that of the Church, had doubled in force until even the universities and schools, like the Church and State, seemed about to be driven into an attempt to educate, though specially forbidden to do it.

Like most of his generation, Adams had taken the word of science that the new unit was as good as found. It would not be an intelligence — probably not even a consciousness — but it would serve. He passed sixty years waiting for it, and at the end of that time, on reviewing the ground, he was led to think that the final synthesis of science and its ultimate triumph was the kinetic theory of gases; which seemed to cover all motion in space, and to furnish the measure of time. So far as he understood it, the theory asserted that any portion of space is occupied by molecules of gas, flying in right lines at velocities varying up to a mile in a second, and colliding with each other at intervals varying up to 17,750,000 times in a second. To this analysis — if one understood it right — all matter whatever was reducible, and the only difference of opinion in science regarded the doubt whether a still deeper analysis would reduce the atom of gas to pure motion.

Thus, unless one mistook the meaning of motion, which might well be, the scientific synthesis commonly called Unity was the scientific analysis commonly called Multiplicity. The two things were the same, all forms being shifting phases of motion. Granting this ocean of colliding atoms, the last hope of humanity, what happened if one dropped the sounder into the abyss — let it go — frankly gave up Unity altogether? What was Unity? Why was one to be forced to affirm it?

Here everybody flatly refused help. Science seemed content with its old phrase of "larger synthesis," which was well enough for science, but meant chaos for man. One would have been glad to stop and ask no more, but the anarchist bomb bade one go on, and the bomb is a powerful persuader. One could not stop, even to enjoy the charms of a perfect gas colliding seventeen million times in a second, much like an automobile in Paris. Science itself had been crowded so close to the edge of the abyss that its attempts to escape were as metaphysical as the leap, while an ignorant old man felt no motive for trying to escape, seeing that the only escape possible lay in the form of *vis a tergo* commonly called Death. He got out his Descartes again; dipped into his Hume and Berkeley; wrestled anew with his Kant; pondered solemnly over his Hegel and Schopenhauer and Hartmann; strayed gaily away with his Greeks — all merely to ask what Unity meant, and what happened when one denied it.

Apparently one never denied it. Every philosopher, whether sane or insane, naturally affirmed it. The utmost flight of anarchy seemed to have stopped with the assertion of two principles, and even these fitted into each other, like good and evil, light and darkness. Pessimism itself, black as it might be painted, had been content to turn the universe of contradictions into the human thought as one Will, and treat it as representation. Metaphysics insisted on treating the universe as one thought or treating thought as one universe; and philosophers agreed, like kinetic gas, that the universe could be known only as motion of mind, and therefore as unity. One could know it only as one's self; it was psychology.

Of all forms of pessimism, the metaphysical form was, for a historian, the least enticing. Of all studies, the one he would rather have avoided was that of his own mind. He knew no tragedy so heartrending as introspection, and the more, because — as Mephistopheles said of Marguerite—he was not the first. Nearly all the highest intelligence known to history had drowned itself in the reflection of its own thought, and the bovine survivors had rudely told the truth about it, without affecting the intelligent. One's own time had not been exempt. Even since 1870 friends by scores had fallen victims to it. Within five-and-twenty years, a new library had grown out of it. Harvard College was a focus of the study; France supported hospitals for it; England published magazines of it. Nothing was easier than to take one's mind in one's hand, and ask one's psychological friends what they made of it, and the more because it mattered so little to either party, since their minds, whatever they were, had pretty nearly ceased to reflect, and let them do what they liked with the small remnant, they could scarcely do anything very new with it. All one asked was to learn what they hoped to do.

Unfortunately the pursuit of ignorance in silence had, by this time, led the weary pilgrim into such mountains of ignorance that he could no longer see any path whatever, and could not even understand a signpost. He failed to fathom the depths of the new psychology, which proved to him that, on that side as on the mathematical side, his power of thought was atrophied, if, indeed, it ever existed. Since he could not fathom the science, he could only ask the simplest of questions: Did the new psychology hold that the ψυχή — soul or mind — was or was not a unit? He gathered from the books that the psychologists had, in a few cases, distinguished several personalities in the same mind, each conscious and constant, individual and exclusive. The fact seemed scarcely surprising, since it had been a habit of mind from earliest recorded time, and equally familiar to the last acquaintance who had taken a drug or caught a fever, or eaten a Welsh rarebit before bed; for surely no one could follow the action of a vivid dream, and still need to be told that the actors evoked by his mind were not himself, but quite unknown to all he had ever recognized as self. The new psychology went further, and seemed convinced that it had actually split personality not only into dualism, but also into complex groups, like telephonic centres and systems, that might be isolated and called up at will, and whose physical action might be occult in the sense of strangeness to any known form of force. Dualism seemed to have become as common as binary stars. Alternating personalities turned up constantly, even among one's friends. The facts seemed certain, or at least as certain as other facts; all they needed was explanation.

This was not the business of the searcher of ignorance, who felt himself in no way responsible for causes. To his mind, the compound υχή took at once the form of a bicycle-rider, mechanically balancing himself by inhibiting all his inferior personalities, and sure to fall into the sub-conscious chaos below, if one of his inferior personalities got on top. The only absolute truth was the sub-conscious chaos below, which every one could feel when he sought it.

Whether the psychologists admitted it or not, mattered little to the student who, by the law of his profession, was engaged in studying his own mind. On him, the effect was surprising. He woke up with a shudder as though he had himself fallen off his bicycle. If his mind were really this sort of magnet, mechanically dispersing its lines of force when it went to sleep, and mechanically orienting them when it woke up — which was normal, the dispersion or orientation? The mind, like the body, kept its unity unless it happened to lose balance, but the professor of physics, who slipped on a pavement and hurt himself, knew no more than an idiot what knocked him down, though he did know — what the idiot could hardly do — that his normal condition was idiocy, or want of balance, and that his sanity was unstable artifice. His normal thought was dispersion, sleep, dream, inconsequence; the simultaneous action of different thought-centres without central control. His artificial balance was acquired habit. He was an acrobat, with a dwarf on his back, crossing a chasm on a slack-rope, and commonly breaking his neck.

By that path of newest science, one saw no unity ahead — nothing but a dissolving mind — and the historian felt himself driven back on thought as one continuous Force, without Race, Sex, School, Country, or Church. This has been always the fate of rigorous thinkers, and has always succeeded in making them famous, as it did Gibbon, Buckle, and Auguste Comte. Their method made what progress the science of history knew, which was little enough, but they did at last fix the law that, if history ever meant to correct the errors she made in detail, she must agree on a scale for the whole. Every local historian might defy this law till history ended, but its necessity would be the same for man as for space or time or force, and without it the historian would always remain a child in science.

Any schoolboy could see that man as a force must be measured by motion, from a fixed point. Psychology helped here by suggesting a unit — the point of history when man held the highest idea of himself as a unit in a unified universe. Eight or ten years of study had led Adams to think he might use the century 1150–1250, expressed in Amiens Cathedral and the Works of Thomas Aquinas, as the unit from which he might measure motion down to his own time, without assuming anything as true or untrue, except relation. The movement might be studied at once in philosophy and mechanics. Setting himself to the task, he began a volume which he mentally knew as "Mont-Saint-Michel and Chartres: a Study of Thirteenth-Century Unity." From that point he proposed to fix a position for himself, which he could label: "The Education of Henry Adams: a Study of Twentieth-Century Multiplicity." With the help of these two points of relation, he hoped to project his lines forward and backward indefinitely, subject to correction from any one who should know better. Thereupon, he sailed for home.

CHAPTER XXXI

The Grammar of Science (1903)

Of all the travels made by man since the voyages of Dante, this new exploration along the shores of Multiplicity and Complexity promised to be the longest, though as yet it had barely touched two familiar regions — race and sex. Even within these narrow seas the navigator lost his bearings and followed the winds as they blew. By chance it happened that Raphael Pumpelly helped the winds; for, being in Washington on his way to Central Asia he fell to talking with Adams about these matters, and said that Willard Gibbs thought he got most help from a book called the "Grammar of Science," by Karl Pearson. To Adams's vision, Willard Gibbs stood on the same plane with the three or four greatest minds of his century, and the idea that a man so incomparably superior should find help anywhere filled him with wonder. He sent for the volume and read it. From the time he sailed for Europe and reached his den on the Avenue du Bois until he took his return steamer at Cherbourg on December 26, he did little but try to find out what Karl Pearson could have taught Willard Gibbs.

Here came in, more than ever, the fatal handicap of ignorance in mathematics. Not so much the actual tool was needed, as the right to judge the product of the tool. Ignorant as one was of the finer values of French or German, and often deceived by the intricacies of thought hidden in the muddiness of the medium, one could sometimes catch a tendency to intelligible meaning even in Kant or Hegel; but one had not the right to a suspicion of error where the tool of thought was algebra. Adams could see in such parts of the "Grammar" as he could understand, little more than an enlargement of Stallo's book already twenty years old. He never found out what it could have taught a master like Willard Gibbs.

Yet the book had a historical value out of all proportion to its science. No such stride had any Englishman before taken in the lines of English thought. The progress of science was measured by the success of the "Grammar," when, for twenty years past, Stallo had been deliberately ignored under the usual conspiracy of silence inevitable to all thought which demands new thought-machinery. Science needs time to reconstruct its instruments, to follow a revolution in space; a certain lag is inevitable; the most active mind cannot instantly swerve from its path; but such revolutions are portentous, and the fall or rise of half-a-dozen empires interested a student of history less than the rise of the "Grammar of Science," the more pressingly because, under the silent influence of Langley, he was prepared to expect it.

For a number of years Langley had published in his Smithsonian Reports the revolutionary papers that foretold the overflow of nineteenth-century dogma, and among the first was the famous address of Sir William Crookes on psychical research, followed by a series of papers on Roentgen and Curie, which had steadily driven the scientific lawgivers of Unity into the open; but Karl Pearson was the first to pen them up for slaughter in the schools. The phrase is not stronger than that with which the "Grammar of Science" challenged the fight: "Anything more hopelessly illogical than the statements with regard to Force and Matter current in elementary textbooks of science, it is difficult to imagine," opened Mr. Pearson, and the responsible author of the "elementary text-book," as he went on to explain, was Lord Kelvin himself. Pearson shut out of science everything which the nineteenth century had brought into it. He told his scholars that they must put up with a fraction of the universe, and a very small fraction at that — the circle reached by the senses, where sequence could be taken for granted — much as the deep-sea fish takes for granted the circle of light which he generates. "Order and reason, beauty and benevolence, are characteristics and conceptions which we find solely associated with the mind of man." The assertion, as a broad truth, left one's mind in some doubt of its bearing, for order and beauty seemed to be associated also in the mind of a crystal, if one's senses were to be admitted as judge; but the historian had no interest in the universal truth of Pearson's or Kelvin's or Newton's laws; he sought only their relative drift or direction, and Pearson went on to say that these conceptions must stop: "Into the chaos beyond sense-impressions we cannot scientifically project them." We cannot even infer them: "In the chaos behind sensations, in the 'beyond' of sense-impressions, we cannot infer necessity, order or routine, for these are concepts formed by the mind of man on this side of sense-impressions"; but we must infer chaos: "Briefly chaos is all that science can logically assert of the supersensuous." The kinetic theory of gas is an assertion of ultimate chaos. In plain words, Chaos was the law of nature; Order was the dream of man.

No one means all he says, and yet very few say all they mean, for words are slippery and thought is viscous; but since Bacon and Newton, English thought had gone on impatiently protesting that no one must try to know the unknowable at the same time that every one went on thinking about it. The result was as chaotic as kinetic gas; but with the thought a historian had nothing to do. He sought only its direction. For himself he knew, that, in spite of all the Englishmen that ever lived, he would be forced to enter supersensual chaos if he meant to find out what became of British science — or indeed of any other science. From Pythagoras to Herbert Spencer, every one had done it, although commonly science had explored an ocean which it preferred to regard as Unity or a Universe, and called Order. Even Hegel, who taught that every notion included its own negation, used the negation only to reach a "larger synthesis," till he reached the universal which thinks itself, contradiction and all. The Church alone had constantly protested that anarchy was not order, that Satan was not God, that pantheism was worse than atheism, and that Unity could not be proved as a contradiction. Karl Pearson seemed to agree with the Church, but every one else, including Newton, Darwin and Clerk Maxwell, had sailed gaily into the supersensual, calling it:—

> "One God, one Law, one Element,
> And one far-off, divine event,
> To which the whole creation moves."

Suddenly, in 1900, science raised its head and denied.

Yet, perhaps, after all, the change had not been so sudden as it seemed. Real and actual, it certainly was, and every newspaper betrayed it, but sequence could scarcely be denied by one who had watched its steady approach, thinking the change far more interesting to history than the thought. When he reflected about it, he recalled that the flow of tide had shown itself at least twenty years before; that it had become marked as early as 1893; and that the man of science must have been sleepy indeed who did not jump from his chair like a scared dog when, in 1898, Mme. Curie threw on his desk the metaphysical bomb she called radium. There remained no hole to hide in. Even metaphysics swept back over science with the green water of the deep-sea ocean and no one could longer hope to bar out the unknowable, for the unknowable was known.

The fact was admitted that the uniformitarians of one's youth had wound about their universe a tangle of contradictions meant only for temporary support to be merged in "larger synthesis," and had waited for the larger synthesis in silence and in vain. They had refused to hear Stallo. They had betrayed little interest in Crookes. At last their universe had been wrecked by rays, and Karl Pearson undertook to cut the wreck loose with an axe, leaving science adrift on a sensual raft in the midst of a supersensual chaos. The confusion seemed, to a mere passenger, worse than that of 1600 when the astronomers upset the world; it resembled rather the convulsion of 310 when the *Civitas Dei* cut itself loose from the *Civitas Romae*, and the Cross took the place of the legions; but the historian accepted it all alike; he knew that his opinion was worthless; only, in this case, he found himself on the raft, personally and economically concerned in its drift.

English thought had always been chaos and multiplicity itself, in which the new step of Karl Pearson marked only a consistent progress; but German thought had affected system, unity, and abstract truth, to a point that fretted the most patient foreigner, and to Germany the voyager in strange seas of thought alone might resort with confident hope of renewing his youth. Turning his back on Karl Pearson and England, he plunged into Germany, and had scarcely crossed the Rhine when he fell into libraries of new works bearing the names of Ostwald, Ernst Mach, Ernst Haeckel, and others less familiar, among whom Haeckel was easiest to approach, not only because of being the oldest and clearest and steadiest spokesman of nineteenth-century mechanical convictions, but also because in 1902 he had published a vehement renewal of his faith. The volume contained only one paragraph that concerned a historian; it was that in which Haeckel sank his voice almost to a religious whisper in avowing with evident effort, that the "proper essence of substance appeared to him more and more marvellous and enigmatic as he penetrated further into the knowledge of its attributes — matter and energy — and as he learned to know their innumerable phenomena and their evolution." Since Haeckel seemed to have begun the voyage into multiplicity that Pearson had forbidden to Englishmen, he should have been a safe pilot to the point, at least, of a "proper essence of substance" in its attributes of matter and energy; but Ernst Mach seemed to go yet one step further, for he rejected matter altogether, and admitted but two processes in nature — change of place and interconversion of forms. Matter was Motion — Motion was Matter — the thing moved.

A student of history had no need to understand these scientific ideas of very great men; he sought only the relation with the ideas of their grandfathers, and their common direction towards the ideas of their grandsons. He had long ago reached, with Hegel, the limits of contradiction; and Ernst Mach scarcely added a shade of variety to the identity of opposites; but both of them seemed to be in agreement with Karl Pearson on the facts of the supersensual universe which could be known only as unknowable.

With a deep sigh of relief, the traveller turned back to France. There he felt safe. No Frenchman except Rabelais and Montaigne had ever taught anarchy other than as path to order. Chaos would be unity in Paris even if child of the guillotine. To make this assurance mathematically sure, the highest scientific authority in France was a great mathematician, M. Poincaré of the Institut, who published in 1902 a small volume called "La Science et l'Hypothèse," which purported to be relatively readable. Trusting to its external appearance, the traveller timidly bought it, and greedily devoured it, without understanding a single consecutive page, but catching here and there a period that startled him to the depths of his ignorance, for they seemed to show that M. Poincaré was troubled by the same historical landmarks which guided or deluded Adams himself: "[In science] we are led," said M. Poincaré, "to act as though a simple law, when other things were equal, must be more probable than a complicated law. Half a century ago one frankly confessed it, and proclaimed that nature loves simplicity. She has since given us too often the lie. To-day this tendency is no longer avowed, and only as much of it is preserved as is indispensable so that science shall not become impossible."

Here at last was a fixed point beyond the chance of confusion with self-suggestion. History and mathematics agreed. Had M. Poincaré shown anarchistic tastes, his evidence would have weighed less heavily; but he seemed to be the only authority in science who felt what a historian felt so strongly — the need of unity in a universe. "Considering everything we have made some approach towards unity. We have not gone as fast as we hoped fifty years ago; we have not always taken the intended road; but definitely we have gained much ground." This was the most clear and convincing evidence of progress yet offered to the navigator of ignorance; but suddenly he fell on another view which seemed to him quite irreconcilable with the first: "Doubtless if our means of investigation should become more and more penetrating, we should discover the simple under the complex; then the complex under the simple; then anew the simple under the complex; and so on without ever being able to foresee the last term."

A mathematical paradise of endless displacement promised eternal bliss to the mathematician, but turned the historian green with horror. Made miserable by the thought that he knew no mathematics, he burned to ask whether M. Poincaré knew any history, since he began by begging the historical question altogether, and assuming that the past showed alternating phases of simple and complex — the precise point that Adams, after fifty years of effort, found himself forced to surrender; and then going on to assume alternating phases for the future which, for the weary Titan of Unity, differed in nothing essential from the kinetic theory of a perfect gas.

Since monkeys first began to chatter in trees, neither man nor beast had ever denied or doubted Multiplicity, Diversity, Complexity, Anarchy, Chaos. Always and everywhere the Complex had been true and the Contradiction had been certain. Thought started by it. Mathematics itself began by counting one — two — three; then imagining their continuity, which M. Poincaré was still exhausting his wits to explain or defend; and this was his explanation: "In short, the mind has the faculty of creating symbols, and it is thus that it has constructed mathematical continuity which is only a particular system of symbols." With the same light touch, more destructive in its artistic measure than the heaviest-handed brutality of Englishmen or Germans, he went on to upset relative truth itself: "How should I answer the question whether Euclidian Geometry is true? It has no sense! . . . Euclidian Geometry is, and will remain, the most convenient."

Chaos was a primary fact even in Paris—especially in Paris—

as it was in the Book of Genesis; but every thinking being in Paris or out of it had exhausted thought in the effort to prove Unity, Continuity, Purpose, Order, Law, Truth, the Universe, God, after having begun by taking it for granted, and discovering, to their profound dismay, that some minds denied it. The direction of mind, as a single force of nature, had been constant since history began. Its own unity had created a universe the essence of which was abstract Truth; the Absolute; God! To Thomas Aquinas, the universe was still a person; to Spinoza, a substance; to Kant, Truth was the essence of the "I"; an innate conviction; a categorical imperative; to Poincaré, it was a convenience; and to Karl Pearson, a medium of exchange.

The historian never stopped repeating to himself that he knew nothing about it; that he was a mere instrument of measure, a barometer, pedometer, radiometer; and that his whole share in the matter was restricted to the measurement of thought-motion as marked by the accepted thinkers. He took their facts for granted. He knew no more than a firefly about rays — or about race — or sex — or ennui — or a bar of music — or a pang of love — or a grain of musk — or of phosphorus — or conscience — or duty — or the force of Euclidian geometry — or non-Euclidian — or heat — or light — or osmosis — or electrolysis — or the magnet — or ether — or *vis inertiae* — or gravitation — or cohesion — or elasticity — or surface tension — or capillary attraction — or Brownian motion — or of some scores, or thousands, or millions of chemical attractions, repulsions or indifferences which were busy within and without him; or, in brief, of Force itself, which, he was credibly informed, bore some dozen definitions in the textbooks, mostly contradictory, and all, as he was assured, beyond his intelligence; but summed up in the dictum of the last and highest science, that Motion seems to be Matter and Matter seems to be Motion, yet "we are probably incapable of discovering" what either is. History had no need to ask what either might be; all it needed to know was the admission of ignorance; the mere fact of multiplicity baffling science. Even as to the fact, science disputed, but radium happened to radiate something that seemed to explode the scientific magazine, bringing thought, for the time, to a standstill; though, in the line of thought-movement in history, radium was merely the next position, familiar and inexplicable since Zeno and his arrow: continuous from the beginning of time, and discontinuous at each successive point. History set it down on the record — pricked its position on the chart — and waited to be led, or misled, once more.

The historian must not try to know what is truth, if he values his honesty; for, if he cares for his truths, he is certain to falsify his facts. The laws of history only repeat the lines of force or thought. Yet though his will be iron, he cannot help now and then resuming his humanity or simianity in face of a fear. The motion of thought had the same value as the motion of a cannon-ball seen approaching the observer on a direct line through the air. One could watch its curve for five thousand years. Its first violent acceleration in historical times had ended in the catastrophe of 310. The next swerve of direction occurred towards 1500. Galileo and Bacon gave a still newer curve to it, which altered its values; but all these changes had never altered the continuity. Only in 1900, the continuity snapped.

Vaguely conscious of the cataclysm, the world sometimes dated it from 1893, by the Roentgen rays, or from 1898, by the Curies' radium; but in 1904, Arthur Balfour announced on the part of British science that the human race without exception had lived and died in a world of illusion until the last year of the century. The date was convenient, and convenience was truth.

The child born in 1900 would, then, be born into a new world which would not be a unity but a multiple. Adams tried to imagine it, and an education that would fit it. He found himself in a land where no one had ever penetrated before; where order was an accidental relation obnoxious to nature; artificial compulsion imposed on motion; against which every free energy of the universe revolted; and which, being merely occasional, resolved itself back into anarchy at last. He could not deny that the law of the new multiverse explained much that had been most obscure, especially the persistently fiendish treatment of man by man; the perpetual effort of society to establish law, and the perpetual revolt of society against the law it had established; the perpetual building up of authority by force, and the perpetual appeal to force to overthrow it; the perpetual symbolism of a higher law, and the perpetual relapse to a lower one; the perpetual victory of the principles of freedom, and their perpetual conversion into principles of power; but the staggering problem was the outlook ahead into the despotism of artificial order which nature abhorred. The physicists had a phrase for it, unintelligible to the vulgar: "All that we win is a battle—lost in advance—with the irreversible phenomena in the background of nature."

All that a historian won was a vehement wish to escape. He saw his education complete, and was sorry he ever began it. As a matter of taste, he greatly preferred his eighteenth-century education when God was a father and nature a mother, and all was for the best in a scientific universe. He repudiated all share in the world as it was to be, and yet he could not detect the point where his responsibility began or ended.

As history unveiled itself in the new order, man's mind had behaved like a young pearl oyster, secreting its universe to suit its conditions until it had built up a shell of *nacre* that embodied all its notions of the perfect. Man knew it was true because he made it, and he loved it for the same reason. He sacrificed millions of lives to acquire his unity, but he achieved it, and justly thought it a work of art. The woman especially did great things, creating her deities on a higher level than the male, and, in the end, compelling the man to accept the Virgin as guardian of the man's God. The man's part in his Universe was secondary, but the woman was at home there, and sacrificed herself without limit to make it habitable, when man permitted it, as sometimes happened for brief intervals of war and famine; but she could not provide protection against forces of nature. She did not think of her universe as a raft to which the limpets stuck for life in the surge of a supersensual chaos; she conceived herself and her family as the centre and flower of an ordered universe which she knew to be unity because she had made it after the image of her own fecundity; and this creation of hers was surrounded by beauties and perfections which she knew to be real because she herself had imagined them.

Even the masculine philosopher admired and loved and celebrated her triumph, and the greatest of them sang it in the noblest of his verses:—

> "Alma Venus, coeli subter labentia signa
> Quae mare navigerum, quae terras frugiferenteis
> Concelebras . . . .
> Quae quoniam rerum naturam sola gubernas,
> Nec sine te quidquam dias in luminis oras
> Exoritur, neque fit laetum neque amabile quidquam;
> Te sociam studeo!"

Neither man nor woman ever wanted to quit this Eden of their own invention, and could no more have done it of their own accord than the pearl oyster could quit its shell; but although the oyster might perhaps assimilate or embalm a grain of sand forced into its aperture, it could only perish in face of the cyclonic hurricane or the volcanic upheaval of its bed. Her supersensual chaos killed her.

Such seemed the theory of history to be imposed by science on the generation born after 1900. For this theory, Adams felt himself in no way responsible. Even as historian he had made it his duty always to speak with respect of everything that had ever been thought respectable — except an occasional statesman; but he had submitted to force all his life, and he meant to accept it for the future as for the past. All his efforts had been turned only to the search for its channel. He never invented his facts; they were furnished him by the only authorities he could find. As for himself, according to Helmholz, Ernst Mach, and Arthur Balfour, he was henceforth to be a conscious ball of vibrating motions, traversed in every direction by infinite lines of rotation or vibration, rolling at the feet of the Virgin at Chartres or of M. Poincaré in an attic at Paris, a centre of supersensual chaos. The discovery did not distress him. A solitary man of sixty-five years or more, alone in a Gothic cathedral or a Paris apartment, need fret himself little about a few illusions more or less. He should have learned his lesson fifty years earlier; the times had long passed when a student could stop before chaos or order; he had no choice but to march with his world.

Nevertheless, he could not pretend that his mind felt flattered by this scientific outlook. Every fabulist has told how the human mind has always struggled like a frightened bird to escape the chaos which caged it; how—appearing suddenly and inexplicably out of some unknown and unimaginable void; passing half its known life in the mental chaos of sleep; victim even when awake, to its own ill-adjustment, to disease, to age, to external suggestion, to nature's compulsion; doubting its sensations, and, in the last resort, trusting only to instruments and averages—after sixty or seventy years of growing astonishment, the mind wakes to find itself looking blankly into the void of death. That it should profess itself pleased by this performance was all that the highest rules of good breeding could ask; but that it should actually be satisfied would prove that it existed only as idiocy.

Satisfied, the future generation could scarcely think itself, for even when the mind existed in a universe of its own creation, it had never been quite at ease. As far as one ventured to interpret actual science, the mind had thus far adjusted itself by an infinite series of infinitely delicate adjustments forced on it by the infinite motion of an infinite chaos of motion; dragged at one moment into the unknowable and unthinkable, then trying to scramble back within its senses and to bar the chaos out, but always assimilating bits of it, until at last, in 1900, a new avalanche of unknown forces had fallen on it, which required new mental powers to control. If this view was correct, the mind could gain nothing by flight or by fight; it must merge in its supersensual multiverse, or succumb to it.

CHAPTER XXXIV

A Law of Acceleration (1904)

Images are not arguments, rarely even lead to proof, but the mind craves them, and, of late more than ever, the keenest experimenters find twenty images better than one, especially if contradictory; since the human mind has already learned to deal in contradictions.

The image needed here is that of a new centre, or preponderating mass, artificially introduced on earth in the midst of a system of attractive forces that previously made their own equilibrium, and constantly induced to accelerate its motion till it shall establish a new equilibrium. A dynamic theory would begin by assuming that all history, terrestrial or cosmic, mechanical or intellectual, would be reducible to this formula if we knew the facts.

For convenience, the most familiar image should come first; and this is probably that of the comet, or meteoric streams, like the Leonids and Perseids; a complex of minute mechanical agencies, reacting within and without, and guided by the sum of forces attracting or deflecting it. Nothing forbids one to assume that the man-meteorite might grow, as an acorn does, absorbing light, heat, electricity—or thought; for, in recent times, such transference of energy has become a familiar idea; but the simplest figure, at first, is that of a perfect comet—say that of 1843—which drops from space, in a straight line, at the regular acceleration of speed, directly into the sun, and after wheeling sharply about it, in heat that ought to dissipate any known substance, turns back unharmed, in defiance of law, by the path on which it came. The mind, by analogy, may figure as such a comet, the better because it also defies law.

Motion is the ultimate object of science, and measures of motion are many; but with thought as with matter, the true measure is mass in its astronomic sense — the sum or difference of attractive forces. Science has quite enough trouble in measuring its material motions without volunteering help to the historian, but the historian needs not much help to measure some kinds of social movement; and especially in the nineteenth century, society by common accord agreed in measuring its progress by the coal-output. The ratio of increase in the volume of coal-power may serve as dynamometer.

The coal-output of the world, speaking roughly, doubled every ten years between 1840 and 1900, in the form of utilized power, for the ton of coal yielded three or four times as much power in 1900 as in 1840. Rapid as this rate of acceleration in volume seems, it may be tested in a thousand ways without greatly reducing it. Perhaps the ocean steamer is nearest unity and easiest to measure, for any one might hire, in 1905, for a small sum of money, the use of 30,000 steam-horse-power to cross the ocean, and by halving this figure every ten years, he got back to 234 horse-power for 1835, which was accuracy enough for his purposes. In truth, his chief trouble came not from the ratio in volume of heat, but from the intensity, since he could get no basis for a ratio there. All ages of history have known high intensities, like the iron-furnace, the burning-glass, the blow-pipe; but no society has ever used high intensities on any large scale till now, nor can a mere by-stander decide what range of temperature is now in common use. Loosely guessing that science controls habitually the whole range from absolute zero to 3000° Centigrade, one might assume, for convenience, that the ten-year ratio for volume could be used temporarily for intensity; and still there remained a ratio to be guessed for other forces than heat. Since 1800 scores of new forces had been discovered; old forces had been raised to higher powers, as could be measured in the navy-gun; great regions of chemistry had been opened up, and connected with other regions of physics. Within ten years a new universe of force had been revealed in radiation. Complexity had extended itself on immense horizons, and arithmetical ratios were useless for any attempt at accuracy. The force evolved seemed more like explosion than gravitation, and followed closely the curve of steam; but, at all events, the ten-year ratio seemed carefully conservative. Unless the calculator was prepared to be instantly overwhelmed by physical force and mental complexity, he must stop there.

Thus, taking the year 1900 as the starting point for carrying back the series, nothing was easier than to assume a ten-year period of retardation as far back as 1820, but beyond that point the statistician failed, and only the mathematician could help. Laplace would have found it child's-play to fix a ratio of progression in mathematical science between Descartes, Leibnitz, Newton, and himself. Watt could have given in pounds the increase of power between Newcomen's engines and his own. Volta and Benjamin Franklin would have stated their progress as absolute creation of power. Dalton could have measured minutely his advance on Boerhaave. Napoleon I must have had a distant notion of his own numerical relation to Louis XIV. No one in 1789 doubted the progress of force, least of all those who were to lose their heads by it.

Pending agreement between these authorities, theory may assume what it likes—say a fifty, or even a five-and-twenty-year period of reduplication for the eighteenth century, for the period matters little until the acceleration itself is admitted. The subject is even more amusing in the seventeenth than in the eighteenth century, because Galileo and Kepler, Descartes, Huygens, and Isaac Newton took vast pains to fix the laws of acceleration for moving bodies, while Lord Bacon and William Harvey were content with showing experimentally the fact of acceleration in knowledge; but from their combined results a historian might be tempted to maintain a similar rate of movement back to 1600, subject to correction from the historians of mathematics.

The mathematicians might carry their calculations back as far as the fourteenth century when algebra seems to have become for the first time the standard measure of mechanical progress in western Europe; for not only Copernicus and Tycho Brahe, but even artists like Leonardo, Michael Angelo, and Albert Dürer worked by mathematical processes, and their testimony would probably give results more exact than that of Montaigne or Shakespeare; but, to save trouble, one might tentatively carry back the same ratio of acceleration, or retardation, to the year 1400, with the help of Columbus and Gutenberg, so taking a uniform rate during the whole four centuries (1400–1800), and leaving to statisticians the task of correcting it.

Or better, one might, for convenience, use the formula of squares to serve for a law of mind. Any other formula would do as well, either of chemical explosion, or electrolysis, or vegetable growth, or of expansion or contraction in innumerable forms; but this happens to be simple and convenient. Its force increases in the direct ratio of its squares. As the human meteoroid approached the sun or centre of attractive force, the attraction of one century squared itself to give the measure of attraction in the next.

Behind the year 1400, the process certainly went on, but the progress became so slight as to be hardly measurable. What was gained in the east or elsewhere, cannot be known; but forces, called loosely Greek fire and gunpowder, came into use in the west in the thirteenth century, as well as instruments like the compass, the blow-pipe, clocks and spectacles, and materials like paper; Arabic notation and algebra were introduced, while metaphysics and theology acted as violent stimulants to mind. An architect might detect a sequence between the Church of St. Peter's at Rome, the Amiens Cathedral, the Duomo at Pisa, San Marco at Venice, Sancta Sofia at Constantinople and the churches at Ravenna. All the historian dares affirm is that a sequence is manifestly there, and he has a right to carry back his ratio, to represent the fact, without assuming its numerical correctness. On the human mind as a moving body, the break in acceleration in the Middle Ages is only apparent; the attraction worked through shifting forms of force, as the sun works by light or heat, electricity, gravitation, or what not, on different organs with different sensibilities, but with invariable law.

The science of prehistoric man has no value except to prove that the law went back into indefinite antiquity. A stone arrowhead is as convincing as a steam-engine. The values were as clear a hundred thousand years ago as now, and extended equally over the whole world. The motion at last became infinitely slight, but cannot be proved to have stopped. The motion of Newton's comet at aphelion may be equally slight. To evolutionists may be left the processes of evolution; to historians the single interest is the law of reaction between force and force — between mind and nature — the law of progress.

The great division of history into phases by Turgot and Comte first affirmed this law in its outlines by asserting the unity of progress, for a mere phase interrupts no growth, and nature shows innumerable such phases. The development of coal-power in the nineteenth century furnished the first means of assigning closer values to the elements; and the appearance of supersensual forces towards 1900 made this calculation a pressing necessity; since the next step became infinitely serious.

A law of acceleration, definite and constant as any law of mechanics, cannot be supposed to relax its energy to suit the convenience of man. No one is likely to suggest a theory that man's convenience had been consulted by Nature at any time, or that Nature has consulted the convenience of any of her creations, except perhaps the *Terebratula*. In every age man has bitterly and justly complained that Nature hurried and hustled him, for inertia almost invariably has ended in tragedy. Resistance is its law, and resistance to superior mass is futile and fatal.

Fifty years ago, science took for granted that the rate of acceleration could not last. The world forgets quickly, but even today the habit remains of founding statistics on the faith that consumption will continue nearly stationary. Two generations, with John Stuart Mill, talked of this stationary period, which was to follow the explosion of new power. All the men who were elderly in the forties died in this faith, and other men grew old nursing the same conviction, and happy in it; while science, for fifty years, permitted, or encouraged, society to think that force would prove to be limited in supply. This mental inertia of science lasted through the eighties before showing signs of breaking up; and nothing short of radium fairly wakened men to the fact, long since evident, that force was inexhaustible. Even then the scientific authorities vehemently resisted.

Nothing so revolutionary had happened since the year 300. Thought had more than once been upset, but never caught and whirled about in the vortex of infinite forces. Power leaped from every atom, and enough of it to supply the stellar universe showed itself running to waste at every pore of matter. Man could no longer hold it off. Forces grasped his wrists and flung him about as though he had hold of a live wire or a runaway automobile; which was very nearly the exact truth for the purposes of an elderly and timid single gentleman in Paris, who never drove down the Champs Élysées without expecting an accident, and commonly witnessing one; or found himself in the neighborhood of an official without calculating the chances of a bomb. So long as the rates of progress held good, these bombs would double in force and number every ten years.

Impossibilities no longer stood in the way. One's life had fattened on impossibilities. Before the boy was six years old, he had seen four impossibilities made actual — the ocean-steamer, the railway, the electric telegraph, and the Daguerreotype; nor could he ever learn which of the four had most hurried others to come. He had seen the coal-output of the United States grow from nothing to three hundred million tons or more. What was far more serious, he had seen the number of minds, engaged in pursuing force — the truest measure of its attraction — increase from a few scores or hundreds, in 1838, to many thousands in 1905, trained to sharpness never before reached, and armed with instruments amounting to new senses of indefinite power and accuracy, while they chased force into hiding-places where Nature herself had never known it to be, making analyses that contradicted being, and syntheses that endangered the elements. No one could say that the social mind now failed to respond to new force, even when the new force annoyed it horribly. Every day Nature violently revolted, causing so-called accidents with enormous destruction of property and life, while plainly laughing at man, who helplessly groaned and shrieked and shuddered, but never for a single instant could stop. The railways alone approached the carnage of war; automobiles and fire-arms ravaged society, until an earthquake became almost a nervous relaxation. An immense volume of force had detached itself from the unknown universe of energy, while still vaster reservoirs, supposed to be infinite, steadily revealed themselves, attracting mankind with more compulsive course than all the Pontic Seas or Gods or Gold that ever existed, and feeling still less of retiring ebb.

In 1850, science would have smiled at such a romance as this, but, in 1900, as far as history could learn, few men of science thought it a laughing matter. If a perplexed but laborious follower could venture to guess their drift, it seemed in their minds a toss-up between anarchy and order. Unless they should be more honest with themselves in the future than ever they were in the past, they would be more astonished than their followers when they reached the end. If Karl Pearson's notions of the universe were sound, men like Galileo, Descartes, Leibnitz, and Newton should have stopped the progress of science before 1700, supposing them to have been honest in the religious convictions they expressed. In 1900 they were plainly forced back on faith in a unity unproved and an order they had themselves disproved. They had reduced their universe to a series of relations to themselves. They had reduced themselves to motion in a universe of motions, with an acceleration, in their own case, of vertiginous violence.

With the correctness of their science, history had no right to meddle, since their science now lay in a plane where scarcely one or two hundred minds in the world could follow its mathematical processes; but bombs educate vigorously, and even wireless telegraphy or airships might require the reconstruction of society. If any analogy whatever existed between the human mind, on one side, and the laws of motion, on the other, the mind had already entered a field of attraction so violent that it must immediately pass beyond, into new equilibrium, like the Comet of Newton, to suffer dissipation altogether, like meteoroids in the earth's atmosphere. If it behaved like an explosive, it must rapidly recover equilibrium; if it behaved like a vegetable, it must reach its limits of growth; and even if it acted like the earlier creations of energy — the saurians and sharks — it must have nearly reached the limits of its expansion. If science were to go on doubling or quadrupling its complexities every ten years, even mathematics would soon succumb. An average mind had succumbed already in 1850; it could no longer understand the problem in 1900.

Fortunately, a student of history had no responsibility for the problem; he took it as science gave it, and waited only to be taught. With science or with society, he had no quarrel and claimed no share of authority. He had never been able to acquire knowledge, still less to impart it; and if he had, at times, felt serious differences with the American of the nineteenth century, he felt none with the American of the twentieth. For this new creation, born since 1900, a historian asked no longer to be teacher or even friend; he asked only to be a pupil, and promised to be docile, for once, even though trodden under foot; for he could see that the new American — the child of incalculable coal-power, chemical power, electric power, and radiating energy, as well as of new forces yet undertermined — must be a sort of God compared with any former creation of nature. At the rate of progress since 1800, every American who lived into the year 2000 would know how to control unlimited power. He would think in complexities unimaginable to an earlier mind. He would deal with problems altogether beyond the range of earlier society. To him the nineteenth century would stand on the same plane with the fourth — equally childlike — and he would only wonder how both of them, knowing so little, and so weak in force, should have done so much. Perhaps even he might go back, in 1964, to sit with Gibbon on the steps of Ara Cœli.

Meanwhile he was getting education. With that, a teacher who had failed to educate even the generation of 1870, dared not interfere. The new forces would educate. History saw few lessons in the past that would be useful in the future; but one, at least, it did see. The attempt of the American of 1800 to educate the American of 1900 had not often been surpassed for folly; and since 1800 the forces and their complications had increased a thousand times or more. The attempt of the American of 1900 to educate the American of 2000, must be even blinder than that of the Congressman of 1800, except so far as he had learned his ignorance. During a million or two of years, every generation in turn had toiled with endless agony to attain and apply power, all the while betraying the deepest alarm and horror at the power they created. The teacher of 1900, if foolhardy, might stimulate; if foolish, might resist; if intelligent, might balance, as wise and foolish have often tried to do from the beginning; but the forces would continue to educate, and the mind would continue to react. All the teacher could hope was to teach it reaction.

Even there his difficulty was extreme. The most elementary books of science betrayed the inadequacy of old implements of thought. Chapter after chapter closed with phrases such as one never met in older literature: "The cause of this phenomenon is not understood"; "science no longer ventures to explain causes"; "the first step towards a causal explanation still remains to be taken"; "opinions are very much divided"; "in spite of the contradictions involved"; "science gets on only by adopting different theories, sometimes contradictory." Evidently the new American would need to think in contradictions, and instead of Kant's famous four antinomies, the new universe would know no law that could not be proved by its anti-law.

To educate — one's self to begin with — had been the effort of one's life for sixty years; and the difficulties of education had gone on doubling with the coal-output, until the prospect of waiting another ten years, in order to face a seventh doubling of complexities, allured one's imagination but slightly. The law of acceleration was definite, and did not require ten years more study except to show whether it held good. No scheme could be suggested to the new American, and no fault needed to be found, or complaint made; but the next great influx of new forces seemed near at hand, and its style of education promised to be violently coercive. The movement from unity into multiplicity, between 1200 and 1900, was unbroken in sequence, and rapid in acceleration. Prolonged one generation longer, it would require a new social mind. As though thought were common salt in indefinite solution it must enter a new phase subject to new laws. Thus far, since five or ten thousand years, the mind had successfully reacted, and nothing yet proved that it would fail to react — but it would need to jump.

PRAYER
TO
THE VIRGIN OF CHARTRES

Gracious Lady: —

SIMPLE as when I asked your aid before;
  Humble as when I prayed for grace in vain
Seven hundred years ago; weak, weary, sore
  In heart and hope, I ask your help again.

You, who remember all, remember me;
  An English scholar of a Norman name,
I was a thousand who then crossed the sea
  To wrangle in the Paris schools for fame.

When your Byzantine portal was still young
  I prayed there with my master Abailard;
When Ave Maris Stella was first sung,
  I helped to sing it here with Saint Bernard.

When Blanche set up your gorgeous Rose of
    France
  I stood among the servants of the Queen;
And when Saint Louis made his penitence,
  I followed barefoot where the King had
    been.

For centuries I brought you all my cares,
And vexed you with the murmurs of a child;
You heard the tedious burden of my prayers;
You could not grant them, but at least you smiled.

If then I left you, it was not my crime,
Or if a crime, it was not mine alone.
All children wander with the truant Time.
Pardon me too! You pardoned once your Son!

For He said to you: — "Wist ye not that I
Must be about my Father's business?" So,
Seeking his Father he pursued his way
Straight to the Cross towards which we all must go.

So I too wandered off among the host
That racked the earth to find the father's clue.
I did not find the Father, but I lost
What now I value more, the Mother, — You!

I thought the fault was yours that foiled my search;
I turned and broke your image on its throne,

Cast down my idol, and resumed my march
   To claim the father's empire for my own.

Crossing the hostile sea, our greedy band
   Saw rising hills and forests in the blue;
Our father's kingdom in the promised land!
   — We seized it, and dethroned the father too.

And now we are the Father, with our brood,
   Ruling the Infinite, not Three but One;
We made our world and saw that it was good;
   Ourselves we worship, and we have no Son.

Yet we have Gods, for even our strong nerve
   Falters before the Energy we own.
Which shall be master? Which of us shall serve?
   Which wears the fetters? Which shall bear the crown?

Brave though we be, we dread to face the Sphinx,
   Or answer the old riddle she still asks.
Strong as we are, our reckless courage shrinks
   To look beyond the piece-work of our tasks.

But when we must, we pray, as in the past
Before the Cross on which your Son was nailed.
Listen, dear lady! You shall hear the last
Of the strange prayers Humanity has wailed.

PRAYER TO THE DYNAMO

Mysterious Power! Gentle Friend!
Despotic Master! Tireless Force!
You and We are near the End.
Either You or We must bend
To bear the martyrs' Cross.

We know ourselves, what we can bear
As men; our strength and weakness too;
Down to the fraction of a hair;
And know that we, with all our care
And knowledge, know not you.

You come in silence, Primal Force,
We know not whence, or when, or why;
You stay a moment in your course
To play; and, lo! you leap across
To Alpha Centauri!

PRAYER TO THE VIRGIN

We know not whether you are kind,
  Or cruel in your fiercer mood;
But be you Matter, be you Mind,
We think we know that you are blind,
  And we alone are good.

We know that prayer is thrown away,
  For you are only force and light;
A shifting current; night and day;
We know this well, and yet we pray,
  For prayer is infinite,

Like you! Within the finite sphere
  That bounds the impotence of thought,
We search an outlet everywhere
But only find that we are here
  And that you are — are not!

What are we then? the lords of space?
  The master-mind whose tasks you do?
Jockey who rides you in the race?
Or are we atoms whirled apace,
  Shaped and controlled by you?

Still silence! Still no end in sight!
  No sound in answer to our cry!
Then, by the God we now hold tight,

Though we destroy soul, life and light,
    Answer you shall — or die!

We are no beggars! What care we
    For hopes or terrors, love or hate?
What for the universe? We see
Only our certain destiny
    And the last word of Fate.

Seize, then, the Atom! rack his joints!
    Tear out of him his secret spring!
Grind him to nothing! — though he points
To us, and his life-blood anoints
    Me — the dead Atom-King!

A CURIOUS prayer, dear lady! is it not?
    Strangely unlike the prayers I prayed to
        you!
Stranger because you find me at this spot,
    Here, at your feet, asking your help anew.

Strangest of all, that I have ceased to strive,
    Ceased even care what new coin fate shall
        strike.
In truth it does not matter. Fate will give
    Some answer; and all answers are alike.

So, while we slowly rack and torture death
And wait for what the final void will show,
Waiting I feel the energy of faith
Not in the future science, but in you!

The man who solves the Infinite, and needs
The force of solar systems for his play,
Will not need me, nor greatly care what deeds
Made me illustrious in the dawn of day.

He will send me, dethroned, to claim my rights,
Fossil survival of an age of stone,
Among the cave-men and the troglodytes
Who carved the mammoth on the mammoth's bone.

He will forget my thought, my acts, my fame,
As we forget the shadows of the dusk,
Or catalogue the echo of a name
As we the scratches on the mammoth's tusk.

But when, like me, he too has trod the track
Which leads him up to power above control,
He too will have no choice but wander back
And sink in helpless hopelessness of soul,

Before your majesty of grace and love,
The purity, the beauty and the faith;
The depth of tenderness beneath; above,
The glory of the life and of the death.

When your Byzantine portal still was young,
I came here with my master Abailard;
When Ave Maris Stella was first sung,
I joined to sing it here with Saint Bernard.

When Blanche set up your glorious Rose of France,
In scholar's robes I waited on the Queen;
When good Saint Louis did his penitence,
My prayer was deep like his: my faith as keen.

What loftier prize seven hundred years shall bring,
What deadlier struggles for a larger air,
What immortality our strength shall wring
From Time and Space, we may — or may not — care;

But years, or ages, or eternity,
Will find me still in thought before your throne,

PRAYER TO THE VIRGIN

Pondering the mystery of Maternity,
Soul within Soul,—Mother and Child in One!

Help me to see! not with my mimic sight—
With yours! which carried radiance, like the sun,
Giving the rays you saw with—light in light—
Tying all suns and stars and worlds in one.

Help me to know! not with my mocking art—
With you, who knew yourself unbound by laws;
Gave God your strength, your life, your sight, your heart,
And took from him the Thought that Is—the Cause.

Help me to feel! not with my insect sense,—
With yours that felt all life alive in you;
Infinite heart beating at your expense;
Infinite passion breathing the breath you drew!

Help me to bear! not my own baby load,
  But yours; who bore the failure of the light,
The strength, the knowledge and the thought
    of God,—
  The futile folly of the Infinite!

8

Aldous Huxley, BRAVE NEW WORLD

H. G. Wells, THE TIME MACHINE

1. Do you see any parallels between the conditioning of children described by Huxley and any aspects of current society? Explain.

2. Implicit in *Brave New World* are at least two sharply different views of technology. Characterize them.

3. *Brave New World* was published in 1932. Has the relationship between technology and society developed in the way Huxley envisioned? Why or why not?

4. What aspects of *Brave New World* still seem far fetched, and what aspects already have come to pass or have been overtaken in the contemporary world?

5. Would you look forward to living in a society like that described in *Brave New World?* Is our society moving in that direction? What can be done either to accelerate or to stop that development?

6. Do you think most people of today would be willing to trade freedom for security? Explain.

7. To what extent do you think that the application of science and technology can overcome current problems of population and food resources, war and peace, and energy?

8. If technology were able to solve the problems of population and food, war and peace, and energy, then what would we have? Compare your views with those of Wells.

Aldous Huxley (1894-1963) was born in England, the son of a physician and grandson of Thomas H. Huxley, and educated at Eton and Oxford. A siege of temporary blindness kept him from going to medical school, and he turned his attention to literature. As journalist, essayist, and novelist, he quickly became a prolific and renowned writer.

In the words of Huxley himself, "*Brave New World* presents a fanciful and somewhat ribald picture of a society, in which the attempt to recreate human beings in the likeness of termites has been pushed almost to the limits of the possible."

BRAVE NEW WORLD

CHAPTER ONE

A SQUAT grey building of only thirty-four stories. Over the main entrance the words, CENTRAL LONDON HATCHERY AND CONDITIONING CENTRE, and, in a shield, the World State's motto, COMMUNITY, IDENTITY, STABILITY.

The enormous room on the ground floor faced towards the north. Cold for all the summer beyond the panes, for all the tropical heat of the room itself, a harsh thin light glared through the windows, hungrily seeking some draped lay figure, some pallid shape of academic goose-flesh, but finding only the glass and nickel and bleakly shining porcelain of a laboratory. Wintriness responded to wintriness. The overalls of the workers were white, their hands gloved with a pale corpse-coloured rubber. The light was frozen, dead, a ghost. Only from the yellow barrels of the microscopes did it borrow a certain rich and living substance, lying along the polished tubes like butter, streak after luscious streak in long recession down the work tables.

"And this," said the Director opening the door, "is the Fertilizing Room."

Bent over their instruments, three hundred Fertilizers were plunged, as the Director of Hatcheries and Conditioning entered the room, in the scarcely breathing silence, the absent-minded, soliloquizing hum or whistle, of absorbed concentration. A troop of newly arrived students, very young, pink and callow, followed nervously, rather abjectly, at the Director's heels. Each of them carried a notebook, in which, whenever the great man spoke, he desperately scribbled. Straight from the horse's mouth. It was a rare privilege. The D.H.C. for Central London always made a point of personally conducting his new students round the various departments.

"Just to give you a general idea," he would explain to them. For of course some sort of general idea they must have, if they were to do their work intelligently—though as little of one, if they were to be good and happy members of society, as possible. For particulars, as every one knows, make for virtue and happiness; generalities are intellectually necessary evils. Not philosophers but fret-sawyers and stamp collectors compose the backbone of society.

"To-morrow," he would add, smiling at them with a slightly menacing geniality, "you'll be settling down to serious work. You won't have time for generalities. Meanwhile . . ."

Meanwhile, it was a privilege. Straight from the horse's mouth into the notebook. The boys scribbled like mad.

Tall and rather thin but upright, the Director advanced into the room. He had a long chin and big, rather prominent teeth, just covered, when he was not talking, by his full, floridly curved lips. Old, young? Thirty? Fifty? Fifty-five? It was hard to say. And anyhow the question didn't arise; in this year of stability, A.F. 632, it didn't occur to you to ask it.

"I shall begin at the beginning " said the D.H.C. and the more zealous students recorded his intention in their notebooks: *Begin at the beginning.* "These," he waved his hand, "are the incubators." And opening an insulated door he showed them racks upon racks of numbered test-tubes. "The week's supply of ova. Kept," he explained, "at blood heat; whereas the male gametes," and here he opened another door, "they have to be kept at thirty-five instead of thirty-seven. Full blood heat sterilizes." Rams wrapped in theremogene beget no lambs.

Still leaning against the incubators he gave them, while the pencils scurried illegibly across the pages, a brief description of the modern fertilizing process; spoke first, of course, of its surgical introduction—"the operation undergone voluntarily for the good of Society, not to mention the fact that it carries a bonus amounting to six months' salary"; continued with some account of the technique for preserving the excised ovary alive and actively developing; passed on to a consideration of optimum temperature, salinity, viscosity; referred to the liquor in which the detached and ripened eggs were kept; and, leading his charges to the work tables, actually showed them how this liquor was drawn off from the test-tubes; how it was let out drop by drop onto the specially warmed slides of the microscopes; how the eggs which it contained were inspected for abnormalities, counted and transferred to a porous receptacle; how (and he now took them to watch the operation) this receptacle was immersed in a warm bouillon containing free-swimming spermatozoa—at a minimum concentration of one hundred thousand per cubic centimetre, he insisted; and how, after ten minutes, the container was lifted out of the liquor and its contents re-examined; how, if any of the eggs remained unfertilized, it was again immersed, and, if necessary, yet again; how the fertilized ova went back to the incubators; where the Alphas and Betas remained until definitely bottled; while the Gammas, Deltas and Epsilons were brought out again, after only thirty-six hours, to undergo Bokanovsky's Process.

"Bokanovsky's Process," repeated the Director, and the students underlined the words in their little notebooks.

One egg, one embryo, one adult—normality. But a bokanovskified egg will bud, will proliferate, will divide. From eight to ninety-six buds, and every bud will grow into a perfectly formed embryo, and every embryo into a full-sized adult. Making ninety-six human beings grow where only one grew before. Progress.

"Essentially," the D.H.C. concluded, "bokanovskification consists of a series of arrests of development. We check the normal growth and, paradoxically enough, the egg responds by budding."

Responds by budding. The pencils were busy.

He pointed. On a very slowly moving band a rack-full of test-tubes was entering a large metal box, another rack-full was emerging. Machinery faintly purred. It took eight minutes for the tubes to go through, he told them. Eight minutes of hard X-rays being about as much as an egg can stand. A few died; of the rest, the least susceptible divided into two; most put out four buds; some eight; all were returned to the incubators, where the buds began to develop; then, after two days, were suddenly chilled, chilled and checked. Two, four, eight, the buds in their turn budded; and having budded were dosed almost to death with alcohol; consequently burgeoned again and having budded—bud out of bud out of bud—were thereafter—further arrest being generally fatal—left to develop in peace. By which time the original egg was in a fair way to becoming anything from eight to ninety-six embryos—a prodigious improvement, you will agree, on nature. Identical twins—but not in piddling twos and threes as in the old viviparous days, when an egg would sometimes accidentally divide; actually by dozens, by scores at a time.

"Scores," the Director repeated and flung out his arms, as though he were distributing largesse. "Scores."

But one of the students was fool enough to ask where the advantage lay.

"My good boy!" The Director wheeled sharply round on him. "Can't you see? Can't you *see?*" He raised a hand; his expression was solemn. "Bokanovsky's Process is one of the major instruments of social stability!"

Major instruments of social stability.

Standard men and women; in uniform batches. The whole of a small factory staffed with the products of a single bokanovskified egg.

"Ninety-six identical twins working ninety-six identical machines!" The voice was almost tremulous with enthusiasm. "You really know where you are. For the first time in history." He quoted the planetary motto. "Community, Identity, Stability." Grand words. "If we could bokanovskify indefinitely the whole problem would be solved."

Solved by standard Gammas, unvarying Deltas, uniform Epsilons. Millions of identical twins. The principle of mass production at last applied to biology.

"But, alas," the Director shook his head, "we *can't* bokanovskify indefinitely."

Ninety-six seemed to be the limit; seventy-two a good average. From the same ovary and with gametes of the same male to manufacture as many batches of identical twins as possible—that was the best (sadly a second best) that they could do. And even that was difficult.

"For in nature it takes thirty years for two hundred eggs to reach maturity. But our business is to stabilize the population at this moment, here and now. Dribbling out twins over a quarter of a century—what would be the use of that?"

Obviously, no use at all. But Podsnap's Technique had immensely accelerated the process of ripening. They could make sure of at least a hundred and fifty mature eggs within two years. Fertilize and bokanovskify—in other words, multiply by seventy-two—and you get an average of nearly eleven thousand brothers and sisters in a hundred and fifty batches of identical twins, all within two years of the same age.

"And in exceptional cases we can make one ovary yield us over fifteen thousand adult individuals."

Beckoning to a fair-haired, ruddy young man who happened to be passing at the moment, "Mr. Foster," he called. The ruddy young man approached. "Can you tell us the record for a single ovary, Mr. Foster?"

"Sixteen thousand and twelve in this Centre," Mr. Foster replied without hesitation. He spoke very quickly, had a vivacious blue eye, and took an evident pleasure in quoting figures. "Sixteen thousand and twelve; in one hundred and eighty-nine batches of identicals. But of course they've done much better," he rattled on, "in some of the tropical Centres. Singapore has often produced over sixteen thousand five hundred; and Mombasa has actually touched the seventeen thousand mark. But then they have unfair advantages. You should see the way a negro ovary responds to pituitary! It's quite astonishing, when you're used to working with European material. Still," he added, with a laugh (but the light of combat was in his eyes and the lift of his chin was challenging), "still, we mean to beat them if we can. I'm working on a wonderful Delta-Minus ovary at this moment. Only just eighteen months old. Over twelve thousand seven hundred children already, either decanted or in embryo. And still going strong. We'll beat them yet."

"That's the spirit I like!" cried the Director, and clapped Mr. Foster on the shoulder. "Come along with us and give these boys the benefit of your expert knowledge."

Mr. Foster smiled modestly. "With pleasure." They went.

In the Bottling Room all was harmonious bustle and ordered activity. Flaps of fresh sow's peritoneum ready cut to the proper size came shooting up in little lifts from the Organ Store in the sub-basement. Whizz and then, click! the lift-hatches flew open; the bottle-liner had only to reach out a hand, take the flap, insert, smooth-down, and before the lined bottle had had time to travel out of reach along the endless band, whizz, click! another flap of peritoneum had shot up from the depths, ready to be slipped into yet another bottle, the next of that slow interminable procession on the band.

Next to the Liners stood the Matriculators. The procession advanced; one by one the eggs were transferred from their test-tubes to the larger containers; deftly the peritoneal lining was slit, the morula dropped into place, the saline solution poured in . . . and already the bottle had passed, and it was the turn of the labellers. Heredity, date of fertilization, membership of Bokanovsky Group—details were transferred from test-tube to bottle. No longer anonymous, but named, identified, the procession marched slowly on; on through an opening in the wall, slowly on into the Social Predestination Room.

"Eighty-eight cubic metres of card-index," said Mr. Foster with relish as they entered.

"Containing *all* the relevant information," added the Director.

"Brought up to date every morning."

"And co-ordinated every afternoon."

"On the basis of which they make their calculations."

"So many individuals, of such and such quality," said Mr. Foster.

"Distributed in such and such quantities."

"The optimum Decanting Rate at any given moment."

"Unforeseen wastages promptly made good."

"Promptly," repeated Mr. Foster. "If you knew the amount of overtime I had to put in after the last Japanese earthquake!" He laughed good-humouredly and shook his head.

"The Predestinators send in their figures to the Fertilizers."

"Who give them the embryos they ask for."

"And the bottles come in here to be predestinated in detail."

"After which they are sent down to the Embryo Store."

"Where we now proceed ourselves."

And opening a door Mr. Foster led the way down a staircase into the basement.

The temperature was still tropical. They descended into a thickening twilight. Two doors and a passage with a double turn insured the cellar against any possible infiltration of the day.

"Embryos are like photograph film," said Mr.

Foster waggishly, as he pushed open the second door. "They can only stand red light."

And in effect the sultry darkness into which the students now followed him was visible and crimson, like the darkness of closed eyes on a summer's afternoon. The bulging flanks of row on receding row and tier above tier of bottles glinted with innumerable rubies, and among the rubies moved the dim red spectres of men and women with purple eyes and all the symptoms of lupus. The hum and rattle of machinery faintly stirred the air.

"Give them a few figures, Mr. Foster," said the Director, who was tired of talking.

Mr. Foster was only too happy to give them a few figures.

Two hundred and twenty metres long, two hundred wide, ten high. He pointed upwards. Like chickens drinking, the students lifted their eyes towards the distant ceiling.

Three tiers of racks: ground floor level, first gallery, second gallery.

The spidery steel-work of gallery above gallery faded away in all directions into the dark. Near them three red ghosts were busily unloading demijohns from a moving staircase.

The escalator from the Social Predestination Room.

Each bottle could be placed on one of fifteen racks, each rack, though you couldn't see it, was a conveyor travelling at the rate of thirty-three and a third centimetres an hour. Two hundred and sixty-seven days at eight metres a day. Two thousand one hundred and thirty-six metres in all. One circuit of the cellar at ground level, one on the first gallery, half on the second, and on the two hundred and sixty-seventh morning, daylight in the Decanting Room. Independent existence—so called.

"But in the interval," Mr. Foster concluded, "we've managed to do a lot to them. Oh, a very great deal." His laugh was knowing and triumphant.

"That's the spirit I like," said the Director once more. "Let's walk round. You tell them everything, Mr. Foster."

Mr. Foster duly told them.

Told them of the growing embryo on its bed of peritoneum. Made them taste the rich blood surrogate on which it fed. Explained why it had to be stimulated with placentin and thyroxin. Told them of the *corpus luteum* extract. Showed them the jets through which at every twelfth metre from zero to 2040 it was automatically injected. Spoke of those gradually increasing doses of pituitary administered during the final ninety-six metres of their course. Described the artificial maternal circulation installed on every bottle at Metre 112; showed them the reservoir of blood-surrogate, the centrifugal pump that kept the liquid moving over the placenta and drove it through the synthetic lung and waste-product filter. Referred to the embryo's troublesome tendency to anæmia, to the massive doses of hog's stomach extract and foetal foal's liver with which, in consequence, it had to be supplied.

Showed them the simple mechanism by means of which, during the last two metres out of every eight, all the embryos were simultaneously shaken into familiarity with movement. Hinted at the gravity of the so-called "trauma of decanting," and enumerated the precautions taken to minimize, by a suitable training of the bottled embryo, that dangerous shock. Told them of the tests for sex carried out in the neighbourhood of metre 200. Explained the system of labelling—a T for the males, a circle for the females and for those who were destined to become freemartins a question mark, black on a white ground.

"For of course," said Mr. Foster, "in the vast majority of cases, fertility is merely a nuisance. One fertile ovary in twelve hundred—that would really be quite sufficient for our purposes. But we want to have a good choice. And of course one must always leave an enormous margin of safety. So we allow as many as thirty per cent. of the female embryos to develop normally. The others get a dose of male sex-hormone every twenty-four metres for the rest of the course. Result: they're decanted as freemartins—structurally quite normal (except," he had to admit, "that they *do* have just the slightest tendency to grow beards), but sterile. Guaranteed sterile. Which brings us at last," continued Mr. Foster, "out of the realm of mere slavish imitation of nature into the much more interesting world of human invention."

He rubbed his hands. For of course, they didn't content themselves with merely hatching out embryos: any cow could do that.

"We also predestine and condition. We decant our babies as socialized human beings, as Alphas or Epsilons, as future sewage workers or future . . ." He was going to say "future World controllers," but correcting himself, said "future Directors of Hatcheries," instead.

The D.H.C. acknowledged the compliment with a smile.

They were passing Metre 320 on rack 11. A young Beta-Minus mechanic was busy with screwdriver and spanner on the blood-surrogate pump of a passing bottle. The hum of the electric motor deepened by fractions of a tone as he turned the nuts. Down, down . . . A final twist, a glance at the revolution counter, and he was done. He moved two paces down the line and began the same process on the next pump.

"Reducing the number of revolutions per minute," Mr. Foster explained. "The surrogate goes round slower; therefore passes through the lung at longer intervals; therefore gives the embryo less oxygen. Nothing like oxygen-shortage for keeping an embryo below par." Again he rubbed his hands.

"But why do you want to keep the embryo below par?" asked an ingenuous student.

"Ass!" said the Director, breaking a long silence. "Hasn't it occurred to you that an Epsilon embryo must have an Epsilon environment as well as an Epsilon heredity?"

It evidently hadn't occurred to him. He was covered with confusion.

"The lower the caste," said Mr. Foster, "the shorter the oxygen." The first organ affected was the brain. After that the skeleton. At seventy per cent. of normal oxygen you got dwarfs. At less than seventy eyeless monsters.

"Who are no use at all," concluded Mr. Foster.

Whereas (his voice became confidential and eager), if they could discover a technique for shortening the period of maturation what a triumph, what a benefaction to Society!

"Consider the horse."

They considered it.

Mature at six; the elephant at ten. While at thirteen a man is not yet sexually mature; and is only full-grown at twenty. Hence, of course, that fruit of delayed development, the human intelligence.

"But in Epsilons," said Mr. Foster very justly, "we don't need human intelligence."

Didn't need and didn't get it. But though the

Epsilon mind was mature at ten, the Epsilon body was not fit to work till eighteen. Long years of superfluous and wasted immaturity. If the physical development could be speeded up till it was as quick, say, as a cow's, what an enormous saving to the Community!

"Enormous!" murmured the students. Mr. Foster's enthusiasm was infectious.

He became rather technical; spoke of the abnormal endocrine co-ordination which made men grow so slowly; postulated a germinal mutation to account for it. Could the effects of this germinal mutation be undone? Could the individual Epsilon embryo be made a revert, by a suitable technique, to the normality of dogs and cows? That was the problem. And it was all but solved.

Pilkington, at Mombasa, had produced individuals who were sexually mature at four and full-grown at six and a half. A scientific triumph. But socially useless. Six-year-old men and women were too stupid to do even Epsilon work. And the process was an all-or-nothing one; either you failed to modify at all, or else you modified the whole way. They were still trying to find the ideal compromise between adults of twenty and adults of six. So far without success. Mr. Foster sighed and shook his head.

Their wanderings through the crimson twilight had brought them to the neighbourhood of Metre

170 on Rack 9. From this point onwards Rack 9 was enclosed and the bottles performed the remainder of their journey in a kind of tunnel, interrupted here and there by openings two or three metres wide.

"Heat conditioning," said Mr. Foster.

Hot tunnels alternated with cool tunnels. Coolness was wedded to discomfort in the form of hard X-rays. By the time they were decanted, the embryos had a horror of cold. They were predestined to emigrate to the tropics, to be miners and acetate silk spinners and steel workers. Later on their minds would be made to endorse the judgment of their bodies. "We condition them to thrive on heat," concluded Mr. Foster. "Our colleagues upstairs will teach them to love it."

"And that," put in the Director sententiously, "that is the secret of happiness and virtue—liking what you've *got* to do. All conditioning aims at that: making people like their unescapable social destiny."

In a gap between two tunnels, a nurse was delicately probing with a long fine syringe into the gelatinous contents of a passing bottle. The students and their guides stood watching her for a few moments in silence.

"Well, Lenina," said Mr. Foster, when at last she withdrew the syringe and straightened herself up.

The girl turned with a start. One could see that, for all the lupus and the purple eyes, she was uncommonly pretty.

"Henry!" Her smile flashed redly at him—a row of coral teeth.

"Charming, charming," murmured the Director and, giving her two or three little pats, received in exchange a rather deferential smile for himself.

"What are you giving them?" asked Mr. Foster, making his tone very professional.

"Oh, the usual typhoid and sleeping sickness."

"Tropical workers start being inoculated at Metre 150," Mr. Foster explained to the students. "The embryos still have gills. We immunize the fish against the future man's diseases." Then, turning back to Lenina, "Ten to five on the roof this afternoon," he said, "as usual."

"Charming," said the Director once more, and, with a final pat, moved away after the others.

On Rack 10 rows of next generation's chemical workers were being trained in the toleration of lead, caustic soda, tar, chlorine. The first of a batch of two hundred and fifty embryonic rocket-plane engineers was just passing the eleven hundred metre mark on Rack 3. A special mechanism kept their containers in constant rotation. "To improve their sense of balance," Mr. Foster explained. "Doing repairs on the outside of a rocket in mid-air is a ticklish job. We slacken off the circulation when they're right way up, so that they're half starved, and double the flow of surrogate when they're upside down. They learn to associate topsy-turvydom with well-being; in fact, they're only truly happy when they're standing on their heads.

"And now," Mr. Foster went on, "I'd like to show you some very interesting conditioning for Alpha Plus Intellectuals. We have a big batch of them on Rack 5. First Gallery level," he called to two boys who had started to go down to the ground floor.

"They're round about Metre 900," he explained. "You can't really do any useful intellectual conditioning till the foetuses have lost their tails. Follow me."

But the Director had looked at his watch. "Ten to three," he said. "No time for the intellectual embryos, I'm afraid. We must go up to the Nurseries before the children have finished their afternoon sleep."

Mr. Foster was disappointed. "At least one glance at the Decanting Room," he pleaded.

"Very well then." The Director smiled indulgently. "Just one glance."

CHAPTER SIXTEEN

THE room into which the three were ushered was the Controller's study.

"His fordship will be down in a moment." The Gamma butler left them to themselves.

Helmholtz laughed aloud.

"It's more like a caffeine-solution party than a trial," he said, and let himself fall into the most luxurious of the pneumatic arm-chairs. "Cheer up, Bernard," he added, catching sight of his friend's green unhappy face. But Bernard would not be cheered; without answering, without even looking at Helmholtz, he went and sat down on the most uncomfortable chair in the room, carefully chosen in the obscure hope of somehow deprecating the wrath of the higher powers.

The Savage meanwhile wandered restlessly round the room, peering with a vague superficial inquisitiveness at the books in the shelves, at the sound-track rolls and the reading machine bobbins in their numbered pigeon-holes. On the table under the window lay a massive volume bound in limp black leather-surrogate, and stamped with large golden T's. He picked it up and opened it. My Life and Work, by Our Ford. The book had been published at Detroit by the Society for the Propagation of Fordian Knowledge. Idly he turned the pages, read a sentence here, a paragraph there, and had just come to the conclusion that the book didn't interest him, when the door opened, and the Resident World Controller for Western Europe walked briskly into the room.

Mustapha Mond shook hands with all three of them; but it was to the Savage that he addressed himself. "So you don't much like civilization, Mr. Savage," he said.

The Savage looked at him. He had been prepared to lie, to bluster, to remain sullenly unresponsive; but, reassured by the good-humoured intelligence of the Controller's face, he decided to tell the truth, straightforwardly. "No." He shook his head.

Bernard started and looked horrified. What would the Controller think? To be labelled as the friend of a man who said that he didn't like civilization—said it openly and, of all people, to the Controller—it was terrible. "But, John," he began. A look from Mustapha Mond reduced him to an abject silence.

"Of course," the Savage went on to admit, "there are some very nice things. All that music in the air, for instance . . ."

"Sometimes a thousand twangling instruments will hum about my ears and sometimes voices."

The Savage's face lit up with a sudden pleasure. "Have you read it too?" he asked. "I thought nobody knew about that book here, in England."

"Almost nobody. I'm one of the very few. It's prohibited, you see. But as I make the laws here, I can also break them. With impunity, Mr. Marx," he added, turning to Bernard. "Which I'm afraid you *can't* do."

Bernard sank into a yet more hopeless misery.

"But why is it prohibited?" asked the Savage. In the excitement of meeting a man who had read Shakespeare he had momentarily forgotten everything else.

The Controller shrugged his shoulders. "Because it's old; that's the chief reason. We haven't any use for old things here."

"Even when they're beautiful?"

"Particularly when they're beautiful. Beauty's attractive, and we don't want people to be attracted by old things. We want them to like the new ones."

"But the new ones are so stupid and horrible. Those plays, where there's nothing but helicopters flying about and you *feel* the people kissing." He made a grimace. "Goats and monkeys!" Only in Othello's words could he find an adequate vehicle for his contempt and hatred.

"Nice tame animals, anyhow," the Controller murmured parenthetically.

"Why don't you let them see *Othello* instead?"

"I've told you; it's old. Besides, they couldn't understand it."

Yes, that was true. He remembered how Helmholtz had laughed at *Romeo and Juliet.* "Well then," he said, after a pause, "something new that's like *Othello*, and that they could understand."

"That's what we've all been wanting to write," said Helmholtz, breaking a long silence.

"And it's what you never will write," said the Controller. "Because, if it were really like *Othello* nobody could understand it, however new it might be. And if it were new, it couldn't possibly be like *Othello.*"

"Why not?"

"Yes, why not?" Helmholtz repeated. He too was forgetting the unpleasant realities of the situation. Green with anxiety and apprehension, only Bernard remembered them; the others ignored him. "Why not?"

"Because our world is not the same as Othello's world. You can't make flivvers without steel—and you can't make tragedies without social instability. The world's stable now. People are happy; they get what they want, and they never want what they can't get. They're well off; they're safe; they're never ill; they're not afraid of death; they're blissfully ignorant of passion and old age; they're plagued with no mothers or fathers; they've got no wives, or children, or lovers to feel strongly about; they're so conditioned that they practically can't help behaving as they ought to behave. And if anything should go wrong, there's *soma*. Which you go and chuck out of the window in the name of liberty, Mr. Savage. *Liberty!*" He laughed. "Expecting Deltas to know what liberty is! And now expecting them to understand *Othello!* My good boy!"

The Savage was silent for a little. "All the same," he insisted obstinately, "*Othello's* good, *Othello's* better than those feelies."

"Of course it is," the Controller agreed. "But that's the price we have to pay for stability. You've got to choose between happiness and what people used to call high art. We've sacrificed the high art. We have the feelies and the scent organ instead."

"But they don't mean anything."

"They mean themselves; they mean a lot of agreeable sensations to the audience."

"But they're . . . they're told by an idiot."

The Controller laughed. "You're not being very polite to your friend, Mr. Watson. One of our most distinguished Emotional Engineers . . ."

"But he's right," said Helmholtz gloomily. "Because it *is* idiotic. Writing when there's nothing to say . . ."

"Precisely. But that requires the most enormous ingenuity. You're making flivvers out of the absolute minimum of steel—works of art out of practically nothing but pure sensation."

The Savage shook his head. "It all seems to me quite horrible."

"Of course it does. Actual happiness always looks pretty squalid in comparison with the over-compensations for misery. And, of course, stability isn't nearly so spectacular as instability. And being contented has none of the glamour of a good fight against misfortune, none of the picturesqueness of a struggle with temptation, or a fatal overthrow by passion or doubt. Happiness is never grand."

"I suppose not," said the Savage after a silence. "But need it be quite so bad as those twins?" He passed his hand over his eyes as though he were trying to wipe away the remembered image of those long rows of identical midgets at the assembling tables, those queued-up twin-herds at the entrance to the Brentford monorail station, those human maggots swarming round Linda's bed of death, the endlessly repeated face of his assailants. He looked at his bandaged left hand and shuddered. "Horrible!"

"But how useful! I see you don't like our Bokanovsky Groups; but, I assure you, they're the foundation on which everything else is built. They're the gyroscope that stabilizes the rocket plane of state on its unswerving course." The deep voice thrillingly vibrated; the gesticulating hand implied all space and the onrush of the irresistible machine. Mustapha Mond's oratory was almost up to synthetic standards.

"I was wondering," said the Savage, "why you had them at all—seeing that you can get whatever you want out of those bottles. Why don't you make everybody an Alpha Double Plus while you're about it?"

Mustapha Mond laughed. "Because we have no wish to have our throats cut," he answered. "We believe in happiness and stability. A society of Alphas couldn't fail to be unstable and miserable. Imagine a factory staffed by Alphas—that is to say by separate and unrelated individuals of good heredity and conditioned so as to be capable (within limits) of making a free choice and assuming responsibilities. Imagine it!" he repeated.

The Savage tried to imagine it, not very successfully.

"It's an absurdity. An Alpha-decanted, Alpha-conditioned man would go mad if he had to do Epsilon Semi-Moron work—go mad, or start smashing things up. Alphas can be completely socialized—but only on condition that you make them do Alpha work. Only an Epsilon can be expected to make Epsilon sacrifices, for the good reason that for him they aren't sacrifices; they're the line of least resistance. His conditioning has laid down rails along which he's got to run. He can't help himself; he's foredoomed. Even after decanting, he's still inside a bottle—an invisible bottle of infantile and embryonic fixations. Each one of us, of course," the Controller meditatively continued, "goes through life inside a bottle. But if we happen to be Alphas, our bottles are, relatively speaking, enormous. We should suffer acutely if we were confined in a narrower space. You cannot pour upper-caste champagne-surrogate into lower-caste bottles. It's obvious theoretically. But it has also been proved in actual practice. The result of the Cyprus experiment was convincing."

"What was that?" asked the Savage.

Mustapha Mond smiled. "Well, you can call it an experiment in rebottling if you like. It began in A.F. 473. The Controllers had the island of Cyprus cleared of all its existing inhabitants and re-colonized with a specially prepared batch of twenty-two thousand Alphas. All agricultural and industrial equipment was handed over to them and they were left to manage their own affairs. The result exactly fulfilled all the theoretical predictions. The land wasn't properly worked; there were strikes in all the factories; the laws were set at naught, orders disobeyed; all the people detailed for a spell of low-grade work were perpetually intriguing for high-grade jobs, and all the people with high-grade jobs were counter-intriguing at all costs to stay where they were. Within six years they were having a first-class civil war. When nineteen out of the twenty-two thousand had been killed, the survivors unanimously petitioned the World Controllers to resume the government of the island. Which they did. And that was the end of the only society of Alphas that the world has ever seen."

The Savage sighed, profoundly.

"The optimum population," said Mustapha Mond, "is modelled on the iceberg—eight-ninths below the water line, one-ninth above."

"And they're happy below the water line?"

"Happier than above it. Happier than your friend here, for example." He pointed.

"In spite of that awful work?"

"Awful? *They* don't find it so. On the contrary, they like it. It's light, it's childishly simple. No strain on the mind or the muscles. Seven and a half hours of mild, unexhausting labour, and then the *soma* ration and games and unrestricted copulation and the feelies. What more can they ask for? True," he added, "they might ask for shorter hours. And of course we could give them shorter hours. Technically, it would be perfectly simple to reduce all lower-caste working hours to three or four a day. But would they be any the happier for that? No, they wouldn't. The experiment was tried, more than a century and a half ago. The whole of Ireland was put on to the four-hour day. What was the result? Unrest and a large increase in the consumption of *soma;* that was all. Those three and a half hours of extra leisure were so far from being a source of happiness, that people felt constrained to take a holiday from them. The Inventions Office is stuffed with plans for labour-saving processes. Thousands of them." Mustapha Mond made a lavish gesture. "And why don't we put them into execution? For the sake of the labourers; it would be sheer cruelty to afflict them with excessive leisure. It's the same with agriculture. We could synthesize every morsel of food, if we wanted to. But we don't. We prefer to keep a third of the population on the land. For their own sakes—because it takes *longer* to get food out of the land than out of a factory. Besides, we have our stability to think of. We don't want to change. Every change is a menace to stability. That's another reason why we're so chary of applying new inventions. Every discovery in pure science is potentially subversive; even science must sometimes be treated as a possible enemy. Yes, even science."

Science? The Savage frowned. He knew the word. But what it exactly signified he could not say. Shakespeare and the old men of the pueblo had never mentioned science, and from Linda he had only gathered the vaguest hints: science was something you made helicopters with, something that caused you to laugh at the Corn Dances, something that prevented you from being wrinkled and losing your teeth. He made a desperate effort to take the Controller's meaning.

"Yes," Mustapha Mond was saying, "that's another item in the cost of stability. It isn't only art that's incompatible with happiness; it's also science. Science is dangerous; we have to keep it most carefully chained and muzzled."

"What?" said Helmholtz, in astonishment. "But we're always saying that science is everything. It's a hypnopædic platitude."

"Three times a week between thirteen and seventeen," put in Bernard.

"And all the science propaganda we do at the College . . ."

"Yes; but what sort of science?" asked Mustapha Mond sarcastically. "You've had no scientific training, so you can't judge." I was a pretty good physicist in my time. Too good—good enough to realize that all our science is just a cookery book, with an orthodox theory of cooking that nobody's allowed to question, and a list of recipes that mustn't be added to except by special permission from the head cook. I'm the head cook now. But I was an inquisitive young scullion once. I started doing a bit of cooking on my own. Unorthodox cooking, illicit cooking. A bit of real science, in fact." He was silent.

"What happened?" asked Helmholtz Watson.

The Controller sighed. "Very nearly what's going

popping all around you? That was when science first began to be controlled—after the Nine Years' War. People were ready to have even their appetites controlled then. Anything for a quiet life. We've gone on controlling ever since. It hasn't been very good for truth, of course. But it's been very good for happiness. One can't have something for nothing. Happiness has got to be paid for. You're paying for it, Mr. Watson—paying because you happen to be too much interested in beauty. I was too much interested in truth; I paid too."

"But *you* didn't go to an island," said the Savage, breaking a long silence.

The Controller smiled. "That's how I paid. By choosing to serve happiness. Other people's—not mine. It's lucky," he added, after a pause, "that there are such a lot of islands in the world. I don't know what we should do without them. Put you all in the lethal chamber, I suppose. By the way, Mr. Watson, would you like a tropical climate? The Marquesas, for example; or Samoa? Or something rather more bracing?"

Helmholtz rose from his pneumatic chair. "I should like a thoroughly bad climate," he answered. "I believe one would write better if the climate were bad. If there were a lot of wind and storms, for example . . ."

The Controller nodded his approbation. "I like your spirit, Mr. Watson. I like it very much indeed.

As much as I officially disapprove of it." He smiled. "What about the Falkland Islands?"

"Yes, I think that will do," Helmholtz answered. "And now, if you don't mind, I'll go and see how poor Bernard's getting on."

CHAPTER SEVENTEEN

ART, science—you seem to have paid a fairly high price for your happiness," said the Savage, when they were alone. "Anything else?"

"Well, religion, of course," replied the Controller. "There used to be something called God—before the Nine Years' War. But I was forgetting; you know all about God, I suppose."

"Well . . ." The Savage hesitated. He would have liked to say something about solitude, about night, about the mesa lying pale under the moon, about the precipice, the plunge into shadowy darkness, about death. He would have liked to speak; but there were no words. Not even in Shakespeare.

H. G. Wells (1866-1946) was said by George Orwell to have had more influence on young people than any other writer between 1900 and 1920. He probably was the most prolific and the most widely read author of his time. A graduate of the South Kensington Normal School for Science in London, Wells was one of the first professional novelists to have had a formal education in science, and a primary concern of his was the rôle of science in society.

In essays and in fiction, he addressed questions of science and technology and the future of mankind, world government, and utopias. His *Outline of History* (1920) became an instant best seller. As early as 1903 he anticipated the tank as a battle weapon; in *War in the Air* (1908), he foresaw a world war with cities laid waste by aerial bombers; in *The World Set Free* (1913), he wrote of a war with atomic bombs to take place in the 1950's.

In these selections from *The Time Machine* (published in 1895), Wells looks at a society in which problems of population, war, and security have been overcome through the application of science and technology.

THE TIME MACHINE

THE TIME traveller (for so it will be convenient to speak of him) was expounding a recondite matter to us. His pale grey eyes shone and twinkled, and his usually pale face was flushed and animated. The fire burnt brightly, and the soft radiance of the incandescent lights in the lilies of silver caught the bubbles that flashed and passed in our glasses. Our chairs, being his patents, embraced and caressed us rather than submitted to being sat upon, and there was that luxurious after-dinner atmosphere, when thought runs gracefully free of the trammels of precision. And he put it to us in this way – marking the points with a lean forefinger – as we sat and lazily admired his earnestness over this new paradox (as we thought it) and his fecundity.

'You must follow me carefully. I shall have to controvert one or two ideas that are almost universally accepted. The geometry, for instance, they taught you at school is founded on a misconception.'

'Is not that rather a large thing to expect us to begin upon?' said Filby, an argumentative person with red hair.

'I do not mean to ask you to accept anything without reasonable ground for it. You will soon admit as much as I need from you. You know of course that a mathematical line, a line of thickness *nil*, has no real existence. They taught you that? Neither has a mathematical plane. These things are mere abstractions.'

'That is all right,' said the Psychologist.

'Nor having only length, breadth, and thickness, can a cube have a real existence.'

'There I object,' said Filby. 'Of course a solid body may exist. All real things—'

'So most people think. But wait a moment. Can an *instantaneous* cube exist?'

'Don't follow you,' said Filby.

'Can a cube that does not last for any time at all, have a real existence?'

Filby became pensive. 'Clearly,' the Time Traveller proceeded, 'any real body must have extension in *four* directions: it must have Length, Breadth, Thickness, and – Duration. But through a natural infirmity of the flesh, which I will explain to you in a moment, we incline to overlook this fact. There are really four dimensions, three which we call the three planes of Space, and a fourth, Time. There is, however, a tendency to draw an unreal distinction between the former three dimensions and the latter, because it happens that our consciousness moves intermittently in one direction along the latter from the beginning to the end of our lives.'

'That,' said a very young man, making spasmodic efforts to relight his cigar over the lamp; 'that . . . very clear indeed.'

'Now, it is very remarkable that this is so extensively overlooked,' continued the Time Traveller, with a slight accession of cheerfulness. 'Really this is what is meant by the Fourth Dimension, though some people who talk about the Fourth Dimension do not know they mean it. It is only another way of looking at Time. *There is no difference between Time and any of the three dimensions of Space except that our consciousness moves along it.* But some foolish people have got hold of the wrong side of that idea. You have all heard what they have to say about this Fourth Dimension?'

'*I* have not,' said the Provincial Mayor.

'It is simply this. That Space, as our mathematicians have it, is spoken of as having three dimensions, which one may call Length, Breadth, and Thickness, and is always definable by reference to three planes, each at right-angles to the others. But some philosophical people have been asking why *three* dimensions particularly – why not another direction at right-angles to the other three? – and have even tried to construct a Four-Dimensional geometry. Professor Simon Newcomb was expounding this to the New York Mathematical Society only a month or so ago. You know how on a flat surface, which has only two dimensions, we can represent a figure of a three-dimensional solid, and similarly they think that by models of three dimensions they could represent one of four – if they could master the perspective of the thing. See?'

'I think so,' murmured the Provincial Mayor; and, knitting his brows, he lapsed into an introspective state, his lips moving as one who repeats mystic words. 'Yes, I think I see it now,' he said after some time, brightening in a quite transitory manner.

'Well, I do not mind telling you I have been at work upon this geometry of Four Dimensions for some time.

Some of my results are curious. For instance, here is a portrait of a man at eight years old, another at fifteen, another at seventeen, another at twenty-three, and so on. All these are evidently sections, as it were, Three-Dimensional representations of his Four-Dimensional being, which is a fixed and unalterable thing.

'Scientific people,' proceeded the Time Traveller, after the pause required for the proper assimilation of this, 'know very well that Time is only a kind of Space. Here is a popular scientific diagram, a weather record. This line I trace with my finger shows the movement of the barometer. Yesterday it was so high, yesterday night it fell, then this morning it rose again, and so gently upward to here. Surely the mercury did not trace this line in any of the dimensions of Space generally recognized? But certainly it traced such a line, and that line, therefore, we must conclude was along the Time-Dimension.'

'But,' said the Medical Man, staring hard at a coal in the fire, 'if Time is really only a fourth dimension of Space, why is it, and why has it always been, regarded as something different? And why cannot we move about in Time as we move about in the other dimensions of Space?'

The Time Traveller smiled. 'Are you so sure we can move freely in Space? Right and left we can go, backward and forward freely enough, and men always have done so. I admit we move freely in two dimensions. But how about up and down? Gravitation limits us there.'

'Not exactly,' said the Medical Man. 'There are balloons.'

'But before the balloons, save for spasmodic jumping

and the inequalities of the surface, man had no freedom of vertical movement.'

'Still they could move a little up and down,' said the Medical Man.

'Easier, far easier down than up.'

'And you cannot move at all in Time, you cannot get away from the present moment.'

'My dear sir, that is just where you are wrong. That is just where the whole world has gone wrong. We are always getting away from the present moment. Our mental existences, which are immaterial and have no dimensions, are passing along the Time-Dimension with a uniform velocity from the cradle to the grave. Just as we should travel *down* if we began our existence fifty miles above the earth's surface.'

'But the great difficulty is this,' interrupted the Psychologist. 'You *can* move about in all directions of Space, but you cannot move about in Time.'

'That is the germ of my great discovery. But you are wrong to say that we cannot move about in Time. For instance, if I am recalling an incident very vividly I go back to the instant of its occurrence: I become absent-minded, as you say. I jump back for a moment. Of course we have no means of staying back for any length of Time, any more than a savage or an animal has of staying six feet above the ground. But a civilized man is better off than the savage in this respect. He can go up against gravitation in a balloon, and why should he not hope that ultimately he may be able to stop or accelerate his drift along the Time-Dimension, or even turn about and travel the other way?'

'Oh, *this*,' began Filby, 'is all—'

'Why not?' said the Time Traveller.

'It's against reason,' said Filby.

'What reason?' said the Time Traveller.

'You can show black is white by argument,' said Filby, 'but you will never convince me.'

'Possibly not,' said the Time Traveller. 'But now you begin to see the object of my investigations into the geometry of the Four Dimensions. Long ago I had a vague inkling of a machine—'

'To travel through Time!' exclaimed the Very Young Man.

'That shall travel indifferently in any direction of Space and Time, as the driver determines.'

Filby contented himself with laughter.

'But I have experimental verification,' said the Time Traveller.

'It would be remarkably convenient for the historian,' the Psychologist suggested. 'One might travel back and verify the accepted account of the Battle of Hastings, for instance!'

'Don't you think you would attract attention?' said the Medical Man. 'Our ancestors had no great tolerance for anachronisms.'

'One might get one's Greek from the very lips of Homer and Plato,' the Very Young Man thought.

'In which case they would certainly plough you for the Little-go. The German scholars have improved Greek so much.'

'Then there is the future,' said the Very Young Man. 'Just think! One might invest all one's money, leave it to accumulate at interest, and hurry on ahead!'

'To discover a society,' said I, 'erected on a strictly communistic basis.'

'Of all the wild extravagant theories!' began the Psychologist.

'Yes, so it seemed to me, and so I never talked of it until—'

'Experimental verification!' cried I. 'You are going to verify *that*?'

'The experiment!' cried Filby, who was getting brain-weary.

'Let's see your experiment anyhow,' said the Psychologist, 'though it's all humbug, you know.'

The Time Traveller smiled round at us. Then, still smiling faintly, and with his hands deep in his trousers pockets, he walked slowly out of the room, and we heard his slippers shuffling down the long passage to his laboratory.

The Psychologist looked at us. 'I wonder what he's got?'

'Some sleight-of-hand trick or other,' said the Medical Man, and Filby tried to tell us about a conjuror he had seen at Burslem, but before he had finished his preface the Time Traveller came back, and Filby's anecdote collapsed.

In the Golden Age

'IN ANOTHER moment we were standing face to face, I and this fragile thing out of futurity. He came straight up to me and laughed into my eyes. The absence from his bearing of any sign of fear struck me at once. Then he turned to the two others who were following him and spoke to them in a strange and very sweet and liquid tongue.

"There were others coming, and presently a little group of perhaps eight or ten of these exquisite creatures were about me. One of them addressed me. It came into my head, oddly enough, that my voice was too harsh and deep for them. So I shook my head, and pointing to my ears, shook it again. He came a step forward, hesitated, and then touched my hand. Then I felt other soft little tentacles upon my back and shoulders. They wanted to make sure I was real. There was nothing in this at all alarming. Indeed, there was something in these pretty little people that inspired confidence – a graceful gentleness, a certain child-like ease. And besides, they looked so frail that I could fancy myself flinging the whole dozen of them about like ninepins. But I made a sudden motion to warn them when I saw their little pink hands feeling at the Time Machine. Happily then, when it was not too late, I thought of a danger I had hitherto forgotten, and reaching over the bars of the machine, I unscrewed the little levers that would set it in motion, and put these in my pocket. Then I turned again to see what I could do in the way of communication.

'And then, looking more nearly into their features, I saw some further peculiarities in their Dresden china type of prettiness. Their hair, which was uniformly curly, came to a sharp end at the neck and cheek; there was not the faintest suggestion of it on the face, and their ears were singularly minute. The mouths were small, with bright red, rather thin lips, and the little chins ran to a point. The eyes were large and mild; and – this may seem egotism on my part – I fancied even then that there was a certain lack of the interest I might have expected in them.

'As they made no effort to communicate with me, but simply stood round me smiling and speaking in soft cooing notes to each other, I began the conversation. I pointed to the Time Machine and to myself. Then, hesitating for a moment how to express Time, I pointed to the sun. At once a quaintly pretty little figure in chequered purple and white followed my gesture, and then astonished me by imitating the sound of thunder.

'For a moment I was staggered, though the import of his gesture was plain enough. The question had come into my mind abruptly: were these creatures fools? You may hardly understand how it took me. You see, I had always anticipated that the people of the year Eight Hundred and Two Thousand odd would be incredibly in front of us in knowledge, art, everything. Then one of them suddenly asked me a question that showed him to be on the intellectual level of one of our five-year-old children – asked me, in fact, if I had come from the sun in a thunderstorm! It let loose the judgment I had suspended upon their clothes, their frail light limbs and fragile features. A flow of disappointment rushed across my mind. For a moment I felt that I had built the Time Machine in vain.

'I nodded, pointed to the sun, and gave them such a vivid rendering of a thunderclap as startled them. They all withdrew a pace or so and bowed. Then came one laughing towards me, carrying a chain of beautiful flowers altogether new to me, and put it about my neck. The idea was received with melodious applause; and presently they were all running to and fro with flowers and laughingly flinging them upon me until I was almost smothered with blossom. You who have never seen the like can scarcely imagine what delicate and wonderful flowers countless years of culture had created. Then someone suggested that their plaything should be exhibited in the nearest building, and so I was led past the sphinx of white marble, which had seemed to watch me all the while with a smile at my astonishment, towards a vast grey edifice of fretted stone. As I went with them the memory of my confident anticipations of a profoundly grave and intellectual posterity came, with irresistible merriment, to my mind.

'The building had a huge entry, and was altogether of colossal dimensions. I was naturally most occupied with the growing crowd of little people, and with the big, open portals that yawned before me shadowy and mysterious. My general impression of the world I saw over their heads was of a tangled waste of beautiful bushes and flowers, a long-neglected and yet weedless garden. I saw a number of tall spikes of strange white flowers, measuring a foot perhaps across the spread of the waxen petals. They grew scattered, as if wild, among the variegated shrubs, but, as I say, I did not examine them closely at this time. The Time Machine was left deserted on the turf among the rhododendrons.

'The arch of the doorway was richly carved, but naturally I did not observe the carving very narrowly, though I fancied I saw suggestions of old Phœnician decorations as I passed through, and it struck me that they were very badly broken and weather-worn. Several more brightly-clad people met me in the doorway, and so we entered, I, dressed in dingy nineteenth-century garments, looking grotesque enough, garlanded with flowers, and surrounded by an eddying mass of bright, soft-coloured robes and shining white limbs, in a melodious whirl of laughter and laughing speech.

'The big doorway opened into a proportionately great hall hung with brown. The roof was in shadow, and the windows, partially glazed with coloured glass and partially unglazed, admitted a tempered light. The floor was made up of huge blocks of some very hard white metal, not plates nor slabs – blocks, and it was so much worn, as I judged by the going to and fro of past generations, as to be deeply channelled along the more frequented ways. Transverse to the length were innumerable tables made of slabs of polished stone, raised, perhaps, a foot from the floor, and upon these were heaps of fruits. Some I recognized as a kind of hypertrophied raspberry and orange, but for the most part they were strange.

'Between the tables was scattered a great number of

cushions. Upon these my conductors seated themselves, signing for me to do likewise. With a pretty absence of ceremony they began to eat the fruit with their hands, flinging peel and stalks, and so forth, into the round openings in the sides of the tables. I was not loth to follow their example, for I felt thirsty and hungry. As I did so I surveyed the hall at my leisure.

'And perhaps the thing that struck me most was its dilapidated look. The stained-glass windows, which displayed only a geometrical pattern, were broken in many places, and the curtains that hung across the lower end were thick with dust. And it caught my eye that the corner of the marble table near me was fractured. Nevertheless, the general effect was extremely rich and picturesque. There were, perhaps, a couple of hundred people dining in the hall, and most of them, seated as near to me as they could come, were watching me with interest, their little eyes shining over the fruit they were eating. All were clad in the same soft, and yet strong, silky material.

'Fruit, by the bye, was all their diet. These people of the remote future were strict vegetarians, and while I was with them, in spite of some carnal cravings, I had to be frugivorous also. Indeed, I found afterwards that horses, cattle, sheep, dogs, had followed the Ichthyosaurus into extinction. But the fruits were very delightful; one, in particular, that seemed to be in season all the time I was there – a floury thing in a three-sided husk – was especially good, and I made it my staple. At first I was puzzled by all these strange fruits, and by the strange flowers I saw, but later I began to perceive their import.

'However, I am telling you of my fruit dinner in the distant future now. So soon as my appetite was a little checked I determined to make a resolute attempt to learn the speech of these new men of mine. Clearly that was the next thing to do. The fruits seemed a convenient thing to begin upon, and holding one of these up I began a series of interrogative sounds and gestures. I had some considerable difficulty in conveying my meaning. At first my efforts met with a stare of surprise or inextinguishable laughter, but presently a fair-haired little creature seemed to grasp my intention and repeated a name. They had to chatter and explain their business at great length to each other, and my first attempts to make their exquisite little sounds of the language caused an immense amount of genuine, if uncivil, amusement. However, I felt like a schoolmaster amidst children, and persisted, and presently I had a score of noun substantives at least at my command; and then I got to demonstrative pronouns, and even the verb 'to eat'. But it was slow work, and the little people soon tired and wanted to get away from my interrogations, so I determined, rather of necessity, to let them give their lessons in little doses when they felt inclined. And very little doses I found they were before long, for I never met people more indolent or more easily fatigued.

The Sunset of Mankind

'A QUEER THING I soon discovered about my little hosts, and that was their lack of interest. They would come to me with eager cries of astonishment, like children, but, like children, they would soon stop examining me, and wander away after some other toy. The dinner and my conversational beginnings ended, I noted for the first time that almost all those who had surrounded me at first were gone. It is odd, too, how speedily I came to disregard these little people. I went out through the portal into the sunlit world again so soon as my hunger was satisfied. I was continually meeting more of these men of the future, who would follow me a little distance, chatter and laugh about me, and, having smiled and gesticulated in a friendly way, leave me again to my own devices.

'The calm of evening was upon the world as I emerged from the great hall, and the scene was lit by the warm glow of the setting sun. At first things were very confusing. Everything was so entirely different from the world I had known – even the flowers. The big building I had left was situate on the slope of a broad river valley, but the Thames had shifted, perhaps, a mile from its present position. I resolved to mount to the summit of a crest, perhaps a mile and a half away, from which I could get a wider view of this our planet in the year Eight Hundred and Two Thousand Seven Hundred and One, A.D. For that, I should explain, was the date the little dials of my machine recorded.

'As I walked I was watchful for every impression that could possibly help to explain the condition of ruinous splendour in which I found the world – for ruinous it was. A little way up the hill, for instance, was a great heap of granite bound together by masses of aluminium, a vast labyrinth of precipitous walls and crumbled heaps, amidst which were thick heaps of very beautiful pagoda-like plants – nettles possibly – but wonderfully tinted with brown about the leaves, and incapable of stinging. It was evidently the derelict remains of some vast structure, to what end built I could not determine. It was here that I was destined, at a later date, to have a very strange experience – the first intimation of a still stranger discovery – but of that I will speak in its proper place.

'Looking round, with a sudden thought, from a terrace on which I rested for a while, I realized that there were no small houses to be seen. Apparently, the single house, and possibly even the household, had vanished. Here and there among the greenery were palace-like buildings, but the house and the cottage, which form such characteristic features of our own English landscape, had disappeared.

' "Communism," said I to myself.

'And on the heels of that came another thought. I looked at the half-dozen little figures that were following me. Then, in a flash, I perceived that all had the same form of costume, the same soft hairless visage, and the same girlish rotundity of limb. It may seem strange, perhaps, that I had not noticed this before. But everything was so strange. Now, I saw the fact plainly enough. In costume, and in all the differences of texture and bearing that now mark off the sexes from each other, these people of the future were alike. And the children seemed to my eyes to be but the miniatures of their parents. I judged then that the children of that time were extremely precocious, physically at least, and I found afterwards abundant verification of my opinion.

'Seeing the ease and security in which these people were living, I felt that this close resemblance of the sexes was after all what one would expect; for the strength of the man and the softness of a woman, the institution of the family, and the differentiation of occupations are mere militant necessities of an age of physical force. Where population is balanced and abundant, much child-bearing becomes an evil rather than a blessing to the State: where violence comes but rarely and offspring are secure, there is less necessity – indeed there is no necessity – of an efficient family, and the specialization of the sexes with reference to their children's needs disappears. We see some beginnings of this even in our own time, and in this future age it was complete. This, I must remind you, was my speculation at the time. Later, I was to appreciate how far it fell short of the reality.

'While I was musing upon these things, my attention was attracted by a pretty little structure, like a well under a cupola. I thought in a transitory way of the oddness of wells still existing, and then resumed the thread of my speculations. There were no large buildings towards the top of the hill, and as my walking powers were evidently miraculous, I was presently left alone for the first time. With a strange sense of freedom and adventure I pushed on up to the crest.

'There I found a seat of some yellow metal that I did not recognize, corroded in places with a kind of pinkish rust and half-smothered in soft moss, the arm-rests cast and filed into the resemblance of griffins' heads. I sat down on it, and I surveyed the broad view of our old world under the sunset of that long day. It was as sweet and fair a view as I have ever seen. The sun had already gone below the horizon and the west was flaming gold, touched with some horizontal bars of purple and crimson. Below was the valley of the Thames, in which the river lay like a band of burnished steel. I have already spoken of the great palaces dotted about among the variegated greenery, some in ruins and some still occupied. Here and there rose a white or silvery figure in the waste garden of the earth, here and there came the sharp vertical line of some cupola or obelisk. There were no hedges, no signs of proprietary rights, no evidences of agriculture; the whole earth had become a garden.

'So watching, I began to put my interpretation upon the things I had seen, and as it shaped itself to me that evening, my interpretation was something in this way. (Afterwards I found I had got only a half truth – or only a glimpse of one facet of the truth):

'It seemed to me that I had happened upon humanity upon the wane. The ruddy sunset set me thinking of the sunset of mankind. For the first time I began to realize an odd consequence of the social effort in which we are at present engaged. And yet, come to think, it is a logical consequence enough. Strength is the outcome of need: security sets a premium on feebleness. The work of ameliorating the conditions of life – the true civilizing process that makes life more and more secure – had gone steadily on to a climax. One triumph of a united humanity over Nature had followed another. Things that are now mere dreams had become projects deliberately put in hand and carried forward. And the harvest was what I saw!

'After all, the sanitation and the agriculture of today are still in the rudimentary stage. The science of our time has attacked but a little department of the field of human disease, but, even so, it spreads its operations very steadily and persistently. Our agriculture and horticulture destroy a weed just here and there and cultivate perhaps a score or so of wholesome plants, leaving the greater number to fight out a balance as they can. We improve our favourite plants and animals – and how few they are – gradually by selective breeding; now a new and better peach, now a seedless grape, now a sweeter and larger flower, now a more convenient breed of cattle. We improve them gradually, because our ideals are vague and tentative, and our knowledge is very limited; because Nature, too, is shy and slow in our clumsy hands. Some day all this will be better organized, and still better. That is the drift of the current in spite of the eddies. The whole world will be intelligent, educated, and co-operating; things will move faster and faster towards the subjugation of Nature. In the end, wisely and carefully we shall readjust the balance of animal and vegetable life to suit our human needs.

'This adjustment, I say, must have been done, and

done well: done indeed for all time, in the space of Time across which my machine had leapt. The air was free from gnats, the earth from weeds or fungi; everywhere were fruits and sweet and delightful flowers; brilliant butterflies flew hither and thither. The ideal of preventive medicine was attained. Disease had been stamped out. I saw no evidence of any contagious diseases during all my stay. And I shall have to tell you later that even the processes of putrefaction and decay had been profoundly affected by these changes.

'Social triumphs, too, had been effected. I saw mankind housed in splendid shelters, gloriously clothed, and as yet I had found them engaged in no toil. There were no signs of struggle, neither social nor economical struggle. The shop, the advertisement, traffic, all that commerce which constitutes the body of our world, was gone. It was natural on that golden evening that I should jump at the idea of a social paradise. The difficulty of increasing population had been met, I guessed, and population had ceased to increase.

'But with this change in condition comes inevitably adaptations to the change. What, unless biological science is a mass of errors, is the cause of human intelligence and vigour? Hardship and freedom: conditions under which the active, strong, and subtle survive and the weaker go to the wall; conditions that put a premium upon the loyal alliance of capable men, upon self-restraint, patience, and decision. And the institution of the family, and the emotions that arise therein, the fierce jealousy, the tenderness for offspring, parental self-devotion, all found their justification and support in the imminent dangers of the young. *Now*, where are these imminent dangers? There is a sentiment arising, and it will grow, against connubial jealousy, against fierce maternity, against passion of all sorts; unnecessary things now, and things that make us uncomfortable, savage survivals, discords in a refined and pleasant life.

'I thought of the physical slightness of the people, their lack of intelligence, and those big abundant ruins, and it strengthened my belief in a perfect conquest of Nature. For after the battle comes Quiet. Humanity had been strong, energetic, and intelligent, and had used all its abundant vitality to alter the conditions under which it lived. And now came the reaction of the altered conditions.

'Under the new conditions of perfect comfort and security, that restless energy that with us is strength would become weakness. Even in our own time certain tendencies and desires, once necessary to survival, are a constant source of failure. Physical courage and the love of battle, for instance, are no great help – may even be hindrances – to a civilized man. And in a state of physical balance and security, power, intellectual as well as physical, would be out of place. For countless years I judged there had been no danger of war or solitary violence, no danger from wild beasts, no wasting disease to require strength of constitution, no need of toil. For such a life, what we should call the weak are as well equipped as the strong, are indeed no longer weak. Better equipped indeed they are, for the strong would be fretted by an energy for which there was no outlet. No doubt the exquisite beauty of the buildings I saw was the outcome of the last surgings of the now purposeless energy of mankind before it settled down into perfect harmony with the conditions under which it lived – the flourish of that triumph which began the last great peace. This has ever been the fate of energy in security; it takes to art and to eroticism, and then come languor and decay.

'Even this artistic impetus would at last die away – had almost died in the Time I saw. To adorn themselves with flowers, to dance, to sing in the sunlight; so much was left of the artistic spirit, and no more. Even that would fade in the end into a contented inactivity. We are kept keen on the grindstone of pain and necessity, and it seemed to me that here was that hateful grindstone broken at last!

'As I stood there in the gathering dark I thought that in this simple explanation I had mastered the problem of the world – mastered the whole secret of these delicious people. Possibly the checks they had devised for the increase of population had succeeded too well, and their numbers had rather diminished than kept stationary. That would account for the abandoned ruins. Very simple was my explanation, and plausible enough – as most wrong theories are!